Advances in Mathematics Education

Series Editors
Gabriele Kaiser, University of Hamburg, Hamburg, Germany
Bharath Sriraman, University of Montana, Missoula, MT, USA

Editorial Board Members
Marcelo C. Borba, São Paulo State University, São Paulo, Brazil
Jinfa Cai, Newark, NJ, USA
Christine Knipping, Bremen, Germany
Oh Nam Kwon, Seoul, Korea (Republic of)
Alan Schoenfeld, University of California, Berkeley, Berkeley, CA, USA

Advances in Mathematics Education is a forward looking monograph series originating from the journal ZDM – Mathematics Education. The book series is the first to synthesize important research advances in mathematics education, and welcomes proposals from the community on topics of interest to the field. Each book contains original work complemented by commentary papers.

Researchers interested in guest editing a monograph should contact the series editors Gabriele Kaiser (gabriele.kaiser@uni-hamburg.de) and Bharath Sriraman (sriramanb@mso.umt.edu) for details on how to submit a proposal.

Ole Skovsmose

Critical Mathematics Education

 Springer

Ole Skovsmose
Aalborg University
Aalborg, Denmark

Universidade Estadual Paulista
São Paulo, Brazil

ISSN 1869-4918　　　　　　ISSN 1869-4926　(electronic)
Advances in Mathematics Education
ISBN 978-3-031-26241-8　　　ISBN 978-3-031-26242-5　(eBook)
https://doi.org/10.1007/978-3-031-26242-5

© The Editor(s) (if applicable) and The Author(s), under exclusive license to Springer Nature Switzerland AG 2023

This work is subject to copyright. All rights are solely and exclusively licensed by the Publisher, whether the whole or part of the material is concerned, specifically the rights of reprinting, reuse of illustrations, recitation, broadcasting, reproduction on microfilms or in any other physical way, and transmission or information storage and retrieval, electronic adaptation, computer software, or by similar or dissimilar methodology now known or hereafter developed.

The use of general descriptive names, registered names, trademarks, service marks, etc. in this publication does not imply, even in the absence of a specific statement, that such names are exempt from the relevant protective laws and regulations and therefore free for general use.

The publisher, the authors, and the editors are safe to assume that the advice and information in this book are believed to be true and accurate at the date of publication. Neither the publisher nor the authors or the editors give a warranty, expressed or implied, with respect to the material contained herein or for any errors or omissions that may have been made. The publisher remains neutral with regard to jurisdictional claims in published maps and institutional affiliations.

This Springer imprint is published by the registered company Springer Nature Switzerland AG
The registered company address is: Gewerbestrasse 11, 6330 Cham, Switzerland

Acknowledgments

Most chapters in this book were written during the period 2019–2022. It was a difficult time. The pandemic haunted the world, and I experienced serious health problems. I went through chemotherapy and surgeries. I found out that one way of relaxing my mind was to work with manuscripts, and so I did. However, without the support from Miriam, my wife, I would not have been able to write anything. She was together with me every day at the hospital. She was with me all the time.

The chapters in this book have the format of articles. Some of them are reworked versions of previously published articles, which I acknowledge in a note following such chapters. I thank the publishers for their kind permission to allow me to use such previous writings.

Many people have helped me with suggestions and advices. I want to thank Denner Dias Barros, Richard Barwell, Denival Biotto Filho, Arindam Bose, Débora Vieira de Souza-Carneiro, Manuella Carrijo, Paul Ernest, Ana Carolina Faustino, Brian Greer, David Kollosche, Jamaal Ince, Marit Johnsen-Høines, Renato Marcone, Amanda Queiroz Moura, João Luiz Muzinatti, Luana Oliveira, Aldo Parra, Lucas Penteao, Miriam Godoy Penteado, Ole Ravn, Célia Roncato, Guilherme Henrique Gomes da Silva, Daniela Alves Soares, and Paola Valero for all their support. I thank Peter Gates for commenting on and helping me improve many of the chapters. And I thank Rosalyn Sword for completing a careful language editing of the whole manuscript.

December 2022   Ole Skovsmose

Introduction

In 1994, I visited São Paulo for the first time in my life. I had brought with me a small Danish travel guide, which recommended taking a walk from Praça da Sé to Praça da República, and I followed the recommendation.

For a couple of days, I was staying in the flat of Ubiratan D'Ambrosio and his wife Maria José and enjoying their hospitality. One morning, D'Ambrosio brought me in his car to Praça de Sé. We had agreed that, in the afternoon at 16:00, he would pick me up close to Praça da República at the corner of Avenida Ipiranga and São João. This corner was clearly shown in my Danish tourist guide. Later, I was told that this corner had become famous due to the song and music "Sampa," composed in 1978 by Caetano Veloso. In the "Sampa," Veloso wanted to express the creative energy that he associated with São Paulo.

I started walking around at Praça da Sé. There were so many people, and music emanated from all directions. With loud vices, street sellers were announcing their special offers. One could buy water, soft drinks, pastries, fruit, small figurines, T-shirts, caps; all kinds of touristy things. Next to the Praça, I saw many fountains where children were playing. Later I came to find out that these fountains attracted many street children, some of them coming from far away to get an opportunity to take a shower. I could have stayed the whole morning around the Praça filled with colors, sounds, and movements, but after a while I started my walk.

The streets seemed packed with people. There was much to look at and also to listen to. Music rang out from many of the shops. The music became mixed with loud voices announcing the shops' special offers. My walk led me across a long pedestrian bridge with street sellers on both sides. Deep below the bridge ran a highway, which I later came to pass many times. I found a place to sit and have a cup of coffee. The Brazilian coffee was strong to me, but soon I came to appreciate precisely this taste of coffee. Later I found a small restaurant and enjoyed my lunch there. In the afternoon I found Praça da República, and D'Ambrosio found me.

Since then I have visited Brazil many times. I fell in love with Miriam there, and we have now been married for almost 20 years. We visited São Paula many times, and now we live there. Over the years, we have walked around many different neighborhoods. We came to like Vila Madalena with its restaurants and galleries.

Once I had an exhibition there showing a collection of my paintings. Many times we have walked along Avenida Paulista, a main avenue framed with huge modern concrete-glass buildings. Occasionally we have seen old, still buildings with stucco and architectural styles that have survived from the past. Looking at them, one can travel back in time and imagine how the whole Avenida was once surrounded by such buildings. Over the last few years, Avenida Paulista has also changed appearance due to the growing number of homeless people settling there in small tents or behind shelters made of cardboard and plastic. Together with my daughter, I have walked down Oscar Freire, which has many fashionable shops and nice coffee places, and now also has its share of homeless people.

Miriam and I live in a flat located in Aclimação, a neighborhood next to Liberdade and close to the hospital where I had my surgeries and am now being treated in. When slavery was abolished in Brazil, many of the liberated people came to live in Liberdade, and this is the reason that the neighborhood got this name. Later, many people with Japanese origins came to live there, and in Liberdade, one finds excellent Japanese restaurants. São Paulo encompasses many different villages; it seems to be composed of a patchwork of townscapes.

One can take many different routes when walking in São Paulo. One can naturally pass the same place on different walks and can come to see a location from different perspectives. One can pass by rather fast, but also take one's time. One can look straight ahead when walking, but also look up and around. Over time, places are changing. Old buildings with history are demolished to make way for new, tall buildings shimmering with glass façades. When walking around, one can see buildings from a way off. One could see the Edifício Copan, designed by Oscar Niemeyer and located in walking distance of Praça da República. Like a huge but all too short snake, Copan seems to have found a comfortable position in between the sharp corners of streets and houses. At night, the tall Altino Arantes Building, also called the Farol Santander, is illuminated in purple colors and can be seen from far away, including from the balcony of our flat in Aclimação. Many neighborhoods in São Paulo remain unknown to me. The metropolis includes huge *favelas*, which I have only seen when driving by.

One can think of critical mathematics education as a metropolis, not made up of streets, places, architectures, and neighborhoods, but by ideas, examples, theories, and practices. In this book, I will present different routes around critical mathematics education through eighteen chapters. These routes, like my walks around São Paulo, will be crossing each other several times, and as a reader, you might come to see the same observation several times, although from different perspectives. Like São Paulo, critical mathematics education has also undergone many changes over time, and we will come to see how some ideas and concerns are rooted back in history.

The chapters in the book take the form of articles. All of the chapters can be read in isolation, or in any order one might prefer. I have numbered the chapters and grouped them in four separate parts, but this is only one suggestion for organizing walks around critical mathematics education.

In *Part I: Learning and Landscapes*, I address the notion of social justice, which is vitally important for formulating the concerns and hopes of critical mathematics education. Landscapes of investigation establish learning environments different from those formed by the school mathematics tradition. Such landscapes make it possible for students and teachers in cooperation to identify and challenge cases of social injustice. This could be the exploitation of workers, patterns of structural racism, or erosions of democracy. Landscapes could also turn climate change into focus. Finally, I bring together a circle of concepts for interpreting learning processes, and in particular for understanding students' possible motives, or lack of motives, for learning mathematics. This cycle tends to capture low achievement in mathematics as a socioeconomic phenomenon rather than as a personal failure.

In *Part II: Crises and Formatting*, I explore the notion of crisis and the formatting power of mathematics. It has been advocated that, by means of mathematics, one can provide reliable pictures or models of reality, and also that mathematics is the language of science, ensuring objectivity and neutrality. This understanding of mathematics is challenged when we recognize the formatting power that can be acted out through mathematics. A crisis might concern economy, politics, security, hunger, sustainability, or pandemics. At times, mathematics provides a formatting of such crises by shaping how we see and react to them. A "banality of mathematical expertise" appears when mathematics is brought into action without considering the possible formatting power of such actions. This banality locates mathematics in an ethical vacuum, which might occur both in mathematics research and mathematics education. It is crucial to critical mathematics education to challenge this banality.

In *Part III: Critique and Dialogue*, a conception of mathematics as glorious, harmless, and innocent is confronted with a conception of mathematics as indefinite. Mathematics is indefinite with respect to concepts, proofs, culture, applications, and power. Mathematics might come to serve any kind of interest, which makes a critique of mathematics necessary. Critique has been conceived of as methodology (as inspired by René Descartes), as hammering (as practised by Friedrich Nietzsche), and as dialogue (as I am going to propose). I want to formulate a conception of critique that does not assume the absolutist features connected to critique as the right methodology without embracing the absolute relativism connected to critique as hammering. Seeing critique as dialogue is crucial for the further development of critical mathematics education.

In *Part IV: Width and Depth*, I illustrate how the significance of critical mathematics education reaches much further than an educational project addressing students at social risks socially. It is important for all groups of students to address cases of social injustice. Critical mathematics education concerns ways of doing research, and it tries deliberately to prevent assuming stereotypes common in mathematics education research. Insight provided by critical mathematics education has relevance not only for critical mathematics education itself, but for any kind of social theorizing. This insight might contribute to an understanding of the scope of social structuration, the emergence of risk society, the formation of life worlds, and characteristics of structural violence. In the final walk around in critical mathematics education, we recognize uncertainty as a universal human condition. This turns critique into a necessary human activity.

Contents

Part I Learning and Landscapes

1 Concerns and Hopes 3
   1.1 Concerns 4
   1.2 Hopes 6
   1.3 Social Justice as Construction 10
   References 12

2 Landscapes and Racism 15
   2.1 Landscapes of Investigation 16
   2.2 Racism 18
   2.3 Media and Racism 22
   2.4 A Bias Index 25
   References 29

3 Democracy and Erosions 33
   3.1 Features of Democracy 34
       3.1.1 Voting 35
       3.1.2 Fairness 36
       3.1.3 Equity 37
       3.1.4 Deliberation 37
   3.2 Erosions of Democracy 38
   3.3 Erosions of Democracy Today 40
       3.3.1 Voting 40
       3.3.2 Fairness 41
       3.3.3 Equity 43
       3.3.4 Deliberation 43
   References 44

4 Sustainability and Risks ... 47
4.1 Risk Society ... 48
4.2 Experimental Forecasting and Climate Changes ... 49
4.3 Reflections on Mathematical Modelling ... 51
4.4 Use of Water ... 53
4.5 Environmental Justice ... 56
References ... 57

5 To Learn or Not to Learn? ... 59
5.1 Learning and Action ... 60
5.2 Action and Intention ... 62
5.3 Intention and Foreground ... 64
5.4 Foreground and Motive ... 66
5.5 Motive and Mathematics ... 68
5.6 Mathematics and Life-Worlds ... 70
5.7 A Conclusion: Life-Worlds and Learning ... 73
References ... 73

Part II Crises and Formatting

6 Crisis and Critique ... 79
6.1 Critical Situation ... 80
6.2 A Hierarchy of Crisis and Critique ... 81
6.3 A Multiplicity of Crises and Critique ... 83
References ... 85

7 Banality of Mathematical Expertise ... 87
7.1 A Role Model ... 89
7.2 A Four-Dimensional Philosophy of Mathematics ... 90
7.3 The Ethical Dimension ... 92
7.4 Banality ... 95
7.5 From the Classroom ... 97
References ... 99

8 Mathematics and Ethics ... 103
8.1 Quantifying ... 105
8.2 Digitalising ... 107
8.3 Serialising ... 109
8.4 Categorising ... 111
8.5 Imagining ... 113
8.6 Banality ... 114
References ... 116

9 Mathematics and Crises ... 119
9.1 Mathematics as Picturing a Crisis ... 121
9.2 Mathematics as Constituting a Crisis ... 123

	9.3 Mathematics as Formatting a Crisis.	126
	9.4 Politics of Crises	128
	References.	130
10	**Picturing or Performing?**	133
	10.1 Mathematics as Picturing.	134
	10.2 Language as Picturing	136
	10.3 Language as Performing	138
	10.4 Mathematics as Performing	140
	References.	142

Part III Critique and Dialogue

11	**Critique of Mathematics.**	145
	11.1 Mathematics as Glorious	146
	11.2 Mathematics as Harmless and Innocent.	148
	11.3 Mathematics as Indefinite	149
	References.	153
12	**Critique of Critique**	155
	12.1 Critique as Methodology	156
	12.2 Critique as Hammering	159
	12.3 Critique as Dialogue	162
	References.	164
13	**Initial Formulations of Critical Mathematic Education**	167
	13.1 Education Lost in a Dream World	168
	13.2 Education Turning Critical.	170
	13.3 Mathematics Education Turning Critical.	172
	13.4 Sociological Imagination and Exemplary Learning.	176
	13.5 Brazil: Land of the Future	177
	References.	180
14	**A Dialogic Theory of Learning Mathematics**	185
	14.1 Theory	186
	14.2 Dialogue and Critique	187
	14.3 Learning Interacts	188
	14.3.1 Getting in Contact	189
	14.3.2 Exploring	189
	14.3.3 Positioning	191
	14.3.4 Foregrounding	192
	14.3.5 Externalising	194
	14.3.6 Doubting	196
	14.4 Was That a Theory?	197
	References.	197

Part IV Width and Depth

15 All Students 201
    15.1   Students at Social Risks................................. 202
    15.2   Students in Comfortable Positions......................... 204
    15.3   University Students in Mathematics 206
    References.. 208

16 Beyond Stereotypes... 211
    16.1   Neat 213
    16.2   Affluent 214
    16.3   Placid 217
    16.4   Descriptive 218
    References.. 221

17 Social Theorising and the Formatting Power of Mathematics 223
    17.1   Structuration... 224
    17.2   Risk Society 226
    17.3   Life-Worlds 227
    17.4   Violence 229
    References.. 231

18 A Philosophy of Critical Mathematics Education 233
    18.1   Social Justice 234
    18.2   Environmental Justice 236
    18.3   Mathematics in Action.................................... 237
    18.4   Foregrounds 240
    18.5   Dialogue and Critique 242
    References.. 244

Composed Reference List... 247

Name Index 265

Subject Index... 271

Part I
Learning and Landscapes

Chapter 1
Concerns and Hopes

Abstract Critical mathematics education is characterised in terms of concerns and hopes: concerns about what is taking place in the classroom, and about the students and their future opportunities in life. Concerns about socio-economic exploitation, racism, sexism, homophobia, xenophobia, and violence. Concerns about our environment. Critical mathematics education is also made up of concerns about mathematics itself, not only as a school subject but also as applied in technology, natural science, social science, economy, and management. Hoping is more than an individual or local undertaking; it also concerns out shared future. Hope represents a socio-political utopia, which might turn hope into political acts. Critical mathematics education engages students in the very formulation of what social justice could mean. I see the articulations of concerns and hopes as driving forces in the construction of conceptions of social justice. Such constructions can take place in the mathematics classroom.

Keywords Concern · Hope · Exploitation · Racism · Social justice · Environment · Social construction · Utopia

What is critical mathematics education? One could try to answer this question by describing how it might be acted out in the classroom; the nature of the problems that are addressed; and the ways students and teachers are invited to interact. One could also try to characterise critical mathematics education through its theoretical references, what philosophical ideas it relates to, and what conceptions of society and of learning it assumes. Previously, I have tried both such approaches, but I have never found either of them quite satisfactory.

Many times I have experienced teachers engaging their students in fascinating projects which I would consider exemplary for a critical mathematics education. However, the teacher might not have heard about critical mathematics education, and might describe the ideas of the projects in quite different terms. This is perfectly fine. I am not interested in trying to ringfence critical mathematics education, but rather in the concerns and hopes that are acted out through educational ideas and practices.

After addressing concerns related to critical mathematics education, I will explore the notion of hope more carefully. Finally, I will interpret the notion of social justice as being under construction and illustrate this with a classroom example.

1.1 Concerns

Critical mathematics education is concerned about *what is taking place in the classroom*. Looking at explicitly formulated aims for mathematics education, everything appears obvious: mathematics education is the principal social institution for establishing mathematical knowledge among students. However, the actual roles of social institutions need not be what they are claimed to be. A paradigmatic recognition of this possibility was expressed by Karl Marx, when he labelled religion as being opium of the people. The church was an institution with a quite different function than commonly assumed. Could mathematics education be a socio-political instrument for making people submissive and easy to dominate? Such a possibility could be associated with the exercise paradigm dominating this education. Engaging students in investigative processes is an attempt to create learning environments that provide mathematics education with a different socio-political role.

Critical mathematics education is concerned about the *students*: not only their learning in the classroom, but their status as future citizens. An education for citizenship could mean preparing students to fit into the given social order, but it could also be an education for autonomy, turning students into critical citizens. It has been common in critical mathematics education to pay particular attention to students facing social risks. This could be students living in poor or in violent environments; it could be students of colour subjected to racist patterns of exclusion; it could be students doomed through caste systems or similar structures of oppression; and it could be immigrant students arriving into hostile environments. It is important for critical mathematics education to operate in solidarity with students at social risks. However, all groups of students are important to critical mathematics education, including students in comfortable positions. With this expression, I refer to students from economically affluent environments, about whom statistics suggest they will most likely go on to further education, obtain good jobs, and become economically well situated. However, it is important also to explore cases of social and economic injustice together with students in comfortable positions. It is important to draw their attention to the extreme inequalities of the world. Another group of students also important for critical mathematics education is that of university students in mathematics, who will come to represent mathematical expertise and to bring mathematics into action. They will come to exercise power through applications of mathematics.[1]

---

[1] See Chap. 15, "All Students", for further elaborations of this point.

1.1 Concerns

Critical mathematics education is concerned about the *exploitation of workers and their families*. This concern was the principle one for the emergence of the critical approach as formulated by Karl Marx. Sharp observations on the exploitation of workers during the initial period of industrialisation were made by Friedrich Engels (1993) in *The Condition of the Working Class in England*, first published in German in 1845. Here he documented, in descriptions and in numbers, the horrible conditions that the English working class was subjected to. In *Germinal*, published in 1885, Émile Zola (1954) provided a literary description of working-class conditions in the coal districts in Northern France. In *The Road to Wigan Pier*, first published in 1937, George Orwell (1937) provided an update of Engels' description of English working-class conditions. The accounts by Engels, Zola, and Orwell are different, but the concerns they raise are shared by many political and social movements, and also by critical mathematics education. Inspired by Marx, focus has been put on the exploitation of workers in industrialised countries. However, in a globalised world exploitations reach much further, also manifesting new forms of colonisation. It is a concern of critical mathematics education to address any form of exploitation.

Critical mathematics education is concerned about *racism*. Colonisation was an explicit and brutal form of exploitation. Colonies were supposed to serve as natural resources for the colonising powers. Colonisation was accompanied by slavery, and racism developed as a pathological justification of slavery. Racist ideologies – through which colonial abuse and cruelty sought legitimisation – normalised slavery. The British Empire was a paradigmatic case of worldwide exploitation, combining violent colonisation with the development of different versions of racism, including orientalism.[2] Colonisation means oppression, which marks the individual. In the two books *Black Skin, White Masks* and *The Wretched on the Earth*, first published in French in 1952 and 1961, Franz Fanon (2004, 2008) shows how the brutality of colonialism structures people's experiences and self-conceptions. The coloniser becomes elevated to a superior level, and this causes a perverse pressure on the colonised. Fanon points out that, in order to be accepted in a colonised world, the colonised needs to imitate the coloniser. This observation is condensed into the title of the book: *Black Skin, White Masks*. Fanon helps us to recognise that colonial exploitation operates at all levels: at the macro socio-political level as well as at the micro personal and experiential level. Racism is a devastating all-inclusive phenomenon. Today, racism has been directly confronted by many different movements, and also by critical mathematics education.

Critical mathematics education is concerned about our *environment*. Traditional critical positions, as directly inspired by Marx, did not express concern about the natural environment. The focus was first of all on the miserable social conditions for the working class in industrialised societies. Nature was considered an infinite resource that needed to be dominated abd used for cresting welfare. A profound preoccupation with the environment, however, developed during the 1960s and 1970s. Since then, a deep concern about our environment has been broadly

---

[2] For a discussion of orientalism, see Said (1979).

articulated. Formerly considered an unrestricted resource, nature has now been recognised as fragile. Human enterprises might have a devastating impact on nature. Pollution is recognised to have a tremendous impact on our living conditions. The oceans, once considered bottomless and used as a global waste bin, now need to be cleaned up. Preoccupation with our environment has become part of critical mathematics education.

Critical mathematics education is concerned about *mathematics* itself, not only as a school subject but as applied is a variety of contexts. Mathematics has been celebrated as a unique phenomenon, providing unquestionable truths. However, it is a concern of critical mathematics education to leave aside any such glorification of mathematics and to formulate a critical conception of it instead. It is a concern to present a critical awareness of the impact of bringing mathematics into action. Mathematics is a powerful tool for creating new technological possibilities. However, such possibilities might come to serve any kind of political and economic interests, also of the most questionable kind.

1.2 Hopes

To be concerned about something means hoping for something to be different. The notions of concern and hope are intrinsically connected, but they are also different. While the notion of concern is part of everyday discourse, hope has been addressed philosophically, and is recently turned into a political conception as well.

When we look through the history of philosophy, we find some notions appearing again and again, like "knowledge", "truth", and "justice". These belong to the crown jewels of philosophy. The notion of hope has also been addressed in philosophy, but it has not played any significant role. In the dialogue *Timaeus*, written 360 years before Christ, Plato points out that "hope easily led astray", as it mixes with irrationality.[3] Throughout his whole philosophy, Plato celebrates knowledge, ranking it much higher than belief, while hope, like belief, is seen as an expression of lack of knowledge. With the emergence of philosophies dedicated to Christianity, the discussion of hope took a new direction. Hope became connected to religious faith; it not only concerned a better life on earth, but first of all salvation and eternal life after death.

René Descartes and several other philosophers, both from the rationalist and the empiricist traditions, make observations about hope. Hope was associated with the notion of belief, and therefore considered less important than knowledge. In *Meditations on First Philosophy*, first published in Latin in 1641, Descartes (1993) concentrates on the principal theme of his philosophy: to establish the foundation of certain knowledge. In this work, one does not find references to the notion of hope, nor to the related notion of faith. However, in *The Passions of the Soul*, completed

---

[3] See Plato, *Timaeus*. The Internet Classics Archive, p. 26.

1.2 Hopes

in 1649, Descartes addresses the notion of hope, although with some reservations. Compared to knowledge, the concept of hope remains in an inferior position.[4]

Through the work of Immanuel Kant, the notion of hope obtained a new position in philosophy. In *Critique of Pure Reason*, first published in German in 1781, Kant (1929) formulates the question "What may I hope?" in parallel with the two questions "What can I know?" and "What ought I do?" (p. 635). This means that a philosophy of hope runs parallel to epistemology and ethics, and so one may consider Kant the founder of a philosophy of hope.

In several places in his authorship, Søren Kierkegaard reflects on the notion of hope, although nowhere does he provide a systematic analysis of the concept. Contrary to Kant, Kierkegaard cuts all possible connections between hope and knowledge. To him, hope does not contain epistemic residues, but first of all religious worries. Hope concerns faith and is a strictly personal affair. According to Kierkegaard, hope of salvation is an entirely irrational affair. This cannot be different; this is a condition of life. By making such a claim, Kierkegaard introduces a new direction for existentialist philosophy.[5]

In radical opposition to Kierkegaard's position, Friedrich Nietzsche claims that religious belief does not make any sense. He sees Christianity as an expression of a deplorable slave morality. Hope associated to faith is not only pointless, it is the cause of human suffering. In *Beyond Good and Evil*, first published in German in 1886, Nietzsche (2009) claims all forms of hope in a "hidden harmony, in future blessedness and justice" to be pointless (§55, p. 60), and in *Human, All Too Human*, Nietzsche (1986) refers to hope as "the most evil of evils because it prolongs man's torment" (§71, p. 58).

In Ernst Bloch's (1995) monumental work *The Principle of Hope*, the philosophy of hope gets its own home ground. Bloch was German and of Jewish origins, and when the Nazis came to power, he and his wife fled from Germany, and via Switzerland, Austria, France, and Czechoslovakia, they finally got to the USA. Here, during the period 1938–1947, he drafted *Das Prinzip Hoffnung*, published in three volumes in 1954, 1955, and 1959, later translated as *The Principle of Hope*.

In the introduction, Bloch states that in the book he has made an extensive attempt "to bring philosophy to hope" (Volume I, p. 6). Contrary to Descartes, Bloch gives the notion of hope his full attention. He shares with Kant the conviction that hope is a crucial notion that needs to be investigated philosophically; however, he disagrees with most of what – until then – philosophy had contributed to its clarification. In particular, Bloch does not make any association between hope and religious ideas. Hope has nothing to do with faith or salvation, but is a completely mundane affair. Bloch's interpretation contrasts directly with the one provided by

---

[4] *The Passions of the Soul* has always been considered a peripheral work in Descartes' philosophy. It was dedicated to the Swedish queen Christina, who in 1649 had brought Descartes to Sweden in order to teach her philosophy. See *The Passions of the Soul*, https://TheVirtualLibrary.org

[5] Careful expositions of Kierkegaard's conception of hope are found in Bernier (2015), Fremstedal (2012), and Sweeney (2016).

Kierkegaard. Bloch might, in fact, share Nietzsche's conviction that hope, in the form of religious faith in salvation after death, is most dangerous.

In the introduction to *The Principle of Hope*, Bloch states that hope concerns the "dreams of a better life" (Volume I, p. 11), and Bloch has a better life *on earth* in mind. Bloch considers himself a Marxist, and throughout *The Principle of Hope* he makes many references to Marx. However, he is far from being an orthodox Marxist, and his work has only been half-heartedly embraced by other Marxists.

To Bloch, hope is formed through political visions and aspirations. Hope concerns what does not exist – at least, not yet. It concerns socio-political utopias. Bloch highlights that hope points towards possible states of affairs that might not be properly grasped in advance; it refers to vaguely anticipated alternatives. Bloch does not claim that, by departing from a Marxist outlook, one becomes able to specify the content of hope. Such content cannot be deduced from any theoretical or political positions. Still, Bloch does not think of hope as a free-floating fantasy. He locates hope in solid distance from fantasies, as well as from economic determinism that, according to orthodox Marxism, leads to a classless society. Instead, Bloch talks about true hope as being "mediated in terms of history and tendency" (Volume 3, p. 1372).

Bloch expresses his ideas in complex formulations. He talks about the act-content of hope, which I find an important concept for articulating the political dimension of hope. In Bloch's words:

> ... the *act-content* of hope is, as a consciously illuminated, knowingly elucidated content, the *positive utopian function*; the *historical content* of hope, first represented in ideas, encyclopaedically explored in real judgements, is *human culture referred to its concrete-utopian horizon*. (Volume I, p. 146. Italics in original).

The act-content of hope is a utopian matter – not a utopia formed by religious fantasies, but by visions formed through "real judgements" elaborated within a "concrete-utopian horizon". In other places, Bloch refers to this as "true hope", which is a political hope concerning a better life on earth, contrasting a "false hope" about an eternal life after death.

One can think of hoping as an individual undertaking. A person can have many hopes about what they may achieve in the future. Hoping is a common human state of mind. Hopes can also be shared. The whole family can hope for the best for the children and the best for the neighbours. However, hope includes much more than individual or local undertakings. Hope also concerns future socio-political states of affairs. One can hope for a better society. Hope represents a shared socio-political utopia.

In *Pedagogy of Hope*, Freire (2014) pays particular attention to the notion of hope. He does not make any reference to Bloch's work, but he seems to be in harmony with Bloch's position: hope forms an integral part of a struggle for a better society.[6] In Freire's words: "I do not understand human existence, and the struggle needed to improve it, apart from hope and dream. Hope is an ontological need"

---

[6] For comments about Freire's and Bloch's conceptions of hope, see Streck et al. (2018).

1.2 Hopes

(p. 8). Freire makes a similar point to Bloch, namely that one cannot rely on any economic determinism. The fulfilment of socio-political hopes cannot be taken as a result of certain economic dynamics. Freire states that "in the struggle to improve the world, as if that struggle could be reduced to calculated acts alone, or a purely scientific approach, is a frivolous illusion" (p. 9). And he adds: "Without a minimum of hope, we cannot so much as start the struggle. But without the struggle, hope, as an ontological need, dissipates, loses its bearings, and turns into hopelessness" (p. 9). Any struggle for a better society needs hope.[7]

For a period, the Brazilian philosopher Mário Sérgio Cortella worked closely with Freire, and Cortella (2015) has condensed Freire's conception of hope in the following way: "It is necessary to have hope, but to have hope from the verb to hope, because there are people who have hope from the verb to wait. And hope of the verb to wait is not hope, it is waiting" (p. 22, my translation).[8] Furthermore, Cortella highlights: "What I learned most from Paulo Freire was the idea of active hope, which is not pure waiting, but hope that searches, constructs, looks for, and knows that teaching, above all, is not just a job, it is a source of life. Teaching is a source of life, in which hope is our refusal of biocide, our refusal of the failure of life and, therefore, our way of existing and hope" (Cortella, 2017, p. 14, my translation).[9] It is important not to let hoping slide into waiting, but to turn hoping into an active search for what to do. Active hoping is important for Freire, for Bloch, and for critical mathematics education.

Daniela Alves Soares (2022) investigates dreams and hopes as formulated by young people living in disadvantaged situations. She reveals the deep complexities of such hopes, which combine personal and social elements. Students are formulating hopes with respect to their own lives, with respect to their families, with respect to their neighbourhoods, with respect to society in general. With her study, Soares illustrates what it could mean to establish hopes within a "concrete-utopian horizon".

The conception of hope as a socio-political force – and as an actual search for what to do – is crucial for critical mathematics education. Hopes concerns what could come to take place in the classroom, but also the direction social changes might take.

---

[7] That hope is part of political activism has been underlined by Solnit (2016). She points out that even when the future looks dark, hope is possible. The title of her book is precisely that: *Hope in the Dark*.

[8] The original Portuguse version is: "É preciso ter esperança, mas ter esperança do verbo esperançar, porque tem gente que tem esperança do verbo esperar. E esperança do verbo esperar não é esperança, é espera."

[9] The original Portuguse version is: "O que mais aprendi com Paulo Freire foi a ideia de esperança ativa, que não é de pura espera, mas é a esperança que procura, constrói, busca e sabe que a atividade docente, acima de tudo, não é somente um emprego, é fonte de vida. A docência é uma fonte de vida, em que a esperança é a nossa recusa ao biocídio, a nossa recusa à falência da Vida e, portanto, o nosso modo de existir e esperançar."

1.3 Social Justice as Construction

In the literature on mathematics education for social justice, we find fascinating examples of classroom activities; however, it is more difficult to find thorough interpretations of the very notion of social justice. To work for social justice seems to be an intrinsic good thing to do. Who would ever start propagating an education for social injustices? Critical mathematics education is an expression of a concern for social justice, and it is relevant to consider more carefully the notion of *social justice*.

Social justice is a contested concept, meaning that it can be interpreted in different, and also contradictory, ways. Only if one assumes that a concept has a kind of genuine meaning that can be discovered, maybe by some hermeneutic approaches, then one might feel comfortable using the notion to designate some desirable states of affairs. But if we assume that the meaning of a concept becomes defined by how it is bought into operation in diverse discourses, the situation is rather different. The notion of social justice will be without a solid semantic kernel.

While the notion of "justice" has a long history in philosophy, the expression "social justice" first appears in the nineteenth century. It was mixed with religious interpretations, indicating that the meaning of social justice could be clarified through careful studies of the *Bible*. The notion of social justice appears in the title of the book *The Constitution Under Social Justice* (*Costituzione Secondo la Giustizia Sociale*) published in Italian in 1848 by Rosmini-Serbati (2007). He relates the notion to a range of political ideas circulating at the time.

In *A Theory of Justice*, first published in 1971, John Rawls (1999) performs a profound philosophical analysis of the conceptions of justice, drawing on inspirations from the liberal tradition, including works by John Stuart Mill, but at the same time he adds his own philosophic ideas to the discussion. At times, Rawls has addressed the question of how to identify justice in society in terms of a thought experiment. To him, such an identification can be reached through a process of dialogue among a group of people if two conditions are fulfilled: First, that everybody from the group is going to live in the society. Second, that nobody from the groups would know in advance what position they are going to obtain: being rich or poor, ill or health, man of woman, black or white, etc. In other words, Rawls presents idealised (as well as impossible) conditions for establishing a constructive process of the concept of justice.

Let us pay attention to the idea that, in order to define social justice, one is *not* going to be engaged in any analytical process of trying to discover the meaning of "justice", but in a process of social construction.[10] I move on from Rawls' thought experiment and consider the formation of a conception of social justice as taking place – not in any imagined congregation – but in real-life contexts. One such context is the classroom. I think of education for social justice in terms of processes that

---

[10] At times I have suggested ethical constructivism: seeing ethical conceptions, including conceptions of social justice, as an expression of social constructions. See Skovsmose (2012, 2018).

1.3 Social Justice as Construction

engage students in the very formulation of what social justice could mean, and not as an education informing students about justices and injustices. A principal step is to engage students in identifying and articulating what they find to be just or unjust, and also to be ready to challenge their opinions.

Looking along the production line for a given product like a T-shirt, it is possible to locate many cases of injustices. In "Sweatshop Accounting", Larry Steele (2013) presents a project involving high school students "with the hope to be future business leaders" (p. 78).[11] The project starts with the students taking a look at the labels in their own clothes: "I was made in Vietnam" (p. 78), one of the students observed. The students marked the location with a yellow post-it note at a world map, and soon the yellow notes are shown in China, the Philippines, Malaysia, Indonesia, and many other countries.

More information was provided for the students in the form of videos, references, and papers. They received more specific information with respect to the production of certain fashion sports shoes, which in the shops at that time did cost 50 USD. At the other end of this production line, one finds the factory workers whose salaries for producing the pair of shoes were 0.40 USD. Between these two extreme points on the production line, we find, for instance, the production cost to be 1.60 USD; the profit of the factory to be 2 USD; and the cost of advertising and publicity to be 13.50 USD. The image one gets is that the total price of the pair of shoes is added up incrementally. Further information is provided with respect to the living conditions of a worker at a garment factory in Nicaragua. The salary at the time is 0.31 USD per hour, which gives them 14.88 USD per week. Costs for transport to the factory adds up to 15.04 USD per week. Costs for necessities – like electricity, food, and wood for an outdoor stove – bring the total costs up to 29.34 USD per week. Many costs, such as buying clothes, have not been incorporated in the calculations. How could a worker and their family possible survive?

By engaging students in a project addressing what takes place along a production line in our globalised world economy, one might identify different forms and degrees of exploitations. The students acquire a space for expressing what to consider reasonable working conditions, reasonable salaries, and reasonable shop prices. It is possible for the students to argue why one finds something to be unjust; one might also come up with counterarguments. Both arguments and counterarguments might find support in graphs, statistics, calculations, and numerical estimations. The construction of conceptions of social justice is far from being only an epistemic endeavour; it also encompasses experiences and emotions. The teachers as well as the students can express their opinions, which is an important component of the constructive process.

In "Sweatshop Accounting", the students had not experienced the deplorable working conditions that they became aware of. However, for formulating conceptions of social justice, it is important that they become engaged, not only in

---

[11] In Chap. 2, "Landscapes and Racism", I address the notion of landscape of investigation, and "Sweatshop Accounting" is an example of such a landscape.

situations they might be familiar with, but also in situations they come to learn about. It is important to act against any normalisation of injustices. Such normalisation might bring about resignations or indifferences: resignations among those who suffer injustices and indifferences among those who are spectators to injustices. In "Sweatshop Accounting", a possible indifference was challenged: the indifference that we all might subscribe to when buying a pair of sport shoes focusses on the price, and ignoring the exploitations and sufferings taking place along its production lines.

I have been talking about conceptions of social justice in plural, as one cannot assume that social constructive processes will converge towards a uniform clarification of social justice. We are dealing with contested constructions, driven forwards by articulations of concerns and of hopes. Such processes can take place in the mathematics classroom.

References

Bernier, M. (2015). *The task of hope in Kierkegaard*. Oxford University Press.
Bloch, E. (1995). *The principle of hope, I–III*. MIT Press.
Cortella, M. S. (2015). *Educação, convivência e ética: Audácia e esperança! (Education, coexistence and ethics: Audacity and hope!)*. Cortez Editora.
Cortella, M. S. (2017). *A escola e o conhecimento: Fundamentos epistemológicos e políticos (School and knowledge: Epistemological and political foundations)*. Cortez Editora.
Descartes, R. (1993). *Meditations on first philosophy*. Routledge.
Engels, F. (1993). *The condition of the working class in England* (Edited with an introduction and notes by Davis Mclellan). Oxford University Press.
Fanon, F. (2004). *The wretched on the earth*. Grove Press.
Fanon, F. (2008). *Black skin, white masks*. Grove Press.
Freire, P. (2014). *Pedagogy of hope: Reliving pedagogy of the oppressed* (With notes by Ana Maria Araújo Freire; translated by Robert R. Barr). Bloomsbury.
Fremstedal, R. (2012). Kierkegaard on the metaphysics of hope. *The Heythrop Journal, 53*(1), 51–60.
Kant, I. (1929). *Critique of pure reason* (Translated by Norman Kemp Smith). Macmillan.
Nietzsche, F. (1986). *Human, all too human: A book for free spirits* (Translated by Marion Faber with Stephen Lehmann). University of Nebraska Press.
Nietzsche, F. (2009). *Beyond good and evil: Prelude to a philosophy of the future* (Translated by Ian Johnson). Richer Resources Publications.
Orwell, G. (1937). *The road to Wigan Pier*. Victor Gollancz.
Rawls, J. (1999). *A theory of justice*. Oxford University Press.
Rosmini-Serbati, A. (2007). *The constitution under social justice* (Translated by Albetto Mingardi). Lexington Books.
Said, E. (1979). *Orientalism*. Vintage books.
Skovsmose, O. (2012). Justice, foregrounds and possibilities. In T. Cotton (Ed.), *Towards an education for social justice: Ethics applied to education* (pp. 41–62). Peter Lang.
Skovsmose, O. (2018). Critical constructivism: Interpreting mathematics education for social justice. *For the Learning of Mathematics, 38*(1), 38–44.
Soares, D. A. (2022). *Sonhos de adolescantes a suas reflexcões sobra as aulas de mathemática. (Dreams of adolescents and their reflections on mathematics classes)*. Doctoral dissertation. Universidade Estadual Paulista (Unesp).

References

Solnit, R. (2016). *Hope in the dark*. Haymarket Books.
Steele, L. (2013). Sweatshop accounting. In E. Gutstein & B. Peterson (Eds.), *Rethinking mathematics: Teaching social justice by the numbers* (2nd ed., pp. 78–94). Rethinking Schools.
Streck, D. R., Redin, E., & Zitkoski, J. J. (2018). *Dicionário Paulo Freire* (4th ed.). Autêntica.
Sweeney, T. (2016). Hope against hope: Søren Kierkegaard on the breath of eternal possibility. *Philosophy and Theology, 28*(1), 165–184.
Zola, É. (1954). *Germinal*. Penguin Books.

Chapter 2
Landscapes and Racism

Abstract It is a concern of critical mathematics education to construct landscapes of investigation. Such landscapes create possibilities for engaging students in investigative processes, for establishing dialogic interactions in the classroom, and for reflecting critically on socio-political issues. It is a concern of critical mathematics education to address racism in all its forms and appearances and to contribute to the general struggle against racism. Different forms of racism will be discussed, including the phenomenon of structural racism. A landscape of investigation will be presented, entitled *Media and Racism*, which addresses everyday features of racism. A Bias Index will be defined, which provides a powerful way of capturing expressions of structural racism. By means of this index, one can also address ways in which sexism might be acted out. We can consider biases with respect to any kind of gender issues. Any such applications of the Bias Index might evolve into rich landscapes of investigation.

Keywords Landscapes of investigation · Prescription readiness · Racism · Structural racism · Degree of visibility · Bias-Index

Mathematics education takes place all over the world: in schools, faculties, universities. It is a well-established social institution with an apparently well-defined function: to make students learn mathematics. Considering all the economic resources put into this activity, it seems impossible to question the social relevance of mathematics education. However, the socio-political role of mathematics education is far from transparent.

On several occasions, Alexandre Pais has highlighted that we cannot forget that mathematics education takes place in a society where capitalist structures and accompanying ideologies permeate whatever is taking place. As an illustration, he considers the broadly celebrated slogan "mathematics for all". According to Pais (2012), this slogan covers up an obscenity of a school system that "year after year throws thousands of people into the garbage bin of society under the official discourse of an inclusionary and democratic school" (p. 58). The crude reality is that "in order for some to succeed others have to fail" (p. 58). The task of identifying those to be thrown into the garbage bin – and the task of performing this very act – becomes concealed beneath respectable looking aims and metaphors like "mathematics for all". This obscenity forms an integral part of the socio-political role of mathematics education.

In *Travelling Through Education* (Skovsmose, 2005), I highlighted that, if one reads aloud all the exercises with which students are presented during their schooling, it would sound like a long sequence of commands. To follow orders is an important functional "quality" of attentive students as well as of an obedient workforce. I have referred to this obedience as *prescription readiness* (Skovsmose, 2008).

For critical mathematics education, it is crucial to challenge the role that mathematics education might play in preparing students to become part of a submissive and prescription-ready workforce, assuming a given political and economic order of things. One step towards doing so is to invite students to engage in investigative processes. A second step is to address controversial socio-political problems as an integral part of learning mathematics. I see the creation of landscapes of investigation as a practical attempt to create learning environments that make such activities possible.

In what follows, I first present more carefully the concept of landscapes of investigation. I then concentrate on the problem of racism. It is a concern of critical mathematics education to confront racism in all its forms and to contribute to the general struggle against racism. I elaborate in more detail on a landscape of investigation called *Media and Racism*, which addresses some everyday features of racism. I conclude the chapter by presenting a Bias Index that might be a useful tool for pointing out cases of structural racism.

2.1 Landscapes of Investigation

By landscapes of investigation, I refer to learning environments that open the learners up to a broader spectrum of problems. This could include problems related to mathematics and its applications. It could include problems concerning social issues, maybe as experienced by the students, maybe as addressed in public discourses, maybe as presented by the teachers. It could include problems concerning economic exploitation, social marginalisation, sexism, pollution, climate changes, any cases of social injustice. The students can explore landscapes in groups and through project work.

Landscapes of investigation also establish patterns of classroom communication different from those patterns characteristic of the school mathematics tradition. Dialogues can be established among students and between students and teachers. This generates resources for critical reflection, both with respect to mathematical issues and with respect to socio-political issues.[1]

---

[1] The concept of landscape of investigation has been developed over a longer period. For a broad presentation of landscapes of investigation, see Penteado and Skovsmose (2022).

2.1 Landscapes of Investigation

An investigative process is open-ended. It can proceed in one direction, but it can easily take a different turn depending on what is observed – or maybe not observed. A process of investigation is sensitive not only to observations but also to arguments, counter-arguments, and new information that might become available. It is sensitive to changes of interests and personal priorities. Students cannot be forced to carry out investigations, but they can be invited to do so.

Several positions in mathematics education have highlighted the importance of students doing investigative work. The problem-solving approach found inspiration in the work of the George Pólya (1957). According to this approach, the learning begins in problems presented to the students or formulated by the students, and not ready-made exercises. Problem solving can turn into fascinating investigative processes. The approach of problem solving, however, cannot be confused with the approach of problem-based learning, which advocates that investigative processes focus on real-life problems. Although different, problem solving and problem-based learning represent two important approaches for establishing learning through investigative processes.

One can draw a distinction between processes of learning and processes of teaching, and in exercise-based mathematics education the two processes appear quite distinct. However, in investigative processes the distinction blurs. When learning becomes structured as an investigative process, so becomes teaching. It becomes important to communicate, to interact, to make proposals, to share ideas, and to challenge. The teacher does not need to "teach", but can serve as a supervisor supporting the process of investigation.

The processes of dialogue achieve a significant role. Dialogue is an open process, and one cannot expect a dialogue to follow any specific pattern. In a dialogue, one reacts to what has been said, making the course of a dialogue unpredictable. The horizon of a dialogue is always changing. In a dialogue, the content of what is said – and not the position of who is saying it – plays the principal role. This means that the role of student and the role of teacher become less significant. In a dialogue, the experience of equity is important.[2] I find investigative processes to be dialogic processes.[3]

A critique can address a diversity of statements, claims, and positions. It can reveal presumptions, preconceptions, and ideologies. A critique can address social states of affairs. I do not associate making a critique with particular methodological procedures, whatever form of critique we have in mind. We are dealing with non-conclusive processes. I see a close relationship between the open-ended nature of a dialogue and the non-conclusive feature of critique.

I see dialogue as a resource for critical activities. Working with landscapes of investigation makes it possible for students and teachers to engage in dialogic processes and to create critical investigations. In the following, I will present a

---

[2] For further characterisations of dialogic processes, see Alrø and Skovsmose (2004); Faustino and Skovsmose (2020); and Milani and Skovsmose (2014).

[3] This point has, for instance, been highlighted by Bohm (1996).

landscape of investigation addressing forms of racism. However, before we get to that I want to indicate the deep complexity of racism, being a principal concern of critical mathematics education.

2.2 Racism

When I was young, I wanted to become a teacher, and I entered the teacher training college in Hjørring, in the north of Jylland where I was born – at that time a city of around 20,000 inhabitants. At the college we were taught all school subjects. Our textbook in geography, used in most teacher training colleges in Denmark, included a chapter introducing physical anthropology, and it showed how the skull index was calculated. The chapter also included a photo of a not-too-smart-looking elderly man, beneath which it was stated that it showed a person belonging to "a primitive Mediterranean race".

According to Nazi-based racism, it makes sense to talk about different human races: the Jewish race was inferior and represented an extreme danger to the Aryan race. If German blood waw mixed with Jewish blood the consequence would be a devastating degeneration of the Aryan race. The Nuremberg Laws, implemented in 1935, stripped the Jewish population in Germany of all human rights. Marriages and sexual relationships between Jewish and German people were forbidden, and the Jewish partner would be punished severely. To the Nazis, the "Jewish problem" needed to be addressed urgently, and as a result the "final solution" was conceptualised and carried out in all its horror. The Second World War has finished, the *Drittes Reich* has been destroyed, and the Nazi racism has been confronted. Yet shockingly, still, in a Danish textbook from the early 1960s, one can find an explicit racist formulation, echoing a celebration of what has been referred to as the Aryan Race.[4]

During the romantic period in Germany, the notions of *Volk* and of *Volkgeist* were accorded profound interpretations. *Volk* means "people", and the romantic idea was that *Volk* refers to much more than a group of people. A *Volk* was united by language, history, folklores, experiences, and nature; it was united by a *Volkgeist*, which is the German word for "spirit of the people". The German *Volkgeist* was considered unique; different from, say, the spirit of the French people. By celebrating the concepts of *Volk* and *Volkgeist*, the romantic movement created a fertile ground for nationalism and chauvinism. Culture obtained an essentialist interpretation, which, for instance, made it impossible for a Jewish person to become German even after converting to, say, Protestantism. The Nazi racism sprouted from such forms of "culturalism".

Slavery has existed throughout all historic periods. There were slaves in Ancient Greece, also at the time when democracy was first conceptualised. The norm was

---

[4] Besides Nazi Germany, the apartheid South Africa and the Jim Crow states in the USA had state-implemented form of racism. See, for instance, Reed (2022).

2.2 Racism

that those on the losing side of a war could be turned into slaves. This form of slavery also existed in Africa before the Portuguese arrived. The Portuguese did not need to enslave black people as they could buy the slaves they wanted. The slave trade turned into a lucrative business, and soon there emerged the need for "legitimising" this trade. Black people came to be characterised as being inferior, naturally born slaves. During the nineteenth century this racist outlook was elaborated in detail, also through studies at university level. Phrenology became an important component of physical anthropology, linking stipulated racial inferiority to measurable physical appearances.[5] Around 1900, racism was a broadly accepted position. Racist views were explicitly presented in the English parliament, as well as at high tables at university colleges. It was simply taken for granted that there exist different human races, and that some were inferior when compared to others. The British Empire was at its zenith, and racist discourses served as justifications of colonialism and imperialism. Inferior races were not able to govern themselves; they were much better when governed by the British. Colonialism and imperialism became stipulated as acts of charity, the idea of which was summarised by Rudyard Kipling in the poem "The White Man's Burden", published in 1899.[6]

There exist many different forms of racism, and Edward Said (1979) presents in terms of orientalism. With this expression, he refers to the conception of people from the East by people from the West. Orientalism developed in parallel with the British colonisation of the East. Orientalism might be different from black-white racism, but it formed an intrinsic part of the "justification" of British imperialism. Times have changed, but orientalism has not faded away. Today, refugees from the Middle East arrive in Europe and are received with reservations, if not with open hostility. We still experience versions of orientalism.

In the Ottoman Empire in 1915, the genocide of the Armenian people started to take place. The Armenians represented a Christian enclave in the otherwise Muslim empire. Most of the Armenian population was forced to leave their homes and to walk to a concentration camp located in the city of Deir ez-Zor in the middle of the Syrian dessert. Only few arrived; most died along the way. It is estimated that around 1,500,000 Armenians died during the genocide. The USA ambassador at that time in Constantinople, Henry Morgenthau, had many conversations with the leaders of the Ottoman Empire. They directly expressed the necessity of eliminating the Armenian "race" from the empire. Morgenthau (1918) was deeply shocked by what he witnessed, and back in the USA he published *A Personal Account of the*

---

[5] In Great Britain, phrenology was studied intensively, and one centre for this research was Edinburgh, where the Edinburgh Phrenological Society was established in 1820, soon to be followed by other such centres around the country.

[6] In 1910 in English parliament, the previous Prime Minister Arthur James Balfour declared that somebody has to "govern the Orientals", and he added: "I think that experience shows that they have got under it far better government than in the whole history of the world they ever had before, and which not only is a benefit to them, but undoubtedly a benefit to the whole of the civilized West" (quoted after Said, 1979, pp. 32–33).

Armenian Genocide in 1918. One can think of the Armenian genocide as being instigated by religious differences; however, racism played its devastating role.

The world has experienced many forms of race-based genocides. The North American genocide of indigenous people is another horrifying example. The number of people killed might be between 9,000,000 and 18,000,000.[7] The Indian caste system is rooted in religious ideas, claiming that there exist basic differences between different groups of people, and that some groups are inferior compared to other groups. The Dalits are seen as being essentially different from the Brahmas.[8] The number of people in India oppressed by the caste system is comparable to the whole population of the USA. Chinese racism in the USA has many ramifications.[9] In China, most people are classified as Han Chinese; however, the Chinese population includes several minority groupings, who over time have suffered different forms of exclusion and oppression. In Africa, one can find racism acted out in the antagonism between different groups of people. Over time, racism has changed and found new forms of expression; racism continues to be a horrific social phenomenon.

In order to condense the different manifestations of racism, I highlight two of its basic features. *First*, racism stipulates that the notion of race refers to some intrinsic characteristics of a group of people. These could include biological differences, which previously were often articulated in terms of differences in "blood". The characteristics could also include differences expressed through versions of culturalism, claiming that some cultural belongings are so deep-rooted that no transition to other cultural settings is conceivable. Whatever terminology is used, it forms part of a racist position that changing from one race to another is impossible. *Second*, racism stipulates that some races are inferior when compared to certain other races. According to Nazi racism, the Jewish people were inferior to the German people, as were Gypsy people. According to the colonial outlook, black people were inferior to white people. According to orientalism, people from the East were inferior to people from the West. According to my Danish geography textbook, people from the South of Europe might belong to a primitive race.

Critical race theory provides an important contribution to the discussion of racism.[10] Its fundamental premise is that there is nothing biological or essential captured by the notion of race. Critical race theory claims that race is a social construct. Social groups referred to as inferior racial groups are constructed as such through processes of exclusion, domination, violence, and labelling. Beyond social

---

[7] The North American genocide has been described in Brown (1970); Cozzens (2016); and Whitt and Clarke (2019). A detailed investigation of the genocide as it took place in California from 1846 to 1873 is presented in Madley (2017).

[8] Wilkerson (2020) relates the conception of caste systems with experiences of racism in the USA.

[9] See, for instance, Chang (2013).

[10] Derrick Bell has been an important inspiration for the formulation of critical race theory, and Delgado and Stefancic (2005), have edited an important *Derrick Bell Reader*; see also Bell (1987, 2002, 2004). Systematic presentations of critical race theory are found in Delgado and Stefancic (2013); Crenshaw et al. (1995); and Bridges (2019).

constructions, there is nothing in the notion of race. By making this claim, critical race theory confronts the basic assumptions of any version of racism.[11]

"Structural racism" is an important notion for addressing the way racism might be acted out.[12] Several times I have been hospitalised in São Paulo, and never have I witnessed any form of racism among the staff: no pointed remarks about a colleague, no forms of microaggression. However, I have never seen a black doctor, and I doubt if I have seen a doctor of colour. I have seen nurses with different skin colours, although the majority were white. A big group of people takes care of the cleaning of the hospital, and they are predominately black people of colour. Racism can be engraved in socio-economic structures; in the unequal distribution of access to further education, and in the unequal distribution of job opportunities. In Brazil (and in many other countries) structural racism has an impact on the formations of our daily life. As pointed out by Manuella Carrijo (2023), one might experience racism without any racists being present.

The principal development of critical race theory has taken place in the USA with explicit reference to black-white racism. This is definitely an important expression of racism, but observations related to a USA context need not be directly applicable to other contexts. In Europe, refugees and immigrants from the Middle East are often met with hostility, which can be a way of expressing xenophobia rooted in orientalism. Over centuries, black people, indigenous people, and white people have been mixing in Brazil, and during the last century many people from Japan have joined the mixing. Racism in Brazil does exist, but it can hardly be reduced to a black-white binary. Critical race theory is an important resource for addressing racism also outside the USA, but like any other critical theories it is only preliminary and tentative; it needs to address its own possible limitations.

When used in critical race theory, the notion of "race" is stripped of any biological components; there is nothing in the notion except social constructions. However, to me the use of "race" remains problematic. It appears in many different discourses, some of which are far removed from critical race theory. In many such cases, the notion of "race" carries along with it some shades of essentialism, some hints of the existence of biological differences. I do not feel it helpful to talk about the Dalits as being a race, of the Brahmas as being a race, of Jewish people as being a race, of Armenian people as being a race, of Mediterranean races, or of coloured people from Brazil as a race, even though we might be dealing with socially constructed groupings that, inspired by critical race theory, could be nominated as being "races".

---

[11] Critical race theory has inspired the formulation of a range of notions, like nicro aggression, important for identifying and addressing cases of racism. For discussions on microaggression, see Davis (2013). See also Silva and Powell (2016); and Solórzano and Huber (2020).

[12] Structural racism is systematically addressed by Almeida (2019). I did not find "structural racism" mentioned in the index of *Critical Race Theory* (Delgado & Stefancic, 2013) nor in *The Derrick Bell Reader* (Delgado & Stefancic, 2005). *Critical Race Theory* (Crenshaw et al., 1995) does not have an index, but nor did I find in this book a reference to structural racism. However, in *Critical Race Theory: A Primer*, Bridges (2019) elaborates carefully on the notion, which she relates to the notion of institutional racism (addressed by those such as Phillips, 2011). She also includes a reference to Powell (2008), who explicitly refers to structural racism.

I see many differences between human beings. This could be due to appearances, religious convictions, political opinions, geographic locations, cultural traditions, or health conditions. However, I do not see differences in terms of "races".

2.3 Media and Racism

The *Handbook of Critical Race Theory and Education* (Lynn & Dixon, 2022) presents a range of examples and reflections on how to address racism in educational contexts. However, the handbook does not refer to any examples from mathematics education. The word "mathematics" is not mentioned in the index. However, in mathematics education one finds many examples of racism being critically addressed. From more recent publications, let me just refer to *Exploring New Ways to Connect: Proceedings of the Eleventh International Mathematics Education and Society Conference* (Kollosche, 2021), where many contributions address problems of racism.

I find that mathematics education has a crucial role to play in the struggle against racism, as it provides particular resources for identifying patterns of structural racism. I will illustrate this by means of a landscape of investigation called *Media and Racism,* presented by Reginaldo Britto (2013, 2022). He formed research groups among his students, who mostly came from poor neighbourhoods in a city within Brazil. In Brazil almost 55% of the population is black, 44% is white, while around 1% belongs to other ethnicities.[13] In the neighbourhood where Britto was working, the majority of the population was black.

The overall idea of *Media and Racism* was to investigate the visibility of black people in magazines circulating in Brazil. Casually leafing through such magazines, one might get the impression that the way different groups of people are portrayed as biased. However, such casual impressions can be investigated more systematically. The projects *Media and Racism* has been completed every year since 2006. The students were divided into groups, and they were given the task to collect photographs of people shown in magazines during a one-week period. This task was carried out by the students, partly at home and partly in the classroom, where collections of magazines were made available.

At first the students observed whether the person was black or white (photos of other ethnicities were not considered). This was an initial step in turning qualitative observations into quantitative measures. When attempting any sort of quantification of visual input, there is a need to simplify what one sees by recognising some

---

[13] In the officially applied Brazilian categories, blacks are composed of *Pardos* (coloured) and *Negros* (blacks). Available at https://agenciadenoticias.ibge.gov.br/agencia-noticias/2012-agencia-de-noticias/noticias/18282-populacao-chega-a-205-5-milhoes-com-menos-brancos-e-mais-pardos-e-pretos

2.3 Media and Racism

features and ignoring others, and the students came to experience this directly. The classification of people as black or white is not as straightforward as it might seem at first. Looking at a particular photograph, one might doubt whether the person is a sunburned white person, or a person of colour that could be classified as black.

As the next step in the investigation, the students were asked to classify the environment in which the person in the photograph was located. The classification could be either "positive" or "negative". A positive environment could mean that the person was located in a prosperous and attractive environment, while a negative environment could show poverty or violence. Clearly the distinction between "positive" or "negative" environments is blurred and subjective, and applying the distinction will come to include some arbitrary stipulations. Sometimes, the negative features in the photo were only hinted at. Both the process of collecting photographs and of classifying them opened up rich possibilities for dialogue, both among students themselves, and between students and the teacher.

The first years in which the landscape *Media and Racism* was explored, the focus was on children appearing in the photos, and for now I will concentrate on this period of the project. (Later on, the focus on children was changed, and the students collected photographs of people of any age.) Of the 41 photos that were collected by one of the groups in the first year of the project, 36 showed white children, while 5 showed black children. Of the white children, 35 appeared in situations that were classified as positive, while only 1 appeared in a negative situation. Of the 5 photos of black children, 3 appeared in positive situations, while 2 appeared negative. In *Media and Racism* this observation was expressed in terms of the construct Degree of Visibility (*DV*). By introducing this term, the project made space for further quantitative investigations of visibility.

The degree of visibility DV_i, where i refers to a particular ethnic group, is a number calculated in the following way:

$$DV_i = \frac{\text{Number of photos with a person from the ethnic group } i}{\text{Number of photos with a person from any ethnic group}}$$

In the present case, the index i could refer to black children or to white children, as pictures of children of other ethnicities were not considered. The total number of appearances of children as collected of all the groups of students for the year 2009, are shown in Table 2.1.

Table 2.1 Appearances of black and white children, according to Reginaldo Britto (2022)

	2009
Black children	115
White children	626
Total	741

For the year 2009, we have the following results, where DV_b refers to the visibility of back children and DV_w to the visibility of black children:

$$DV_b = 115/741 = 16\%$$

$$DV_w = 626/741 = 84\%$$

The numbers highlighted what the students felt: The visibility of white children was much higher than the visibility of black children.

The number DV_i provides the degree of visibility, whether the situation is positive or negative. However, it is also possible to consider the quality of the environment presented in the photo, and in *Media and Racism* the Degree of Negative Visibility of an ethnic group i, DNV_i, was defined in the following way:

$$DNV_i = \frac{\text{Number of appearances with negative context of the ethnic group } i}{\text{Number of appearances in total of the ethnic group } i}$$

In a similar way, the Degree of Positive Visibility of an ethnic group i, DPV_i, was defined as:

$$DPV_i = \frac{\text{Number of appearances with positive context of the ethnic group } i}{\text{Number of appearances in total of the ethnic group } i}$$

Based on the classification of positive and negative appearances, the students from the first year the project could calculate:

$$DPV_b = 3/5 = 0.60$$

$$DPV_w = 34/35 = 0.97$$

Naturally, such numbers are not the final words in a discussion; rather, they provide a starting point for more profound clarifications and critical observations.

Media and Racism puts on the agenda critical issues which call for further discussions: Why are some groups of people more visible in the magazines than others? How do such priorities occur? What do such priorities imply for perceptions of black people? Such questions allude to the phenomenon of structural racism. We might not be dealing with open forms of racist aggressions, but with racism acted out through implicit priorities, embedded preconceptions, commercial interests, and aesthetic assumptions. The landscape *Media and Racism* illustrates that mathematics can be an important tool for identifying patterns of structural racism, and furthermore that processes of investigation, dialogue, and critique are connected and entwined.

2.4 A Bias Index

A discussion of visibility is closely related to a broader discussion of *representativity*.[14] This may concern how the diversity of political positions in society is represented in parliament. One can compare the number of women and men in society with the number of women and men in parliament. One can compare the number of blacks and whites in society with the number of representatives in parliament. However, the issue of representativity does not only concern political representations; it concerns representations in any positively – as well as negatively – selected groupings. It concerns the numbers of people undergoing further education, getting well-paid jobs, and having long life expectations, but also the number of people put in jail, living in poor conditions, and being victims of violence. The issue of representativity may concern any group of people in any social situation.

Degrees of representation can be explored further through quantification. We can consider a group of people that we refer to as A, and the number of people in this group is referred to as *A* (where *A* is put in italics). The number of the total population P, we refer to as *P*. We then consider a particular subdivision S of P (it could be members of parliament), and we refer to the number of people in this subdivision as *S*. The number of people from A in the subdivision S, we refer to as *As*. Are we dealing with a fair representation of A in this subsection? One could claim that this would be the case if, and only if, the proportionality of A in S is the same as the proportionality of A in the total population P:

$$\frac{As}{S} = \frac{A}{P}$$

If *As/S* is bigger than *A/P*, A is overrepresented in S; if *As/S* is smaller than *A/P*, A is underrepresented. (We assume that $S \neq 0$ and that $P \neq 0$, and in the following we make similar assumptions.) We define the *level of representation Ras* of A in S as:

$$Ras = \frac{As/S}{A/P}$$

A fair representation of A in S would be indicated when *Ras* = 1. Overrepresentation occurs when *Ras* is bigger than 1, and underrepresentation when *Ras* is smaller than 1.

Let us consider another subgroup B in society with the number of people being *B*. The level of representation of B in S, *Rbs* can be calculated as:

$$Rbs = \frac{Bs/S}{B/P}$$

---

[14] The problem of representativity has been carefully addressed by Barros (2021).

We can compare the level of representation in the subdivision S between these two subgroups A and B. The *Bias Index, BI*, shows the bias in representation between these two subgroups A and B defined as:

$$BIabs = \frac{Ras}{Rbs} = \frac{As/S}{A/P} \bigg/ \frac{Bs/S}{B/P} = \frac{AsB}{ABs}$$

If *BIabs* is bigger than 1, the subgroup A is overrepresented in the subdivision S compared to the subgroup B. If *BIabs* is smaller than 1, the subgroup A is underrepresented compared to B, and in the case where *BIabs* is equal to 1, the two groups have the same degree of representation. Let me illustrate the calculation of the Bias Index with some examples.

Example 2.1 We assume that a population is composed of 90 black people (B) and 50 white people (W), in total 140 people. We consider a congregation S of 20 people, where 5 are black and 15 are white. In this case we have $B = 90$, $Bs = 5$, $W = 50$, and $Ws = 15$. The Bias Index will be:

$$BIbws = \frac{BsW}{BWs} = \frac{5 \times 50}{90 \times 15} = 0.19$$

Instead of calculating the degree of underrepresentation of black people in the subdivision *s*, one could calculate the overrepresentation of white people and get:

$$BIwbs = \frac{WsB}{WBs} = \frac{15 \times 90}{50 \times 5} = 5.4$$

We can observe that $BIbws \times BIwbs = 0.19 \times 5.4 = 1.03$, which is not any surprise as we have:

$$BIabs \times BIbas = \frac{AsB}{ABs} \times \frac{BsA}{BAs} = 1$$

The possible underrepresentation of one group is the inverse of the overrepresentation of the other group.

Example 2.2 We assume that a population is composed of 70 black people (B) and 70 white people (W), in total 140. We consider a congregation S of 20 people, where 10 are black and 10 are white. In this case we have $Bs = 10$ and $B = 70$; and $Ws = 10$, and $W = 70$. The Bias Index will be:

$$BIbws = \frac{BsW}{BWs} = \frac{10 \times 70}{70 \times 10} = 1.00$$

2.4 A Bias Index

If we do not know the numbers A and B of people in each ethnic group, but merely the size of the groups in percentages, then we can use these percentages as the values of A and B in the calculations. The same applies if we know the percentages of each ethnic groups in the subdivision S.[15] In both cases, the degree of bias between the two subgroups A and B in the subdivision S can be calculated as:

$$BIabs = \frac{AsB}{ABs}$$

Example 2.3 We consider a situation where the population is composed of 60% black people (B) and 40% white people (W), while in the congregation it is composed of 20% black people and 80% white people. In this case we have $Bs = 20$, and $B = 60$; and $Ws = 80$, and $W = 40$. The Bias Index will be:

$$BIbws = \frac{BsW}{BWs} = \frac{20 \times 40}{60 \times 80} = 0.17$$

The Bias Index $BIabs$ is a relative measure comparing the level of representation of two subgroups A and B in a subdivision S, and we might be dealing with two underrepresented groups.

Example 2.4 We consider populations where 50% are black (B), 30% are coloured (C), and 20% are white (W). The congregation is composed of 10% black, 25% coloured, and 60% are white. In this case we can calculate the bias-index with respect to the representation of black and coloured people based on the follow numbers: $Bs = 10$, and $B = 50$; $Cs = 25$, and $C = 30$. The Bias Index will be:

$$BIbws = \frac{BsC}{BCs} = \frac{10 \times 30}{50 \times 25} = 0.24$$

Example 2.5 We consider a country where 54% of the population is black and 46% is white; in the same country 52% are women and 48% are men. A congregation of 200 people is composed 24 black women, 81 black men, 74 white women and 21 white men. We can calculate the Bias Index with respect to the representation of black (B) and white (W) people in the congregation, and we get:

---

[15] The mathematical reason for this is:

$$\frac{\frac{As}{S}100 \cdot \frac{B}{P}100}{\frac{A}{P}100 \cdot \frac{Bs}{S}100} = \frac{AsB}{ABs} = BIabs$$

$$BIbws = \frac{BsW}{BWs} = \frac{105 \times 46}{54 \times 95} = 0.94$$

The bias-index with respect to the representation of women (W) and men (M) will be:

$$BIwms = \frac{WsM}{WMs} = \frac{98 \times 48}{52 \times 102} = 0.89$$

The two numbers might indicate that we are not dealing with a profound racist or sexist situation. However, we can also calculate the Bias Index with respect to the representation of black women (B) and white women (W), and then we get:

$$BIbws = \frac{BsW}{BWs} = \frac{24 \times 46}{54 \times 74} = 0.28$$

(In making this calculation we assumed that 54% of the women population is black, and that 46% is white.) This number clearly indicates that black women might be subjected to both racist and sexist treatments. Thus the calculation of the Bias Index with respect to particular groupings might reveal problems that need to be further investigated.

In Brazil, the number of people killed by the police is extremely high. In 2017 the number was 5179, and in 2019 the number was 6220. In the year 2017, the distribution of the killed people was 75.4% black people, 24.4% white, and 0.2% others, while the Brazilian population this year was composed of 55.8% black people, 43.1% white and 1.1% others. If *BIbws* refers the degree of bias between the number of black and white people in the subdivision S, killed by the police, in the year 2017, we can calculate:

$$BIbws = \frac{BsW}{BWs} = \frac{75.4 \times 43.1}{55.8 \times 24.4} = 2.39$$

The degree of bias is strong, even considering that there are more black people living in Brazil than white.

A Bias Index can be defined in more general terms. We consider three subsets A, B, and S of a universal set P, the number of which are A, B, S, and P (see the Fig. 2.1).

The bias of A related to B within the subset S, referred to as *BIabs*, is defined as:

$$BIabs = \frac{AsB}{ABs}$$

where As is the number of elements from A being in S, and Bs the number of elements from B being in S. We assume that $A \neq 0$, and that $Bs \neq 0$.

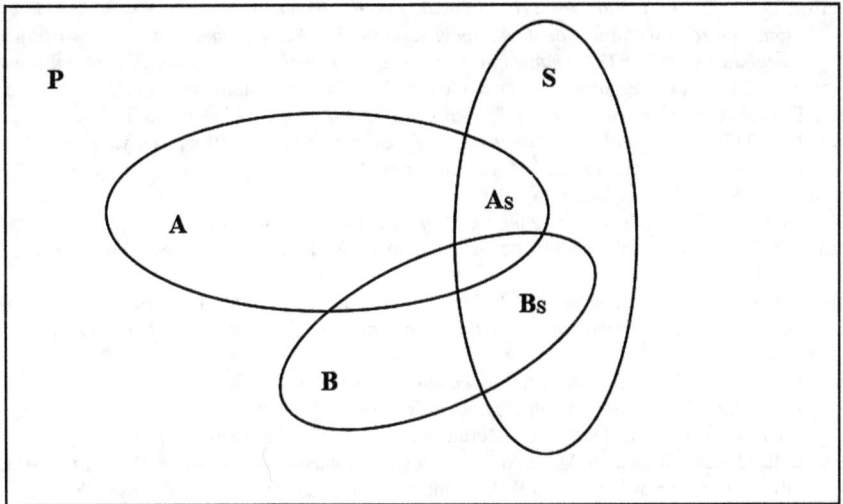

Fig. 2.1 The parameters of the Bias Index

In the previously presented examples, we might have assumed that the sets A and B are disjoint, but this is not a necessary assumption. We can, for instance, consider a possible bias between the number of black women in congress and the number of rich women in congress, knowing that some black women are also rich women.

Mathematics provides a powerful way of addressing structural racism. By calculating the Bias Index with respect to different situations, racism acted out through social structures, procedures, and traditions can be spelled out in numbers. Using the same mathematical approach, we can address how sexism might be acted out, for instance with respect to the representativity of women in leading positions, in parliament, in governments, in low-paid jobs, in jail. One can identify the development of a bias during a certain time period, and in this way open up a discussion of whether, say, structural racism has been increasing or decreasing. Any issue with respect to possible biases can turn into rich landscapes of investigation. Such landscapes might provide students with opportunities for interpreting quantitative data through an operationalisation of ratio, proportions, and indices. Working with such landscapes might also bring students to recognise that any observations of social phenomena expressed by numbers do not bring about concluding observations, but rather are indicators of issues that need further qualitative exploration.

References

Almeida, S. (2019). *Racismo estrutural (Structural racism)*. Editora Jandaíra.
Alrø, H., & Skovsmose, O. (2004). *Dialogue and learning in mathematics education: Intention, reflection, critique*. Kluwer Academic Publishers.

Barros, D. D. (2021). *Leitura e escrita de mundo com a matemática e a comunidade LGBT+: As lutas e a representatividade de um movimento social (Reading and writing the world with mathematics and LGBT+ commmunity: The struggles and representation of a social movement)*. Doctoral Dissertation. Universidade Estadual Paulista (Unesp).
Bell, D. (1987). *And we are not saved: The elusive ques for racial justice*. Basic Books.
Bell, D. (2002). *Ethical ambition: Livign a life of meaning and worth*. Bloomsbury.
Bell, D. (2004). *Silent covenenats*. Oxford University Press.
Bohm, D. (1996). *On dialogue*. Routledge.
Bridges, K. M. (2019). *Critical race theory: A primer*. Foundation Press.
Britto, R. (2013). Educação matemática e democracia: Mídia e racismo. *Anais do VII CIBEM*, 3355–3362.
Britto, R. (2022). Media and racism. In M. G. Penteado & O. Skovsmose (Eds.), *Landscapes of investigation: Contributions to critical mathematics education* (pp. 39–56). Open Book Publishers.
Brown, D. (1970). *Bury my heart at wounded knee*. Henry Holt and Company.
Carrijo, M. H. S. (2023). *Beyond borders and racism: Towards an inclusive mathematics education for immigrant students*. Doctoral Dissertation. Universidade Estadual Paulista (Unesp).
Chang, R. (2013). Toward an Asian American legal scholarship: Critical race theory, poststructuralism, and narrative space. In R. Delgado & J. Stefancic (Eds.), *Critical race theory: The cutting edge* (3rd ed., pp. 466–478). Tempe University Press.
Cozzens, P. (2016). *The earth is weeping: The epic story of the Indian wars for the American West*. Vintage Books.
Crenshaw, K., Gotanda, N., Peller, G., & Thomas, K. (Eds.). (1995). *Critical race theory: The key writing that formed the movement*. The New Press.
Davis, P. C. (2013). Law of microaggression. In R. Delgado & J. Stefancic (Eds.), *Critical race theory: The cutting edge* (pp. 187–196). Tempe University Press.
Delgado, R., & Stefancic, J. (Eds.). (2005). *The Derrick Bell reader*. New York University Press.
Delgado, R., & Stefancic, J. (Eds.). (2013). *Critical race theory: The cutting edge* (3rd ed.). Tempe University Press.
Faustino, A. C., & Skovsmose, O. (2020). Dialogic and non-dialogic acts in learning mathematics. *For the Learning of Mathematics, 40*(1), 9–14.
Kollosche, D. (Ed.). (2021). *Exploring new ways to connect: Proceedings of the Eleventh International Mathematics Education and Society Conference. 3 Volumes*. University of Klagenfurt.
Lynn, M., & Dixon, A. D. (Eds.). (2022). *Handbook of critical race theory in education* (2nd ed.). Routledge.
Madley, B. (2017). *An American genocide: The United States and the California Indian catastrophe, 1846–1873*. Yale University Press.
Milani, R., & Skovsmose, O. (2014). Inquiry gestures. In O. Skovsmose (Ed.), *Critique as uncertainty* (pp. 45–56). Information Age Publishing.
Morgenthau, H. (1918). *Ambassador Morgenthau's story: A personal account of the Armenian genocide*. Doubleday, Page and Company.
Pais, A. (2012). A critical approach to equity in mathematics education. In O. Skovsmose & B. Greer (Eds.), *Opening the cage: critique and politics of mathematics education* (pp. 49–91). Sense Publishers.
Penteado, M. G., & Skovsmose, O. (Eds.). (2022). *Landscapes of investigation: Contributions to critical mathematics education*. Open Book Publishers.
Phillips, C. (2011). Institutional racism and ethnic inequalities: An expanded multilevel framework. *International Journal of Sociology and Social Policy, 40*(1), 173–192.
Pólya, G. (1957). *How to solve it*. Doubleday.
Powell, J. A. (2008). Structural racism: Building upon the insights of John Calmore. *North Carolina Law Review, 86*, 791–816.
Reed, A. L. (2022). *The south: Jim Crow and its afterlives*. Verso.

References

Said, E. (1979). *Orientalism*. Vintage books.

Silva, G. H. G., & Powell, A. B. (2016). Microagressoes no ensino superior nas vias da Educacão Matemática (Microaggressions in higher education occurring in mathematics education). *Revista Latinoamericana de Etnomatemática: Perspectivas Socioculturales de la Educacion Matemática, 9*(3), 44–76.

Skovsmose, O. (2005). *Travelling through education: Uncertainty, mathematics, responsibility*. Sense Publishers.

Skovsmose, O. (2008). Mathematics education in a knowledge market. In E. de Freitas & K. Nolan (Eds.), *Opening the research text: Critical insights and in(ter)ventions into mathematics education* (pp. 159–174). Springer.

Solórzano, D. G., & Huber, L. P. (2020). *Racial microaggressions: Using critical race theory to respond to everyday racism*. Teachers College Press.

Whitt, L., & Clarke, A. W. (2019). *North American genocides: Indigenous nations, settler colonialism, and international law*. Cambridge University Press.

Wilkerson, I. (2020). *Caste: The origins of our discontents*. Random House.

Chapter 3
Democracy and Erosions

Abstract Features of democracy are elaborated upon in terms of voting, fairness, equity, and deliberation. In the struggle for democracy, the right to vote has played a crucial role. One can imagine democracies in both rich and poor countries, but it is difficult to imagine a functioning democracy in a country with an extreme unfair distribution of economic resources. In a democracy, one cannot legitimately differentiate with respect to gender, ethnicity, religion, richness, sexuality, poverty, caste, or any other such categories. It is assumed that, in a democracy, a principle of equity must be put into operation. Democracy must include the processes of reaching decisions, and in such processes, deliberation plays a crucial role. Erosions of democracy have taken place in the past – for instance, during the Weimer Republic – and erosions are again taking place today. I illustrate how one can address such erosions by means of mathematics. Mathematics is important in order to identify undemocratic features of different voting systems. Mathematics can be used for clarifying degrees of economic inequalities and for pointing out any lack of equity with respect to political representations. Finally, I highlight that democracy also concerns life in the mathematics classroom and that deliberations with respect to both form and content contribute to turning democracy into a reality.

Keywords Democracy · Erosion of democracy · Voting · Fairness · Equity · Deliberation · Gini Index · Lorenz Curve

Critical mathematics education does not operate with a transparent and well-defined conception of democracy. However, it is important to make space for students to engage in discussions of what democracy might include, and of what to consider to be erosions of democracy. Democracy concerns decision-making at all levels of society: in parliaments, in companies, in local communities, in families, and in the classroom. Democracy might represent a way of living.

A range of questions are important to raise with respect to democracy: What kind of anti-democratic forces are developing in today's society? How might a totalitarianism influence people's everyday lives? How does totalitarianism bring new, popular discourses into circulation? Does totalitarianism establish itself in a new informational format, different from previous formats? Does liberal democracy represent the only form of a democratic way of being? What limitations might capitalist structures put on democratic ideals? What other conceptions of democracy can be formulated? What does a democratic way of living mean for educational processes? What does it mean for mathematics education?

As no straightforward definition of democracy exists, I will restrict myself to outlining some features that I find relevant to address for conceptualising what democracy could mean. Then I present a historical case of an erosion of democracy by referring to the short and sad history of the Weimar Republic. Finally, I point out possibilities for critical mathematics education to engage in discussions about democracy and possible erosions of democracy.

3.1 Features of Democracy

Throughout history, we have seen different totalitarian regimes and movements, Nazi Germany being one example, and fascism another. Totalitarian regimes and movements belong to the past, but most unfortunately also to the present. We are living through a period where totalitarianism seems to be again proliferating.[1]

One may assume that one should confront totalitarianism by advocating for democracy, but the very notion of democracy is crammed with preconceptions. The Greek word *demos* means "people", and *cratos* means "rule". So "ruled by people" seems to be a direct interpretation of democracy. In Ancient Greece the ruling was only for "free men", meaning that women and slaves had no voice in that democracy. In *Two Treatises of Government* from 1689, democratic ideals were advocated by John Locke (1988). However, democracy was first of all considered an instrument for the dominant class to gain some control over the ruling monarch. That women could constitute part of a democratic life was not considered. Locke elaborated on many topics that formed the liberal tradition with respect to democracy, including strict protection of private property. He also addressed slavery, which he found himself able to justify.

In *The Social Contract*, first published in 1762 in French, Jean-Jacques Rousseau (1968) raised the question: Who is the sovereign? In a monarchy, the answer is simple: the king. However, according to Rousseau, the answer is just as simple when we consider a democracy: the people are the sovereign. The question then is, how could the people come to exercise their sovereignty? Rousseau claimed that, in a democracy, it should be possible for people to actually participate in the governing. In this way, he shared ideas with the Antique Greek model of direct democracy.

---

[1] Arendt (1951) analyses totalitarian positions as they have emerged previously.

3.1 Features of Democracy

The very notion of democracy is dynamic, and here I will consider some features of democracy, namely *voting, fairness, equity*, and *deliberation*.[2] I am not trying to reach out for any proper definition of democracy, but simply aiming to show the complexities of the notion.

3.1.1 Voting

In the struggle for democracy, the right to vote has played a crucial role. As part of its principal struggle against the Apartheid regime, the African National Congress (ANC) insisted that everybody – blacks, whites, Indians, coloured – must have the right to vote: one person, one vote! During the nineteenth century, women's right to vote was a highly controversial issue. People went to jail for propagating such ideas. Moving a bit further back in time, the idea that only landowners should have the right to vote was a dominant position. As part of the struggle for democracy, the right to vote has always been an issue.

When we do not have the conditions for establishing a direct democracy (wherein decisions of the people can be taken in plenary), voting becomes a crucial act. This is an act through which one allows somebody else to speak in one's name. Voting is the formal procedure for creating a congregation that executes the shared will of the whole population.

In *Capitalism, Socialism and Democracy*, first published in 1943, Joseph Schumpeter (1987) provides a critique of previous interpretations of democracy that he finds to be based on a most dubious assumption, namely that people operate in a rational way. Schumpeter considers the act of voting to be crucial, and he highlights: "The democratic method is that institutional arrangement for arriving at political decisions in which individuals acquire the power to decide by means of a competitive struggle for the people's vote" (p. 269). Schumpeter does not pay much attention to what can be referred to as the "democratic life". He concentrates on the implementation of the "democratic method", which defines the procedures for deciding "who the leading man shall be" (p. 269). Schumpeter's characteristic of democracy turns the procedure for electing the president of the USA into an exemplary "democratic method": it is a competitive struggle for the people's vote. *Capitalism, Socialism and Democracy* was published during the Second World War, and it might not be surprising that it includes a celebration of the political life in the USA compared to what had taken place Germany and in Austria. Schumpeter was born in Austria, but in 1932 he emigrated to the USA.

---

[2] Previously (Skovsmose, 1994), I have highlighted some ideals through which I tried to characterise democracy. I am going to draw on these formulations here, but I also provide reformulations and elaborations. See also Part 3, "Democracy as a Challenge", in Skovsmose (2014), and Skovsmose and Penteado (2016).

However, a "democratic method" defined through voting procedures does not ensure a functioning democracy. As an illustration, one can imagine a small community consisting of seven people, four of them coming from the North and three from the South. Every decision in this community is based on voting, so apparently we are dealing with an example of direct democracy. However, the four from the North vote together; they always do so. It has to be decided who is going to do the manual work in the community. With four votes against three, it is decided that people from the South have to do it. Who is going to pay taxes? With four votes against three, it is decided that people from the South have to do it. In the end, it is decided with four votes against three that the three people from the South should serve as slaves in the community.

My point with this small fable is that although voting is crucial for democracy, voting procedures do not, as such, ensure democracy. We also have to consider some less well-defined conceptions such as fairness, equity, and deliberation.

3.1.2 Fairness

One can imagine democracies in both rich and poor countries. It is more difficult to imagine a functioning democracy in a country with extreme differences between rich and poor. Decisions in a democracy are expected to be expressions of a shared will, but if extreme economic differences are maintained, one might suspect that, one way or another, the shared will has become manipulated and substituted by the will of some selected groups.

The classic liberal interpretation of democracy does not highlight the point about fair distribution of welfare. However, to me material conditions also make up a defining part of a democracy. For in creating huge economic differences, processes of exploitation might be brought into operation. Maybe they are explicit and brutal; maybe they are hidden and sophisticated. Whatever pattern of exploitation we are dealing with, I see it as an obstruction for a democratic life. In a democracy, a majority of voters cannot decide that a certain group of people should become slaves. In a similar way, in a democracy a majority of voters cannot decide that a certain group of people should stay in poverty. Systemic poverty is a sign that the sovereignty of the people has been bypassed.

By specifying the "democratic method" in terms of voting procedures, Schumpeter tends to provide a definition of democracy – if not in strictly quantitative terms, then at least through some well-defined procedures. I find such an approach problematic, as I see democracy as being an open notion that includes many features that escape clear definitions, such as fairness. By stressing the importance of economic and material fairness, I want to highlight that it is contradictory to claim that a democracy is operating in a situation of exploitation and oppression.

3.1.3 Equity

It is generally assumed that, in a democracy, one cannot differentiate with respect to gender, ethnicity, religion, richness, sexuality, poverty, caste, or any other such categories. Equity must be put into operation.

One can consider in what sense, then, the so-called "old democracies" as in England, France, and Switzerland in fact deserve this label. In England women got the right to vote in 1928, in France in 1944, and in Switzerland in 1971. The apartheid regime in South Africa was not considered democratic. However, the apartheid regulations in the USA were only abolished by the Civil Rights Act in 1964. So how are we to think of the USA as being an old democracy? The antique democracy in Athens was not for women and slaves. The history of democracy is filled with examples of exclusions and brutal marginalisations.

Equity not only concerns the right to vote, but the relationships to the laws in general. In a democracy, one cannot operate with laws applying differently to different groups of people. Women have been prohibited from entering universities, even in countries which during that time claimed to be democratic. Black people have been blocked from further studies. When formal rules manifest different rights to different groups of people, we are dealing with a ruthless cancellation of the principle of equity. However, such cancellations need not be explicit and formal, but can operate via traditions, economic resources, and preconceptions. It makes perfectly good sense to talk about structural inequalities, considering structural racism as one manifest example. Equity, with all its different features, is an ongoing preoccupation in discussions of what may count as being democratic.

3.1.4 Deliberation

Democracy includes many processes for reaching decisions. Such processes cannot be cut short by dictatorial decisions, and in many cases not even by voting procedures. In order to reach democratic decisions, one needs to explore a range of perspective, including the perspectives of minorities. The process of moving towards a democratic decision is complex, and this process I refer to as a deliberation. The issue of freedom of speech concerns possibilities for expressing oneself's views about political, economic, religious, or cultural issues – or any issues for that matter. Freedom of speech is crucial for the classic liberal conception of democracy; it is also an important feature of a deliberation.[3]

As part of his ethnomathematical studies, Aldo Parra (2018) worked with the Nasa People, an Indian population located in Colombia. He lived with the people and participated in many of their activities. He observed an event that played an

---

[3] The notion of deliberative democracy has been elaborated upon in great detail; see, for instance, Bohman and Rehg (1997).

important role in the life of the people, namely a *barter*. It is a getting together by different groups of the Nasa people. A barter can have the purpose of settling some sort of disagreement and of reaching clarifications, agreements, and decisions. It can also have the purpose of doing something together, such as harvesting or building a new house. For joining a barter, one could bring extra food, some presents, and also tools for working. A barter includes many features of deliberation, as well as of cooperation and support.

My point in referring to the *barter* is that the very practice of democracy is not a Western invention, anticipated by the Ancient Greeks. Democracy formed through deliberation can be found in very many different cultural contexts, both when looking back in history and looking around the world today. To me, it is extremely important to clarify the non-European roots of democracy, as this would help us to understand better what the very notion of democracy might include.

3.2 Erosions of Democracy

After the defeat in the First World War and the escape of the Kaiser Wilhelm II to the Netherlands, Germany was shattered. Everything had to be rebuilt, including a new constitution. The Weimar Republic was formed in 1919, when this new constitution was decided upon in the city of Weimar. According to the general conception at the time, the Weimar Republic was a democracy defined through the principles engraved in the new constitution. However, its erosion started right from the beginning.

Using the word "erosion" in this context seems to presuppose that the original socio-political order was democratic. I chose to refer to the Weimar Republic as a democracy, but certainly it was not any lived-through democracy. It was rather defined as such through some formal stipulations. Instead of talking about erosions of democracy, I could have talked about obstructions of democracy or about some further distancing from democratic ideals. However, let me stick to the expression of erosion, without making strong claims about what becomes eroded.

Partly due to the conditions set by the Treaty of Versailles, the German economy collapsed due to hyperinflation. Whatever money people might have saved became nullified. In the end, people selling potatoes in the street brought another basket as well as their basket of potatoes, more or less of the same size, to collect the money paid for the potatoes. It was a time of fear. For the majority, securing the means necessary to go about daily life was a severe struggle.

Street violence became common, and the Nazi movement appeared to be well-organised in undertaking physical attacks on political opponents. Campaigns of hate were launched first of all with Jewish citizens as their target. Such campaigns served as a form of legitimation for more violence, and one could be attacked or humiliated on the streets for appearing Jewish. It is a basic principle in a democracy that everybody is equal before the law. Formally speaking, this was the case during the Weimar Republic, but reality was different.

3.2 Erosions of Democracy

The Nazi party was in the ascendency. Adolf Hitler's first strategy for getting to power was to overthrow the existing democracy with a coup. In 1923, he tried to take power in Munich, and together with General Ludendorff, he and many Nazi supporters marched towards the Marienplatz in the centre of the city. There, however, the police confronted the march, and the coup ended in a fiasco. Hitler was sentenced to 5 years in jail, but was released in less than 1 year.

Hitler changed strategy. In order to come to power, he decided to use formal possibilities included in the democratic system and tried to get votes for Parliament. Hitler was a captivating speaker, and the Nazi party demonstrated a formidable ability in making propaganda. In the general election in 1928, the Nazi party got 12 seats in Parliament. In the election in 1930, this number became 107, and in the election in 1932, the party got 230 seats and became the biggest party in Parliament.[4]

In January 1933, Adolf Hitler became appointed chancellor by the president of the Republic, Paul von Hindenburg, everything apparently in accordance with the democratic rules inscribed in the Weimar constitution. Hitler formed a government composed of two parties, namely NSDAP (*Nationalsozialistische Deutsche Arbeiterpartei*), his own Nazi Party, and DNVP (*Deutschnationale Volkspartei*), a conservative party whose leader, Franz von Papen, became nominated as vice-chancellor. The conservative party DNVP was put under pressure by Hitler, and a few months later it dissolved itself. From then on, the government consisted of only Nazi members. Two months after Hitler came to power, the Enabling Act was implemented, meaning that the government acquired the right to make laws and decisions without consulting Parliament. In July 1933, the Nazi Party was declared the only legal political party, and from that day on, Parliament was just a decoration. The Nazi movement dominated the new media, the radio, which became the main component of the Nazi propaganda machinery. Hitler's voice could be heard by everybody in *Das Dritte Reich*.

In 1935 the Laws for the Protection of German Blood and German Honour, also referred to as the Nuremberg Laws, were implemented. Here, it was stated that *marriages between Jews and citizens of German or related blood are forbidden*, and also that *extramarital relations between Jews and citizens of German or related blood are forbidden*.[5] The laws reflect directly the racist assumption that the German people are "clean" and "superior", while the Jewish people are "unclean" and "inferior". The way of maintaining the cleanness of the German people was to prevent any "mixing of blood".

The Nuremberg Laws included a law about the *Reichsbürger* (state citizens), representing the deep violence of democratic ideals. The definition of a *Reichsbürger* is the following: *A Reich citizen is a subject of the state who is of German or related blood, and proves by his conduct that he is willing and fit to faithfully serve the German people and Reich*. However, there are groups of people that do not fit into

---

[4] An insightful presentation of this development is found in Neumann (2009).

[5] For this and the following quotations, see Nuremberg Laws https://en.wikipedia.org/wiki/Nuremberg_Laws

this category: *People belonging to the Jewish race and communities are not approved to have Reich citizenships.* People who did not meet the conditions of being a *Reichsbürger* were considered to be "state subjects" without any civil rights. The Nuremberg Laws represent the ultimate collapse of Weimar democracy.

3.3 Erosions of Democracy Today

To what degree do erosions of democracy take place today? Today one finds extreme right-wing movements in progress all over the world. From 2016 to 2020, Donald Trump was the president of the USA, but what power Trumpism might gain in the future is uncertain. From 2018–2022, Jair Bolsonaro was president of Brazil, and with him a range of right-wing and fascist opinions was circulated.

Democratic ideals might not only be threatened by explicit totalitarian movements. They might also be wiped away by the capitalist economic order, as it operates in proclaimed democratic societies. Democratic ideals might be demolished by the way poverty and unequal distributions of welfare are ignored and subsequently maintained. They might be threatened by the way different groups of people are excluded; for instance, due to caste systems or policies for dealing with refugees.

During the Weimar Republic, one did not find any proposals for education confronting the emerging totalitarian positions. Education was rather considered an apolitical task focussing on the child. The very notions of critical education and of critical mathematics education were not conceptualised.[6] However, critical mathematics education is now an articulated educational approach assuming a sociopolitical role. But how to play this role? I do not have definite answers in mind, but I want to illustrate that by means of mathematics it is possible to address erosions of democracy today.

3.3.1 Voting

Hans-Georg Steiner (1988) exposes a range of phenomena related to different voting systems. He does not only consider voting for Parliament, but any form of voting. For instance, he addresses carefully the voting used by the European Economic Community from 1958 composed of six nations with the following distribution of votes: France 4 votes, Germany 4 votes, Italy 4 votes, Belgium 2 votes, The Netherlands 2 votes, and Luxembourg 1 vote.

By studying such particular cases by means of mathematics, Steiner identifies different particular phenomena related to voting systems: a voter or a group of voters might be "powerless", being a "dictator", or having "veto power". When the four

---

[6] For an outline of the emergence of these educational positions, see Chap. 13, "Initial formulations of Critical Mathematics Education".

people from the North, referring to the example of the small community of seven people, vote together, they form a dictatorship, as their decisions become the decisions for everybody. A group of voters might have the power of veto, which means that although they are without the power to make decisions alone, they have the power to block decisions made by others. The three people coming from the south did not have any veto power; they were simply powerless, even with almost 43% of the votes. In real life situations, when members of Parliament vote together party for party, one might find situations where a party becomes powerless, acts as a dictator, or obtains veto power.

Steiner provides formal definitions of a voting system which he refers to as a voting body, and drawing on such definitions he provides formal definitions of the notions of being "powerless", being a "dictator", and having "veto power". Building upon such definitions, he formulates some theorems concerning voting systems and provides them with formal proofs. We are dealing with a mathematisation of procedures for voting. Through this process, we gain a deeper understanding of why the existence of voting procedures does not ensure the functioning a democracy. In fact, certain voting systems might be revealed as undemocratic.[7]

Steiner's presentation highlights two important things, namely that mathematics can help to provide critical investigations of voting systems, and that such investigations can be conducted in mathematics education. Steiner worked with upper secondary students who seemed ready to engage in sophisticated mathematical considerations. I find it relevant also to consider possibilities for addressing the voting system in secondary schools in general, and in elementary education, using less advanced mathematical reasoning.

3.3.2 Fairness

As highlighted, a fair distribution of wealth is a premise for a functioning democracy. There are different mathematical measurements of such distributions; one is referred to as the Gini Index, named after statistician Corrado Gini.

The mathematical idea of the Gini Index is illustrated in Fig. 3.1. The x-axis indicates the percentage of households organised according to income, starting with the lowest income (we assume that there is nothing called negative income). The y-axis shows the cumulative share of the total income, also in percentage. A point (x_0, y_0) at the Lorenz curve, named after the economist Max Otto Lorenz, indicates that the $x_0\%$ of the households with the lowest income earns $y_0\%$ of the total income. The Gini Index, GI, is defined as $GI = 100A/(A + B)$, where A is the area between the equality line and the Lorenz curve, and B is the area below the Lorenz curve. $GI = 0$ represents complete equality where every household earns the same; in this case the Lorenz curve is identical with the equality line, and $A = 0$. Extreme inequality occurs when

---

[7] For further discussions of mathematics and voting systems, see, for instance, Brams (2008) and Wallis (2014).

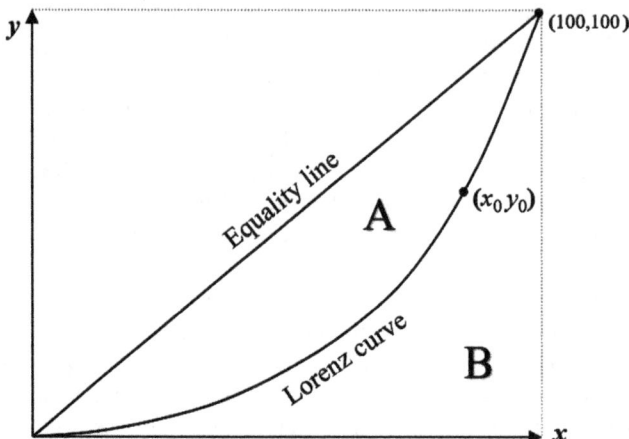

Fig. 3.1 The Gini index and the Lorenz curve

one household earns everything, meaning that B = 0. In this case we have GI = 100. In 2020, the Gini Index for Denmark was 28.5; for the United Kingdom 33.12; for Brazil 52.44; and on top of the list appeared South Africa with 62.73.

Working with the Gini Index can bring students deep into mathematics, as the very definition of the index is based on calculus, and the further exploration of the Lorenz curve leads us deeper into that topic. However, one can also explore the interpretation of the index with fewer mathematical means, and students can, for instance, investigate how the index for a particular country has developed over time. This development can be related to the different governments that have been in power during the corresponding time periods.

There are, however, many other ways of addressing fairness by drawing on mathematical insight. In the article "When Equal Isn't Fair", Aguirre and Zavala (2013) describe how they presented students with an imagined situation: three children planned to celebrate their father's 70th birthday with a surprise party, which would, however, cost a lot of money. How should they divide the expenses? In three equal portions? This would look, on the surface, like a fair solution. However, one could also consider the children's very different monthly incomes. Mathematics could be brought into action in order to elaborate on conceptions of proportionality and, in this way, illuminate further what could be considered fair and unfair with respect to the payment of the costs for the surprise party. One could move on and consider in which cases it might be more fair to look at not the monthly, but the yearly income. Or instead of income, should one consider the three children's fortunes? This apparently simple example provides an entrance to deep and complex discussions of fairness, where conceptions of proportionality might play an important role.[8]

---

[8] This example was discussed in a course for Master's and PhD students, organised by Miriam Godoy Penteado, Amanda Queiroz Moura, and myself as part of the Port-Graduate Programme in Mathematics Education at Universidade Estadual Paulista (Unesp), Rio Claro. My comments here are inspired by the many reflections formulated during the course.

3.3 Erosions of Democracy Today

Fairness concerns the distribution of costs, not only within a family, but within society in general. Fairness is a defining feature of democracy.

3.3.3 Equity

In a democracy, one should not be able to identify differences in the administration of juridical procedures with respect to different groups of people. The application fir legal procedures should be independent of gender, ethnicity, religion, wealth, sexuality, poverty, caste, or any other such categories. Everybody has to be treated equally with respect to the law.

Several issues concerning rights and obligations can be expressed in terms of problems of representativity.[9] This concerns the election of a Congress where the political diversities of the whole population are supposed to be reflected adequately. However, the problem of representativity reaches much further. It also concerns the degree to which different social groups are properly represented in Congress. One can consider the number of women and men in society, and compare this with the number of women and men in Congress. One can consider the number of black and white people in society, and compare with the representation in Congress. One can consider representativity with respect to any group of people and any type of representation. The problem of representativity also concerns the number of black and white people killed by the police, the number of black and white people put in jail, the number of black and white people living in extreme poverty, and the number of black and white people getting a medical degree, compared with the number of black and white people living in society. Any kind of implicit or explicit acted-out preconceptions can be addressed in terms of problems of representativity.

Numbers can reveal problems related to representativity. The existence of such problems shows that the principle of equity might have been violated. However, the principle makes integral part of the conception of democracy. While equity is an open concept with no clear definition, the notion of representativity seems more straightforward to handle, and mathematics is important for addressing issues of representativity.[10]

3.3.4 Deliberation

It is possible to use mathematics as an analytical tool for addressing the conditions for and processes of deliberation. One can, for instance, consider to what degree a certain TV channel provides different degrees of visibility of different political

---

[9] The issue of representativity has been addressed by Barros (2021).

[10] The Bias Index, as presented in Chap. 2, "Landscapes and Racism", can be used for addressing representativity as well as lack of representativity in all kinds of context.

opinions. However, here I want to pay attention to another relationship between mathematics education and democracy.

It has been highlighted that it is an important educational task to ensure that students experience what democracy could mean, and that the mathematics classroom prepares them for a democratic life in society. Democracy in a mathematics classroom might concern organisational issues, implying that students and teachers together discuss what format the lessons should follow. Should they be centred on the teacher's presentation of new topics? Should they be strictly related to the textbook? Should group work be tried out? And what about project work? Should such project work be rooted in mathematical issues, or rather in real-life problems? Such questions can set an agenda for deliberations in the classroom.

Democracy in a mathematics classroom can also be acted out with respect to content-matter issues; for instance, with respect to how the very mathematical investigations become conducted. Instead of first of all focussing on ready-made mathematical knowledge, it is possible to concentrate on the formulation ideas and proposals, and to make space for arguments, counter-arguments, examples, and counter-examples.[11] Democracy in a mathematics classroom also concerns the formulation of socio-political visions. An important feature of mathematics education for social justice is to create space for discussions of what social justice might mean, and for pointing out cases of social injustices. The mathematics classroom can turn into a space for students to show indignation (or maybe indifference) with respect to economic inequalities.[12] It is a space where political positions can be expressed and challenged. The deliberation of political visions constitutes an important element of a democratic life in a mathematics classroom.

Through mathematics, it is possible to address issues related to voting, fairness, equity, and deliberation. This is a concern of critical mathematics education.

References

Aguirre, J. M., & Zavala, M. R. (2013). When equal isn't fair. In E. Gutstein & B. Peterson (Eds.), *Rethinking mathematics: Teaching social justice by the numbers* (2nd ed., pp. 115–121). Rethinking Schools.

Arendt, H. (1951). *The origins of totalitarianism*. Schocken Books.

Barros, D. D. (2021). *Leitura e escrita de mundo com a matemática e a comunidade LGBT+: As lutas e a representatividade de um movimento social (Reading and writing the world with mathematics and LGBT+ commmunity: The struggles and representation of a social movement)*. Doctoral dissertation. Universidade Estadual Paulista (Unesp).

Bohman, J., & Rehg, W. (Eds.). (1997). *Deliberative democracy: Essays on reason and politics*. The MIT Press.

---

[11] See, for instance, Jaworski (2006). See also Hannaford (1998).

[12] In Chap. 1, "Concerns and Hopes", I refer to the project "Sweatshop Accounting", which creates space for such discussions.

References

Brams, S. J. (2008). *Mathematics and democracy: Designing better voting and fair-division procedures*. Princeton University Press.

Hannaford, C. (1998). Mathematics teaching *is* democratic education. *Zentralblatt für Didaktik der Mathematik, 98*, 181–187.

Jaworski, B. (2006). Theory and practice in mathematics teaching development: Critical inquiry as a mode of learning in teaching. *Journal of Mathematics Teacher Education, 9*(2), 187–211.

Locke, J. (1988). In P. Laslett (Ed.), *Two treatises of government*. Cambridge University Press.

Neumann, F. (2009). *Behemoth: The structure and practice of National Socialism*. Roman and Littlefield.

Parra, A. I. S. (2018). *Curupira's walk: Prowling ethnomathematics theory through decoloniality. Doctoral dissertation*. Aalborg University Press.

Rousseau, J.-J. (1968). *The social contract*. Penguin Books.

Schumpeter, J. A. (1987). *Capitalism, socialism and democracy*. Unwin Paperbacks.

Skovsmose, O. (1994). *Towards a philosophy of critical mathematics education*. Kluwer Academic Publishers.

Skovsmose, O. (2014). *Critique as uncertainty*. Information Age Publishing.

Skovsmose, O., & Penteado, M. G. (2016). Mathematics education and democracy: An open landscape of tensions, uncertainties, and challenges. In L. D. English & D. Kirshner (Eds.), *Handbook of international research in mathematics education* (3rd ed., pp. 359–373). Routledge.

Steiner, H.-G. (1988). Mathematization of voting systems: Some classroom experiences. *International Journal of Mathematical Education in Science and Technology, 19*(2), 199–213.

Wallis, W. D. (2014). *The mathematics of elections and voting*. Springer.

Chapter 4
Sustainability and Risks

Abstract Social justice concerns first of all social relationships, while fears for the environment and the topics of sustainability first of all concern our relationship with nature. In order to explore further the concern about nature, I present the notion of risk society. This notion captures the idea that social development might not follow long-term tendencies, but rather might be determined by critical events. With reference to the discussion of climate change, I show what reflections with respect to mathematical modelling might include. This possibility is illustrated further with a classroom example, *Use of Water*, before concluding that both social justice and environmental justice are concerns that should be addressed by critical mathematics education.

Keywords Social justice · Sustainability · Risk society · Climate changes · Reading and writing of the world with mathematics · Environmental justice

Critical education has focused on socio-political issues and challenged a range of social injustices. This focus has deep historical roots. The critical work by Karl Marx was fully concentrated on the socio-economic exploitation of workers. Classic Marxism does not include considerations with respect to environmental issues or sustainability. In this respect, Marxism shared the outlook of Modernity by considering nature an infinite resource that has to be used for ensuring human welfare. The further development of critical approaches through Critical Theory widened the scope of critique. Critique of art, music, and literature became integrated in the overall socio-political perspectives, and many more features of social life were addressed. Still, even within this broader critical stance, environmental issues continued to be ignored.

It is urgent for critical mathematics education also to address environmental issues. When doing so, it is important to remember that there is no sharp distinction between environmental and socio-political issues. Such issues are integrated. One can, for instance, think of the global hunger problem. One might try to interpret this as an environmental issue, provoked by droughts and bad conditions for farming.

However, another important factor is the capitalist-based and profit-oriented production and distribution of food, which adds to the global inequalities.

In the following, I present the notion of *risk society*. Then I elaborate on the concept of experimental forecasting, with particular reference to the discussion of climate change. I consider different issues for reflection with respect to mathematical modelling. I refer to a classroom example, where students around 10 years old work on the project *Use of Water*. Finally, I consider the relevance of operating with a notion of environmental justice.

4.1 Risk Society

The notion of risk society was coined by Ulrich Beck (1992), when he in 1986 published the book *Risk Society* in German. This was the very same year that the Chernobyl disaster took place and provided a most unwelcome illustration of what the notion could mean.

Nature may cause catastrophes due to floods, droughts, earthquakes, and hurricanes. Such risks have accompanied human history, and both in the *Bible* and in the *Epic of Gilgamesh* one reads about catastrophes caused by nature. What is new and what Beck highlights is that today risks are *also* caused by human beings, particularly due to recent technological development. We might be entering a new historic epoch where human-fabricated risks set the conditions for social life.

A paradigmatic case considered by Beck is the risks accompanying the construction of atomic power plans. In the so-called *Rasmussen Report* first published in 1975, the odds against a serious atomic power-plant accident were estimated to be 10,000 against 1.[1] It appears to be a small likelihood, but not an impossibility. This point illustrates the very circumstances of entering a risk society. Most likely nothing will happen – but accidents might happen, and in such cases the consequences might be catastrophic. One might wonder how the 10,000-against-1 estimation was calculated and how questionable this estimation might have been right from the beginning. In any case, accompanied by such estimations, reliable or not, we enter the risk society.

A more general phenomenon that brings us deep into the risk society is captured by Beck (1998) when stating that society "has become a laboratory where there is absolutely nobody in charge" (p. 9). An experiment is normally conducted in a laboratory, where it is somehow under control. What is happening today is that society has turned into a laboratory without anybody in control. The possible consequences of implementing new technological inventions seem impossible to identify in advance through some small-scale experiment. There is no existing experimental observation through which a 10,000-against-1 estimation can be corroborated. What

---

[1] The title of the report is *Reactor Safety Study: An Assessment of the Accident Risk in the U.S. Commercial Nuclear Power Plants*. The *Executive Summary* is available at https://www.osti.gov/servlets/purl/7134131

will happen in the long run when the production of food becomes more and more gen-manipulated? What impact will such production have on ecological systems? Will some new forms of health problems emerge? We do not know, and we cannot come to know in advance by means of specific laboratory experiments. Later we might come to experience the consequences, as we all participate in a worldwide experiment with gen-manipulated food production. Naturally, gen-manipulation need not bring about disasters, but it may. This is the same for many other technological inventions. Human-created natural risks create the risk society that we are entering.

According to Beck, the risk society is a human construction based on science and technology. According to the modern outlook, science and technology were reliable drivers of progress, but this image is questioned by identifying science and technology as driving forces for bringing us into the risk society. To this observation I want to add that mathematics – by playing a crucial role in scientific and technological development – also contributes to the formation of dramatic risk structures. Mathematics is a crucial factor in the formation of the risk society.[2]

4.2 Experimental Forecasting and Climate Changes

By referring to the Anthropocene, Alf Coles (2016) highlights the importance of addressing environmental issues in mathematics education.[3] The Anthropocene refers to the recent historical period where the human influence on the climate of the planet is significant. We do not know the consequences of this influence, and this uncertainty is part of the ongoing configuration of our risk society.

Weather forecasts have been created in all historical periods. For the farmers, the prediction of whether it is going to rain or not is important. For the fishermen, the anticipation of storms and bad weather is crucial. Throughout history there have been many ways of reading the sky. Today, weather forecasting has turned into a highly developed mathematical discipline. This also applies when the forecast does not concern the weather of tomorrow, but the future climate of the planet.

According to Dana McKenzie (2007), a climate model in its simplest form includes four components dealing with the atmosphere, the ocean, the land, and the ice. The atmosphere model tries to grasp the dynamics of the air, and it can include several differential equations. The ocean model tries to comprehend the dynamics of the sea, and is also composed of a range of equations. One important feature of water dynamics is the degree of salinity, and this parameter has to be included in the model. The structure of land is relatively stable, as the continents are not rapidly changing positions. This is, however, the amount of ice. The ice model shows

---

[2] In Chap. 7, "Mathematics and Crises", I address more carefully the relationship between mathematics and crises.

[3] See also Coles et al. (2013).

obvious changes which have an impact on, for instance, the dynamics captured by the ocean model. In fact, the models for atmosphere, ocean, land, and ice need to be carefully interrelated in order to establish a reliable climate model.[4]

There is no way of doing real experimentation with respect to the future climate of the planet, but by means of a mathematical climate model one becomes able to make what I will refer to as *experimental forecasting*. In order to do so, one can start stipulating the values of the parameters of the model; it could be the values as one assumes them to be today. Based on these stipulations, one can make a forecast of what will happen to the climate in, say, 30 years, if no further initiatives are taken. Then one can change the values of some of the parameters, and see what then becomes the output of the model. By making such experimental forecasting in a systematic way, one might become able to identify which parameters have a particular significance, say, with respect to global warming. In this way, experimental forecasting becomes crucial for defining political measures for coping with climate changes.[5]

All this sounds rather straightforward, but the use of mathematics for conducting experimental forecasting is associated with many problems. In the article "The Mathematical Formatting of Climate Change: Critical Mathematics Education and Post-Normal Science," Richard Barwell (2013) talks about the "mathematical formatting of climate change". This formulation appears strange. Climate change appears to be a natural and physical phenomenon, which is independent of possible mathematical descriptions of this phenomenon. How could it make sense to talk about a mathematical formatting of climate change? Barwell observes that any mathematics-based identification of climate change reflects not only some possible physical phenomena, but also the very nature of the mathematics being used. Mathematics shapes a particular reading of climate change, making space for a selective set of possible actions. In Barwell's formulation: "Mathematics also formats how we interact with the climate. Through the mathematised, model-based perspective prevalent in climate research, the climate is constructed in particular ways: as, for example, measurable, predictable, technical and controllable by humans, rather like the temperature in a high-tech sensor controlled greenhouse" (p. 2). This is the point: mathematics formats how we interact with the climate.[6]

Mathematics does not provide any neutral reading of possible climate changes; rather, it establishes a particular perspective that may serve particular political, economic, or industrial interests – for instance, by portraying climate change as being manageable. As an example of such a reading, I can refer to a study presented in *The Sceptical Environmentalist: Measuring the Real State of the World* by Bjørn Lomborg (2001). His bases his argumentation on mathematical studies, and reaches

---

[4] See also Coiffier (2011); and Warner (2011).

[5] Experimental forecasting is an integral part of the construction of the North Sea Model, which was elaborated with the purpose of identifying the best possible fishing policy for the North Sea. For a presentation of the model, see Chap. 18, "Reflective Knowledge: Its Relation to the Mathematical Modeling Process", in Skovsmose (2014, pp. 263–280).

[6] See Chap. 9, "Mathematics and Crises", or further comments on Barwell (2013).

the conclusion that global warming is a natural, rather than a human, phenomenon. Lomborg confronts the whole conception of the Anthropocene as an emerging and threatening historical period. His conclusion was welcomed by industries as a justification for continuing as usual. Naturally, there were other mathematics-based studies that pointed towards other conclusions; there were also other readings of material that Lomborg did refer to. In all cases, mathematics operates in powerful ways by forming opinions, decisions and actions.

Any mathematical climate model will include simplifications, distortions, and misinterpretations; still it is applied for making environmental forecasting. Such forecasting might be misleading, not adequately grasping the implications of what we are doing. Decisions and actions based on mathematical modelling might create new problems; they might have disastrous consequences. Mathematical climate models might bring us deeper into the risk society. I see the title of Barwell's article as making perfectly good sense: mathematics may format how we are dealing with climate changes.

4.3 Reflections on Mathematical Modelling

A mathematical climate model need not "look like" the climate. It can be very different from what it is assumed to describe. This observation also applies to many other mathematical models used in experimental forecasting.

I will talk about the existence of a *similarity gap* between a mathematical model and what it is assumed to describe, acknowledging that the use of the word "gap" is metaphorical. A similarity gap need not be a gap between two well-defined phenomena. It need not be a gap in terms of degrees. A similarity gap may take the form of huge and fundamental differences. It could be between two entities, a mathematical and a non-mathematical entity, that cannot be compared. It might be a gap between a world of numbers and a world of real-life phenomena.

Many of the risks that constitute our risk society enter through similarity gaps provoked by applications of mathematical models. It could be a climate model; it could be a model trying to capture the security levels at an atomic power plant; it could be a model through which one tries to grasp the long-term implications of gen-manipulation used in farming; it could be a model of economic risk-taking. Whatever mathematical models we are dealing with, they are accompanied by similarity gaps. What the model presents might be radically different from the reality that one assumes that it portrays.

I will refer to such phenomena as a mathematics-based creation of a parallel world. When we are making any analysis of a situation based on a mathematical model, we are not analysing the situation as such. We are analysing the mathematical representation of the situation, and this representation is located in a parallel world made up of numbers and equations. When operating within this parallel world, one is easily captured by mathematics-based illusions of predictability. We assume that we are able to make real-life predictions, but they are only parallel-world predictions.

I find that a principal cause for bringing us into a risk society is the mathematics-based formation of parallel worlds, creating illusions of predictability. In order not to be lost in parallel worlds and not to be ensnared by illusions of predictability, one needs to engage in a broad set of reflections. This applies to any situation where mathematics is applied. It applies to any educational context, and now I concentrate on such a context.[7]

A first set of questions that students and teachers might raise when engaging in any form of mathematical modelling could be: *Have we done the calculation correctly? Are there different ways of performing the calculation? Have we followed the algorithm accurately?* Such questions address the technical aspects of a calculation. A second set of questions is also technical, but they have a broader scope: *Have we used the right algorithm? Is it possible to choose between different algorithms?* Whenever a choice of technical approach has to be made, it is important to reflect on this choice. A third set of questions concerns the reliability of the solution. Even if the calculations are done correctly and the choice of the algorithm is appropriate, the result produced by means of the modelling need not be trusted. It is important to ask questions like: *Can we rely on the result from the calculations? Even if we have calculated correctly and used an appropriate algorithm, do we get a relevant result?* While the first and second sets of questions concern the more technical aspects of the mathematical modelling, this third set of questions directs the reflections towards the contexts of the modelling.

Such broader questioning leads us to a fourth set of questions such as: *Could we have found an answer without using mathematics? Is it appropriate to use a mathematical technique in the present situation? Do we need mathematics?* These questions draw attention to the fact that formal techniques need not be necessary for achieving a certain aim. In some cases, a non-formal common-sense-based way of handling a problem may be more appropriate. When applying mathematical modelling, one always need to reflect on both the possibility and the necessity of using mathematics.

A fifth set of questions concerns the broader consequences of using mathematical techniques for addressing a problem: *How does the application of an algorithm affect our conception of the problem? How do our realities become formed through applications of mathematics? What kind of risk might be fabricated through such applications?* Naturally, it might be rather complex to handle such questions in an educational context, but whenever one is engaged in modelling processes, it is important to consider what could be the impact of doing so. A sixth set of questions concerns the relationship between mathematics and power: *What political interests might be acted out through mathematical modelling? What economic priorities might be served? What ideological and metaphysical conceptions might become*

---

[7] In the following presentation of different forms of reflections, I draw on Skovsmose (1994). For a careful discussion of the notion of reflection, see also Hauge et al. (2017).

propagated through a mathematical modelling? It is important to reflect on any form of mathematics-power integration.

Critical reflections include self-reflections, and a seventh set of questions concern the way one has reflected on the mathematical modelling process. Critical reflections must address their own status. Some questions could be: *Are our critical reflections limited by some sets of preunderstanding and preconceptions? Are there other questions that could be raised with respect to the use of mathematics? What might we have overlooked?*

4.4 Use of Water

It is a concern of critical mathematics education to bring issues about climate change and sustainability into the classroom, and let me present the landscape of investigation *Use of Water* as described by Ana Carolina Faustino (2018).[8] As part of her doctoral study, Faustino worked with a class of 26 students around 10 years old in a public school in a city within the São Paulo State. The students were working in groups, and each group had to choose an issue concerning sustainability for further exploration. The groups were asked to produce a video that showed the results of their work.

One of the groups concentrated on the question: How much water does a family use during a day? The group – consisting of Carla, Júlia, Denis, Luiza, and Pedro (all pseudonyms) – were very dedicated to the project, and they engaged in different kinds of calculations and estimations. Over a weekend, they met in Julia's home to conclude the project and to plan how to organise the video. The video became divided into different scenes which were recorded, edited, and saved on a pen drive. For doing the editing, the students used the programme *Kinemaster,* with which they were already familiar.

On the day of the presentation, the classroom was filled with people, not only from the students' own class, but also from another class engaged in a similar sustainability project.[9] Furthermore, there were invited guests, one of them with almost white hair and a beard. Each group of students had brought their pen drive, but here we concentrate on the group of Carla, Júlia, Denis, Luiza, and Pedro.

The first scene in their video is from the bathroom, and the word "Bath" appears on the screen. Then we see a person opening the door to the shower. The person enters, closes the door, and starts humming. The camera focusses on the door to the shower, and the humming continues now mixing with the sound of water. Then the words "Ten minutes later" appear, and we see the person getting out of the shower.

---

[8] Steffensen (2021) carefully relates the discussion of climate changes to the concerns of critical mathematics education. See also Abtahi et al. (2017), and Barwell (2018).

[9] The day of presentation is described in Faustino (2018), as well as in Faustino (2021).

Then the commentator appears; it is Carla, who points out: "In one minute one uses 9 litres of water, in 10 minutes it gets up to 90 litres".

The second scene has the title "The Pavement". Here we see Ryan looking at the pavement in front of his house, and he remarks: "The pavement is dirty, I am going to clean it". He opens the gate, takes a hose, and starts spurting the pavement with water. Then, in the corner of the screen, appears a person making the ringtone of a telephone: "Trilim, trilim, trilim". Ryan lifts his hand to his ear and answers the call:

– Hello!
– Hello, who is it?
– It's Ryan.
– How are you, Ryan?
– I am fine. Are you spending a lot of water when you wash your driveway?
– No.
– That is good, because the water of the world is ending. We have to save a lot of water so that by 2070 we have water to drink and live. Thank you. That is all I wanted to tell you.
– That is good! Bye, bye.

The scene is concluded when the commentator Carla appears and points out: "During these five minutes, when Ryan was cleaning the driveway, he used 279 litres of water".

The third scene has the title "Washing Clothes". Júlia is in the laundry room and puts some dirty clothes into the washing machine, adds a little softener, closes the machine, and says: "Today I'm going to wash these clothes". Then Júlia repeats the procedure and says: "Today I will have to wash these clothes again". Then the commentator appears and states: "A machine uses a hundred and twenty litres of water. If Júlia does all the washing, it will use eight hundred and forty litres per week."

The video from the group continues with two more scenes, and at the end of each scene one hears applause from the audience. The other groups show their videos, and receive much applause as well. When all the videos have been shown, the attention turns towards the visitors; one of them is Miriam Godoy Penteado, a mathematics educator from the university who is familiar with the school and the teachers. The other one is me, married to Miriam and supervisor of Ana Carolina. We are going to make comments on the videos. We are impressed: the videos demonstrate a professionalism both with respect to presenting a story and with respect to technical features that we had not imagined.

In her comments, Miriam concentrates on issues of sustainability, and she points out the many important issues addressed in the videos. Sometimes she goes into detail, and the students get the opportunity to elaborate further on their calculations. Sometimes Miriam points out some principal issues regarding sustainability and the importance of everybody having access to clean water. The students eagerly follow up with more comments. The concern about sustainability is elaborated upon as a shared concern, not only among the students and the audience in the classroom, but as a general global concern. In my comments, I concentrate on the aesthetic and technical aspects of the videos: how the cuts are made, slowly or fast; how the

4.4 Use of Water

camera perspective is used, fixed or moving; how the sound effects have been composed; how the colour contrasts are used, light and darkness, or green and red. As a conclusion, Miriam and I congratulate all the students on the remarkable work they had done.

The use of mathematics is crucial for addressing climate change and issues about sustainability, and the students exploring the landscape *Use of Water* also experienced the usefulness of mathematics for underlining observations. It became clear, for instance, that there is an important difference between stating that a large quantity of water is used for cleaning a pavement and stating that 279 litres is used. In this direct way, the students experienced the power of using mathematical terminology.

Let us reconsider the different sets of questions through which one can stimulate reflections on mathematical modelling. The first group of questions can be represented by the question: *Have we done the calculation correctly?* Such a question appears rather straightforward, and it clearly makes sense – for instance, with respect to the calculations of the use of 279 litres of water. The second group of questions can be represented by: *Have we used the right algorithm?* This is also a question that will make direct sense in a classroom setting. It concerns issues like: Does one have to multiply? Or to add? Does one have to add before multiplying? And so on. As in the first set of questions, we are dealing with reflections addressing the technical aspect of mathematics.

The third group of questions concerns: *Can we rely on the results from the calculations?* This question stimulates reflections that move beyond technicalities. For instance, one could take a more careful look at the number 279. How was it identified? One could imagine that one has misestimated how much water Ryan used per minute. How could one try to provide a more reliable estimation of this quantity? And for how long did Ryan clean the pavement? There are many ways that one can address the reliability of observations put in numbers. The fourth group of questions can be represented by: *Could we find an answer without using mathematics?* This is a critical issue when applying mathematics. With respect to the discussion of climate change, mathematics appears a necessary tool. However, in many situations the relevance of mathematics can be questioned. The question was not explicitly raised in the classroom, but it would have been interesting to hear the students' opinions about that.

A fifth and sixth group of questions concern raises the general issue about the social consequences of using mathematics. However, it might be possible to address such questions only with older students. The same might be the case with respect to the seventh groups of questions concerning the way the reflections themselves have been conducted.

By relating the reflective questions to the project *Use of Water*, I wanted to show what reflections with respect to mathematical modelling are not only crucial when one is dealing with advanced applications of mathematics, but also relevant with respect to classroom activities.

4.5 Environmental Justice

In 1848, when the notion of social justice was coined, there was no awareness of ecological issues. Nature is an infinite resource that we human beings need to control and dominate in order to extract its resources for improving human life. This perspective on nature has changed, and now nature is recognised as being both limited and fragile. As a consequence, I want to introduce the notion of *environmental justice* in parallel with the notion of *social justice*.

The concepts of social justice and environmental justice might be overlapping. Apparently, social justice concerns relationships between different groups of people, while environmental justice concerns relationships between people and nature. However, environmental justice also concerns relationships between people. Natural resources become very unequal distributed around the world. Some groups of people benefit much more than other groups of people. Some nations use many more resources per citizen than other nations. Pollution is a huge problem, but it is not the same for everybody. It has become common to construct polluting industries in poorer neighbourhoods or in poorer countries in the world. Mountains of waste are now exported from rich to poor countries.

On Brazilian TV, one often sees recommendations for ways families can save water. What is seldom mentioned is that, in Brazil, the private use of water adds up to only 10% of the total use of water; the rest is used by agriculture and industry. Precautions for using less water are implemented in many cities, including São Paulo. In some neighbourhoods, the water supply simply gets cut during the night: no water comes out of the taps. Such precautions, however, are only implemented in poor neighbourhoods, while other neighbourhoods continue to have water supply 24 h per day. Many examples of environmental injustices can be pointed out, not only with respect to water but with respect to any kinds of resources becoming more and more limited. Most often, such limitations hit poor and rich people very differently.

While social justice concerns actual relationships between groups of people, environmental justice also concerns relationships with the coming generations. The coming generations become prevented from access to natural resources of which the earth is being drained today. We might belong to the generations that make use of the last of several kinds of natural resources, leaving only little (of oil, for example) for the next generations. The initiative *Skolstrejk för klimatet* (school strike for the climate), as guided and inspired by Greta Thunberg, is a strong reminder that environmental justice also concerns future generations.

Like social justice, environmental justice is a constructed and contested concept. Constructions can take place in the mathematics classroom, where students are invited to form their own opinions about environmental issues. The project *Use of Water* illustrates what I have in mind.

References

Abtahi, Y., Gøtze, P., Steffensen, L., Hauge, K. H., & Barwell, R. (2017). Teaching climate change in mathematics classroom: An ethical responsibility. *Philosophy of Mathematics Education Journal, 32*, 1–18.

Andersson, A., & Barwell, R. (Eds.). (2021). *Appling critical mathematics education*. Brill.

Barwell, R. (2013). The mathematical formatting of climate change: Critical mathematics education and post-normal science. *Research in Mathematics Education, 15*(1), 1–16.

Barwell, R. (2018). Some thoughts on a mathematics education for environmental sustainability. In P. Ernest (Ed.), *The philosophy of mathematics education today* (pp. 145–160). Springer.

Barwell, R., & Hauge, K. H. (2021). A critical mathematics education for climate change. In A. Andersson & R. Barwell (Eds.), *Appling critical mathematics education* (pp. 166–184). Brill.

Beck, U. (1992). *Risk society: Towards a new modernity*. Sage Publications.

Beck, U. (1998). Politics of risk society. In J. Franklin (Ed.), *The politics of risk society* (pp. 9–22). Polity Press.

Beck, U. (1999). *World risk society*. Polity Press.

Coiffier, J. (2011). *Fundamentals of numerical weather prediction*. Cambridge University Press.

Coles, A. (2016). *Mathematics education in the anthropocene. Open access*. University of Bristol.

Coles, A., Barwell, R., Cotton, T., Winter, J., & Brown, L. (2013). *Teaching secondary mathematics as if the planet matters*. Routledge.

Faustino, A. C. (2018). *"Como você chegou a esse resultado?": O processo de dialogar nas aulas de matemática dos anos iniciais do Ensino Fundamental. ("How did you get to this result?": The dialogue in mathematics classes of the early years of Elementary School)*. Doctoral dissertation. Universidade Estadual Paulista (Unesp).

Faustino, A. C. (2021). Trabalho com projetos nos anos inicias do ensino fundamental: Dialogando para ler e escrever o mundo com matemática (Work with projects in the early years of elementary school: Dialogue in order to read and write the world with mathematics). In G. H. G. Silva, I. M. S. Lima, & F. A. G. Rodríguez (Eds.), *Educação matemática crítica e a (in)justiça social: Práticas pedagógicas e formação de professores (Critical mathematics education and social (in)justice: Pedagogical practices and teacher education)* (pp. 259–292). Mercado de Letras.

Hauge, K. H., Gøtze, P., Hansen, R., & Lisa Steffensen, L. (2017). Categories of critical mathematics based reflection on climate change. In A. Chronaki (Ed.), *Mathematics education and life at times of crisis: Proceedings of the ninth international mathematics education and society conference* (pp. 524–532).

Lomborg, B. (2001). *The sceptical environmentalist: Measuring the real state of the world*. Cambridge University Press.

McKenzie, D. (2007). *Mathematics of climate change: A new discipline for an uncertain century*. Mathematical Sciences Research Institute.

Skovsmose, O. (1994). *Towards a philosophy of critical mathematics education*. Kluwer Academic Publishers.

Skovsmose, O. (2014). *Critique as uncertainty*. Information Age Publishing.

Steffensen, L. (2021). *Critical mathematics education and climate change: A teaching and research partnership in lower-secondary school*. Doctoral dissertation. Western Norway University of Applied Sciences.

Steffensens, L., Herheim, R., & Rangnes, T. E. (2021). The mathematical formatting of how climate change is perceived. In A. Andersson & R. Barwell (Eds.), *Appling critical mathematics education* (pp. 185–209). Brill.

Warner, T. (2011). *Numerical weather and climate prediction*. Cambridge University Press.

Chapter 5
To Learn or Not to Learn?

Abstract I present a perspective on learning phenomena by relating the concepts of learning, action, intention, foreground, motive, mathematics, and life-world. According to my interpretation, life-worlds are formed by social factors, be they economic, political, cultural, religious, or discursive. Life-worlds are social constructions. They are composed of different domains, one being the students' foregrounds. To interpret students' motives for learning mathematics, I pay particular attention to how mathematics might appear in their foregrounds. It could be that, to students in marginalised positions, mathematics only appears in opaque format in their visions about their own future possibilities in life. Marginalised students' foregrounds might be ruined due to the very socio-political processes of marginalisation. Ruined foregrounds might be a principal factor in turning groups of students into low performers in mathematics.

Keywords Learning · Action · Intentionality · Foreground · Motive · Mathematics · Life-world

What can we expect from a philosophy? Philosophical investigations are different from empirical studies, in particular with respect to the use of empirical data. In an empirical study, the data material is in focus: it is produced, described, and analysed. The analysis takes the form of a meeting between a conceptual structure and a data material. As a result, one formulates observations, conclusions, suggestions, and new questions. A philosophical investigation might also take inspiration from a meeting between conceptual structures and data material. However, such investigations pay particular attention to bringing concepts together in new constellations. To characterise the following discussion as a philosophical investigation might be an exaggeration; it is better to think of it as some philosophical remarks.

In Ludwig Wittgenstein's (1953) book *Philosophical Investigations,* we find references to empirical situations. These are invented examples serving as illustrations of several of Wittgenstein's points – for instance, with respect to the notion of "language game". The examples are not treated as data material to be properly analysed.

In the following philosophical remarks, I am also going to refer to examples but contrary to Wittgenstein, I will refer to real-life examples.

The notions that I am going to relate are: learning, action, intention, foreground, motive, mathematics, and life-world. I think of them as forming a conceptual circle. With reference to this circle, I will address questions like: How could it be that certain groups of students apparently turn into low achievers in mathematics? What causes students' reluctance towards learning mathematics? How could it be that, at times, students are not interested in addressing apparently relevant real-life problems? Can low achievement in mathematics be considered as not a personal, but a socio-political fabrication? How are students' motives for learning mathematics formatted by social, political, and economic factors?

5.1 Learning and Action

Why should we see learning as action? Learning takes place in many different situations. One can think of the very young child starting to learn their mother tongue; one can think of soldiers being drilled to march in time; and one can think of tourists learning to find their way around new places. One need not expect to be able to identify general characteristics of learning valid to all such situations. However, I am focussing on learning as it takes place in school, concentrating on the mathematics classroom in particular. To interpret what might take place here, I find it useful to see learning as action.[1]

A particular episode that I witnessed a long time ago in a mathematics classroom inspired me to see learning as action.[2] There was a nice, warm, and welcoming atmosphere in the classroom; the teacher was kind to all the students, who were around 7 years old. The lessons themselves, however, followed a traditional pattern. At the beginning of each lesson, the teacher carefully explained the new tasks to the students. It could, for instance, be with respect to the addition of two-digit numbers. After the teacher had explained and answered questions, the students had to do exercises from the textbook. I was sitting next to one of the students, Peter (pseudonym), and when the teacher asked the students to start doing the exercises, I was ready to engage Peter in a conversation about how he would do the exercises. However, he ignored my questions and seemed completely occupied in copying the exercises from the textbook and solving them. Looking around the classroom, I noticed that several other of the students had also started working, and were in deep concentration. I soon realised the reason.

After each student had completed the exercises, they went to the teacher's desk. There, the teacher was sitting, receiving the students with a big smile. He looked at

---

[1] For previous suggestions about interpreting learning as action, see Skovsmose (1994, 2005a, b).

[2] For a previous presentation of this example, see, for instance, Chapter 5, "Foregrounds and the Politics of Learning Obstacles", in Skovsmose (2014b, pp. 59–72).

5.1 Learning and Action

their solutions, which he approved, and told them that they had done a great job. However, a different agenda was also operating in the classroom; namely a competition among the students, at least among some of them, to become the first in line next to the teacher's desk.

I also observed a group of girls who seemed to ignore this whole rush sweeping through the classroom. They did work on the exercises, but at their own pace and with many small interruptions. They might have realised that, whatever they were doing, they would never be among the first in line. It was common knowledge that sometimes Peter was the first, sometimes Anna (pseudonym), sometimes Maria (pseudonym), and a few times a fourth member of the class. The winner of the game was to be found among a small group of students.

If one was not participating in the game, what could one do better than ignore it and occupy oneself with one's own business? It might well be that the group of girls did start copying the numbers from the first exercise in the textbook, but perhaps copied a number incorrectly and had to erase it. In order to do so, they needed to choose the proper eraser. I noted that the group of girls had many different erasers to choose from. The erasers had different colours, different sizes, and also different smells. It was not always easy to choose the proper eraser. Several times it was necessary to ask for one of the other's erasers, maybe one with a smell of strawberries. Pencils also had to be sharpened, smiles had to be exchanged, and time passed. The group of girls created their own private room, and by doing so they enjoyed being in the mathematics classroom. Had they struggled to become first in line – lesson after lesson after lesson – and failed, their self-esteem might have been affected. To me, their approach looked like a healthy form of self-protection.

Some of the children, like Peter, decided to engage in the competition to become first in line. Naturally, one can imagine different motives for doing so, including motives that the children might not be aware of. I see the children that were participating in the competition as actors in doing so. In a similar way, I see the group of girls that became silent and focused on other activities as actors in doing so. We are dealing with decisions about whether to participate or not to participate. Naturally, these might not be explicit decisions; they might be implicit and unarticulated. We are dealing with driving forces with respect to learning or not learning.

Partly inspired by this example, I came to see learning as action. By doing so, I distance myself from several other positions in learning theory. According to behaviourism, learning can be interpreted as a agglomerate of stimulus-response processes.[3] However, I am not following a conception according to which learning is something that the learner becomes subject to. Rather, I see learning as something performed by the learner.

I also distance myself from a perspective on learning as suggested by Jean Piaget. In terms of a genetic epistemology, he highlights two phenomena as being crucial to

---

[3] See Watson (1913) for an initial presentation of the behaviourist position. Skinner (1957) provided a principal inspiration for formulating learning theories according to this position.

any learning process, namely assimilation and accommodation.[4] Assimilation refers to the process of incorporating new information in already formatted schemes of understanding, while accommodation refers to the adjustment of schemes in order to grasp some new information that does not fit into the existing schemes.

As a diabetic, I take my insulin as part of my daily routine. I may not pay much attention to it; at times I might also forget about it. Either way, I consider taking my insulin to be an action. Before I got diabetes, my pancreas ensured that insulin was added to my blood in the proper doses. It was a process that I was not aware of, and I do not consider the functioning of my pancreas as being *my* action. It was a biological phenomenon taking place independent of my will. The processes of assimilation and accommodation are presented by Piaget as being similar to biological processes taking place within the learner.[5]

By seeing learning as action, I highlight that learning is not a biological-like phenomenon. Learning, as it takes place in a classroom context, is a complex interactive process. The students might want to participate in this process, but they might also have reasons for withdrawing from it. In order to interpret human actions, one needs to draw on a complex set of notions and ideas, different from those that are important for interpreting biological phenomena. One needs to draw on notions and ideas similar to those that are relevant for understanding human actions in general.

5.2 Action and Intention

How can we characterise an action? One attempt is to relate action and *intention*.[6] Several times the notion of *intentionality* has also been applied, and I will not make an attempt to distinguish between intention and intentionality, even though some philosophical trends have made a point of differentiating between the two notions.

For centuries, the notion of intentionality was part of philosophical controversies over how to explain physical phenomena. Assuming an Aristotelian outlook, one tried to explain why a stone will fall towards the ground by pointing out that the stone does the falling due to some intrinsic tendencies in the stone. A principal feature of the so-called scientific revolution was to eliminate any references to purposes or intentionalities from explanations of physical phenomena: one should only

---

[4] For a presentation for a genetic epistemology, see Piaget (1970). For an elaborated presentation of his conception of learning mathematics, see Part 2, written by Piager, in Beth and Piaget (1966).

[5] Piaget's generic epistemology was reworked by Glasersfeld into what is referred to as radical constructivism (see Glasersfeld, 1991, 1995). This constructivism sees the learner as actually constructing their knowledge. Glasersfeld's notion of construction is closely related to Piaget's conceptions of assimilation and accommodation, presenting the construction of knowledge as being somehow similar to the growth of natural phenomena, rather than being something similar to human actions.

[6] For discussion of the notion of action, see, for instance, Paul (2021), Searle (1983), and Wilson (1989).

5.2 Action and Intention

refer to something taking place *before* the event, and not to something following *after* the event.

An important extension of the scope of such cause-effect explanations was provided by Charles Darwin, when he proposed the notion of natural selection. Until then, biological phenomena had been interpreted in terms of purposes or intentionalities: the rabbit was brown so that it could better hide on the ground; the giraffe's neck was long so that it could eat leaves from the trees; and the kangaroo had a pouch so that it could carry around its baby. However, the theory of natural selection made it possible to provide cause-effect explanations of any such biological phenomena.

The next step came with respect to the explanations of human phenomena. It seems straightforward to use intentional explanations for interpreting human actions: one does something in order to obtain something, one does things with a purpose, one does something due to one's intentions. However, the behaviourist approach tried to conquer the human domain for cause-effect explanations.[7] Sigmund Freud was also deeply inspired by the pattern of cause-effect explanations, and tried to interpret features of human behaviour by referring to the unconsciousness, which he considered a cause of what we do.[8]

Considering this broad trend towards applying cause-effect explanations, it seems it would be swimming against the current to pay particular attention to the notion of intentionality. This is, however, what I am going to do, and let me start by referring to the work of Franz Brentano (1995a, b). In 1874, he published the original German version of *Psychology From an Empirical Standpoint*. Until then, psychology had taken the form of rather philosophical speculations about the human mind, but Brentano wanted to provide psychology with a solid empirical basis. However, what he had in mind was quite different from what one today might associate with "empirical". Brentano did not advocate any form of experimental psychology, but instead an introspective psychology. According to Brentano, mental phenomena can be observed through an "inner consciousness", and furthermore they can *only* be observed this way. Inner consciousness provides us with the only possible access to psychological phenomena by addressing our own consciousness. This is in fact the only consciousness we have access to. Inner consciousness provides us with the only possible empirical basis for psychology.[9]

---

[7] For a systematic exposition of patterns of scientific explanation from the perspective of logical positivism, see Hempel (1965).

[8] See, for instance, the final section of *The Interpretations of Dreams*, first published in German in 1899, where Freud (1997) presents his conception of psychoanalytic explanations and assigns a role to the unconsciousness similar to the role of "reality" in natural science explanations.

[9] In *Principles of Physiological Psychology*, first published in German in 1974, the same year as Brentano published *Psychology From an Empirical Standpoint*, Wundt (1904) suggested an empirical basis for psychology, quite different from the one suggested by Brentano. By basing psychology on physiological observations, Wundt wanted to establish an empirical basis for psychology within the paradigm of natural sciences.

One important observation to be made by "inner consciousness" is that the mind is always directing itself towards something. We always have something in mind, which Brentano refers to as "intentional in-existence". He finds that intentional in-existence is characteristic of mental phenomena and that no physical phenomenon shows any similar characteristics. This means that Brentano can define mental phenomena as "those phenomena which contain an object intentionally within themselves" (see Brentano, 1995a, p. 89). Mental phenomena always demonstrate intentionalities according to Brentano.[10]

I am not trying to follow Brentano's understanding of psychology, nor to follow its further development into a general phenomenological outlook. However, I feel deeply inspired by the attention he pays to intentionalities. I find that, when trying to interpret human phenomena, the notion of intentionality does play an important role. In particular, I find that *the notion of intentionality is crucial for interpreting human actions*.

An action is directed towards something. It is guided by visions about the future and what could be achieved through the action. An acting person has something in mind. Naturally, one should not assume that actions are guided by transparent visions. The intentions driving an action could include concealed features; they might be opaque. Still, I find that identifying such intentions is crucial when trying to understand an action.

When seeing learning as action, one assumes that, in order to interpret students' engagement in learning processes, one needs to consider the possible intentions that might be driving their learning actions. One needs to consider, as well, that students might withdraw their intentions from such processes. The struggle to be first in line next to the teacher's desk is an example of some students putting their intentions into their learning actions. The formation of the group of silent girls is an example of other students withdrawing their intentions from the "official" learning processes and turning them in other directions. In order to interpret what is taking place in the mathematics classroom, I find it important to address how students' intentions for learning mathematics might be strengthened, changed, modified, weakened, withdrawn, and maybe redirected.

5.3 Intention and Foreground

Can intentions be social constructions? As originally presented by Brentano – and further elaborated by Edmund Husserl into a phenomenological outlook – intentionalities refer to a principal feature of the human mind. As an ever-running stream, consciousness directs itself towards something. Brentano and Husserl present intentionalities as operating *a priori* to any social formation of the mind. My conception

---

[10] For further discussions of Brentano's ideas, see, for instance, Albertazzi (2006), Jacquette (2004), and Smith (1994). See also Skovsmose (2014a) for a presentation of Brentano's ideas.

5.3 Intention and Foreground

of intentionalities is different, as I find that we are dealing with phenomena that are deeply influenced by social factors.

Let me illustrate by an example. What kinds of intentions might lie behind a child wishing to become a scientist? This depends on the child's conception of what a scientist is. Such conceptions have been explored by studying children's drawings of scientists.[11] I remember once being told about a study revealing a marked difference between drawings made by children from Europe and drawings made by children from Africa. Among European children, drawings of crazy-looking scientists with rolling eyes and small planets circling their heads were common, while drawings of scientists appearing as medical doctors giving an injection were common among African children. A child's conception of being a scientist depends on many inputs. It could be based on what they experience in school, it could be based on films or commercials, or it could be based on personal experiences. Conceptions of scientists – and, therefore, of the possible intentions behind becoming a scientist – are complex social constructions. In a similar way, I see children's possible intentions to engage, or not, in any kind of learning processes as being complex social constructions.

In order to further conceptualise the social formation of intentionalities, I find the notion of *foreground* to be useful.[12] By using the term in relation to a student, I understand this to mean the possibilities that are available for that student. Very different foregrounds can be spread out in front of a student, depending on economic, political, culture, religious, and discursive factors. Such differences can be directly exposed by means of statistics. For instance, for the year 2013, statistics show that, among the students enrolled in medical courses at the University of São Paulo (USP), 7% were black, even though more than half of the population in Brazil can be classified as being black.[13] This, together with many other statistics, reveals that the prospects in life for black students are very different from that of white students. Their foregrounds are structured differently. This structuration concerns the likelihood of progressing to further education, of moving to an affluent neighbourhood, of becoming a victim of a crime, or of ending up in jail, and so on.

When we consider historical developments, we can identify huge differences with respect to the formation of foregrounds. One can, for instance, consider how the foregrounds of girls and boys in a country like Denmark were laid out a hundred years ago, and compare this with the situation today. One can study the number of girls and boys enrolled in medical courses, in engineering courses, or in mathematical courses. Across time periods, different roles have been assigned to men and women. Such roles might have changed, but we are not in a situation where such differences can be ignored.

---

[11] See, for instance, Jane et al. (2007).

[12] The notion of foreground was introduced in Skovsmose (1994). Since then, it has been developed further in many steps. See, for instance, Skovsmose (2007, 2014a, 2018). See also Biotto Filho (2015), and Roncato (2021).

[13] For a presentation of this and other statistics, see Silva and Skovsmose (2019).

Foregrounds are social constructions. Simultaneously, foregrounds are formed by personal factors. Thus, students located in apparently the same social situation might experience different possibilities in life. They might encounter the same difficulties, for instance with respect to preconceptions, but they might deal with such preconceptions in different ways. One student might be more optimistic than the other, and one might have different levels of support from their parents and family than the other. In this sense, foregrounds are not only social but also personal constructions. Foregrounds are part of who we are going to become, how we perceive possibilities, and how we tend to act.

Foregrounds influence how we direct and redirect our intentionalities. They influence what possibilities, risks, and dangers we might perceive or ignore. Intentionalities are directed towards something, and this whole landscape of "somethings" is part of our foregrounds. A foreground might appear rich with possibilities, but it might also appear devoid of them. The notion of possibility seems to connote something to look forward to; however, whenever I talk about possibilities, I also have in mind those which appear alarming or frightening. Possibilities might include risks and dangers, and one's intentionalities might be directed so as to prevent such risks and dangers from turning real. Foregrounds guide our intentionalities, which are the driving forces of any kind of action, and also of learning activities.

5.4 Foreground and Motive

How could foregrounds structure motives for learning? Before trying to clarify this question, let me refer to two examples, one where the teachers felt frustrated and one where a teacher was surprised.

On several occasions, teachers have come to tell me about their frustrations. They have been teachers dedicated to critical mathematics education, who have been concerned about addressing real-life problems experienced by their students. They tell me about projects that they found to be relevant for their students living in precarious situations. They tell me about projects addressing unemployment, water pollution, or inflation. However, the teachers also tell me that, at times, the students did not seem very interested in such projects. They did not demonstrate any particular motives for engaging in the projects related to their own lived-through realities. How could that be?

A teacher was planning some project work, and she suggested several possible topics, but was also fully aware of the importance of the students themselves deciding on the topic.[14] They discussed the possibilities, and in the end the students voted for their own suggestion: "surfing". The teacher was surprised. How could such a project make any sense to the students? The students, around 10–11 years old, were all living in a poor neighbourhood in a city inside the São Paulo state. It was a long

---

[14] This example I have presented previously; see Skovsmose (2016).

5.4 Foreground and Motive

distance away from the ocean, and most likely none of the students had ever seen the ocean. How could a project about surfing make any sense to them? The project was not related to the students' everyday lives, and for economic reasons it would be impossible to include an excursion to the sea as part of the project.

Nevertheless, the students had voted for "surfing", and they put their full energy into it. They were really motivated. There was much to address with respect to surfing, even so far away from the sea. One could find out about the different types of surfboards, about their shapes and sizes. The prices were also a huge topic of discussion, and the students could explore the remarkable price differences, and also the fact that it might be possible to use homemade surfboards. Where could they go surfing? Even if the students were not leaving the classroom, they could consider where to surf. Which beaches could one find the proper waves at? What would be the price for travelling there? If one cannot do any real travelling, one can still imagine travelling. Many famous surfers come from Brazil, and where do they do their training? Where do they join the international competitions? There are many questions to be addressed, and the students engaged in all of them with enthusiasm.

How can we interpret these two examples? According to common sense, students should get motives for learning mathematics when the topics can be related to their daily life experiences. "Being familiar with" has been associated with "getting motives for". This appears to be a plausible association, but it need not to be taken as a given. Other factors might also create motives for learning mathematics. I agree that having motives for learning can be stimulated by our experiences or relationships, but these need not be relationships between the learning content and the students' everyday experiences. *Relationships between the learning content and features of the students' foregrounds might establish strong motives for learning.*

This observation is a consequence of seeing learning as action. In order to understand actions, one needs to consider the intentionalities that might be brought into the actions, as intentionalities provide the driving force for actions. Intentionalities are directed towards features of the students' foregrounds. In order to understand their possible motives for learning – and also their possible lack of such motives – one needs to consider their foregrounds.

It need not be a surprise that students living in precarious situations might not find motives for addressing some of the everyday problems they have experienced. Like any students, they might be interested in directing their learning activities towards features of their foregrounds that they find fascinating. Instead of simply associating "having motives for" with "being familiar with", I suggest that "having motives for" rather becomes associated with "hoping for".[15] Hopes and visions provide energy for actions, and also for engaging in learning mathematics.[16]

---

[15] By making this emphasis, I distance myself from many interpretations of motivation in mathematics education. See, for instance, Moddleton (2020).

[16] Freire (2014) highlights the importance of hope as making integral part of an educational activity, and Soares (2022) points out the importance of dreams in forming students' motives for learning.

I prefer to talk about *motives* for learning rather than about *motivations* for learning. In everyday discourses, motives and motivations normally operate as synonyms, but they include very different theoretical connotations. The notion of motivation has been elaborated in great detail with reference to a behaviourist outlook, and the notion is part of several attempts to provide cause-effect explanations of human behaviour and actions.[17] The notion of motives I relate to seeing actions as driven by intentions directed towards features of foregrounds. Consequently, I prefer to talk about motives for learning rather than about motivations for learning.

5.5 Motive and Mathematics

How can we locate students' motives for learning mathematics? My suggestion is to look into students' foregrounds, but how might mathematics appear in these foregrounds? Let me indicate some possible answers.

Students might find mathematics to be necessary for doing well in school. They may come to master the curriculum and pass tests and examinations. I can refer to a remark made by Godfrey Hardy (1967) in *A Mathematician's Apology*, where he states that, when he was young, he had no clear vision of becoming a mathematician, and his motive for doing mathematics was rather "to beat the other boys" (p. 144). He was thinking of mathematics as a game where he could be the winner. He could be first in class, in the same way as Peter could be first in line next to the teacher's desk. Such instrumental motives might be driving forces for some students' learning of mathematics.[18]

With a French context in mind, Pierre Bourdieu (1996) uses the expression of "state nobility" as a metaphor for people in high positions in administration and public services. He highlights that mathematics education is part of the nomination of this "nobility". The mathematical knowledge that is required for passing the mathematical tests might be of little relevance for the actual work content of the state nobility. However, what *is* relevant is the act of passing the required tests, which need to be considered difficult. The passing of such tests provides the state nobility with its authority: they were the first in line. Such kinds of instrumentality might also be part of students' foregrounds and have an impact on their motives for learning mathematics.

Students might be interested in doing, say, engineering, and school mathematics might be relevant for getting access to such study programmes. As an example, I can refer to a study where indigenous students living in Brazil were asked about their motives for learning mathematics.[19] One of the students, Matheus (pseudonym),

---

[17] See Madsen (1974) for an overview of theories of motivation resonating with a behaviourist conception of learning.

[18] Mellin-Olsen (1981) explores the notion of instrumentalism.

[19] See Skovsmose, Alrø and Valero in collaboration with Silvério and Scandiuzzi (2007).

found that mathematics was relevant for life in the community. He referred to the measuring of land and to the adding and division of crops as mathematical activities that were relevant. Another student, Patrick (pseudonym), told the researchers that he wanted to study medicine. He also highlighted the relevance of mathematics for doing so, but added that he had no idea how mathematics could be relevant for studying medicine.

In his studies, Sikunder Baber (2007, 2012) interviewed students from Pakistani families living in Denmark. It was generally recognised by both students and parents that, in order for the child of an immigrant family to have a respectable life in Denmark, it was crucial for them to complete further education. To many, mathematics was recognised as being crucial for doing so, and a lot of the students' energy – and also their parents' energy – was directed towards obtaining the best possible marks in mathematics. In a sense, their actions were directed towards being included in the Danish "state nobility".

The Algebra Project, as directed by Bob Moses, paid particular attention to black students living in poor neighbourhoods in the USA.[20] The aim was to secure black students' access to further education. Such access was profoundly obstructed by black students' low achievements in mathematics, and Moses wanted to eliminate this obstruction. The Algebra Project concentrated on mathematical topics, and it aimed at making students master them. The project recognised that passing tests in mathematics plays a crucial role in the formation of black students' foregrounds.

Inspired by Paulo Freire and by Marilyn Frankenstein, Eric Gutstein (2006) has talked about reading and writing the world with mathematics. By "reading the world", he refers to an interpretation of the world with the aim of pointing out cases of injustice, oppression, and exploitation. By "writing the world", he refers to ways of changing the world. He points out that mathematics can play an important role with respect to both reading and writing activities. Through his educational practice, Gutstein shows how mathematics can be relevant for addressing living conditions in poor neighbourhoods, for addressing processes of gentrification, and for pointing out different features of racism and sexism. By doing so, he connects educational processes with activist approaches.[21]

Activities of reading and writing the world might form students' visions about the future, and establish new motives for learning mathematics. Lupes, one of Gutstein's students, expressed this in the following way:

> With every single thing about math that I learned came something else. Sometimes I learned more of other things instead of math. I learned to think of fairness, injustices and so forth everywhere I see numbers distorted in the world. Now my mind is opened to so many new things. I'm more independent and aware. I have learned to be strong in every way you can think of it. (Lupes, quoted from Gutstein, 2003, p. 37)

Students' motives for doing critical readings and writings of the world might be formed through the act of doing so. Motives for learning mathematics should not be

---

[20] See Moses and Cobb (2001).
[21] See Gutstein (2009, 2012, 2016, 2018) and also Frankenstein (1983, 1989, 1998, 2012).

thought of as pre-existing structures, but as being developed and formed through the very process of learning mathematics. Motives are constructed through dynamic interactions between students' foregrounds, backgrounds, as well as the very processes of learning. Motives are social constructions.

Motives for learning mathematics might be formed, not due to any experienced or imagined political or utilitarian value of mathematics, but due to features of mathematics itself. Mathematics might fascinate some students. Mathematical problems might capture some students' imaginations and establish their motives for learning and exploring mathematics. This happened to Hardy, and this might happen to many other students as well. There is, however, also an inverse phenomenon that needs to be considered. Some students express a fear of mathematics and consequently abandon it. This phenomenon has been studied by David Kollosche (2017), who talks about students' auto-exclusions from mathematics. Auto-exclusion can be expressed directly by the students when declaring that mathematics is not for them, that they are not able to learn mathematics, and that they are not interested in doing so. One can imagine that the group of silent girls later on might articulate such auto-exclusion – a kind of self-protection against possible failures in mathematics. Auto-exclusion is a declaration of not having a motive for learning mathematics.

Foregrounds are always in transition, and so are the possible appearances of mathematics. As foregrounds are social constructions, so become students' motives for learning mathematics. These motives are formed, elaborated, intensified, moderated, and maybe eliminated through complex social processes.

5.6 Mathematics and Life-Worlds

How can we locate mathematics in our life-worlds? In 1936, the notion of life-world was introduced by Edmund Husserl (1970), when the original German version of *The Crisis of European Sciences and Transcendental Phenomenology* was published.[22] Husserl considers a life-world as representing our immediate daily life experiences, as they appear before they are moulded by more systematic insight or scientific knowledge. To me, the notion of life-world is extremely useful, but I use it in a different way than Husserl and phenomenology in general. I see life-worlds as complex social constructions, reflecting economic, political, religious, cultural, and discursive structures. No life-world exists *a priori* to such structuration. Furthermore, life-worlds become reconfigured through personal beliefs, conceptions, misconceptions, and idiosyncrasies. Thus, my interpretation of life-world is

---

[22] For a general elaboration of his phenomenological outlook, see Husserl (1998, 1999, 2001). For a presentation of Husserl's philosophy, see, for instance, Moran (2005) and Smith (2007). For further elaborations of the notion of life-worlds within a phenomenological outlook, see Schutz and Luckmann (1973). For an elaboration of the notion of life-world beyond the phenomenological outlook, see Habermas (1984, 1987).

similar to that of foreground: we are dealing with personal elaborations of social constructions.

I see a person's foreground as being a province of the person's life-world. One can also consider a person's background and actual situation as being such provinces. While the foreground represents the person's future possibilities and difficulties in life, the background represents the person's life history. The actual situation refers to that province of the life-world where foregrounds turn into backgrounds. Even though backgrounds represent already-lived-through realities, they are not fixed. Experiences of the past might change over time. A person might come to see his or her past from a new perspective. The person might come to realise that something that was previously taken as a given could have been challenged, and maybe changed. A person might come to see that he or she had suffered some kind of oppression, which at that time was not recognised as such.

How can we locate mathematics in our life-worlds? Answering this apparently straightforward question has turned out to be quite complicated due to two interconnected processes: mathematisation and demathematisation. The significance of these two processes has been pointed out by Eva Jablonka and Uwe Gellert (2007).[23] One can illustrate what is meant by mathematisation and demathematisation by considering the impact that mathematics-based digitalisation has on our daily lives. When buying at the supermarket, we might pay with a credit card. We are dealing with an apparently simple operation: we just have to insert the credit card in the slot machine, press a few numbers, and we have paid for what we have bought. However, beneath this surface of simplicity, a huge amount of mathematics is brought into operation. The processes of digitalisation, coding, decoding, and the very economic transactions are made possible only due to a huge amount of carefully elaborated mathematical algorithms. Not only at the supermarket, but in a broad range of daily life situations, mathematics is operating beneath a surface of apparently simple routines. One can think of our use of mobile phones, searching on the internet, and adding friends on Facebook. Our life-worlds become formed by a huge amounts of mathematical algorithms brought in operation.

Processes of mathematisation are accompanied by processes of demathematisation. By demathematisation, I refer to the phenomenon that mathematisation does not appear to be part of daily life practices. The act of paying with a credit card is based on mathematisations of underlying processes, while the very act of paying with the credit card appears to be a non-mathematical act. Previously, the act of paying for what one is buying included some mathematical calculations like the adding up of prices, as well as subtractions (in order to verify what to receive in return). No such calculations seem relevant any longer: the new practice of paying with the credit card appears demathematised. Similar processes of demathematisation occur when we are dealing with any other process of mathematisation. Nobody associates

---

[23] See also Penteado and Skovsmose (2015) for a further discussion of the relationships between mathematisation and demathematisation.

mathematics with performing a search on the internet, doing a Skype meeting, or looking at Facebook.

The extensive processes of mathematisation and demathematisation make it difficult to locate mathematics in our life-worlds. Massive mathematical procedures provide a deep structuring of our life-worlds, which, however, become covered by a surface of simplicity. Like an act of magic, mathematics has disappeared. This magic has an impact on the visibility, or rather lack of visibility, of mathematics in students' foregrounds, and consequently on the formation of their motives for learning mathematics. Possible motives might be annihilated due to the mathematisation taking place "behind the stage" that brings about a demathematisation taking place "on stage".

That the appearance of mathematics within students' foreground is opaque has been observed several times. I have already referred to Patrick, who did acknowledge that mathematics is important for studying medicine, even though he could not see precisely why. Observations made by Daniela Alves Soares (2022) point in the same direction. She has explored the dreams and aspirations of students coming from poorer neighbourhoods in both Brazil and Colombia. She finds that mathematics does not appear in any manifest way in their visions about their own future, but only in fuzzy formats. In addition, she points out that most students assume the belief that "mathematics is everywhere", although it appears to be "nowhere". Most students do not see mathematics as something important in their lives, neither now nor in the future. Similar observations have been made by Manuella Carrijo (2023) in her studies of immigrant students living in Brazil. To these students, mathematics also appears only in opaque forms in their visions about their own futures.[24]

It might be that, to students in marginalised positions, mathematics only appears in an opaque format in their foregrounds. This might not only be due to processes of demathematisation, but also to the fact that the daily life problems that they experience are so urgent that it becomes difficult, if not impossible, to visualise radical new possibilities in life. *Marginalised students' foregrounds might be ruined due to the very processes of marginalisation.* That groups of marginalised students could end up being classified as low performers in mathematics might first of all be due to their ruined foregrounds. I see a ruined foreground as being a devastating socially fabricated learning obstacle.[25]

---

[24] In Skovsmose (2014a), I have presented interviews with students living in a poor neighbourhood in a city in the interior of the São Paulo State. They also expressed opaque visions of mathematics as a feature of their foregrounds.

[25] In his studies of African-American students in the USA, Jamaal Ince (n.d., in progress) takes inspiration from this observation. Simultaneously, he shows how processes of marginalisation might be confronted through mathematics education.

5.7 A Conclusion: Life-Worlds and Learning

Can learning be a political act? Can learning *mathematics* be a political act? Students' motives for learning mathematics are formed by their foregrounds, which in turn are formed by economic, political, cultural, religious, and discursive factors. Let us now consider how such factors themselves might be changed. Could learning mathematics change structural features of foregrounds and life-worlds?

I will refer to changing social structures as *political acts*. Social movements have been driving forces in implementing political actions, like feminist movements, antiracist movements, ecological movements, and anti-war movements. However, political actions can also be associated to, say, right-wing movements, totalitarian outlooks, and anti-democratic projects. The implementation of the Nuremberg Laws was an example of such an act, which ruined the life-worlds of millions of people as well as that of Husserl. When talking about political action, we need to be fully aware that we are dealing with a contested concept. This can be interpreted in radically different ways, where such differences might be of a deep political nature. Contested concepts include clashes between contrasting world-views.

When interpreting learning as action, it becomes consequential to ask: Can learning change social structures? Can learning mathematics form or reform foregrounds and life-worlds? As mentioned, the background for the Algebra Project was that many black students experienced the foreshortening of their foregrounds, also with respect to moving on to further education. The explicit aim of the project was to add to the students' foregrounds.

Gutstein engaged his students in political actions by reading and writing the world with mathematics. He wanted to engage them in identifying cases of social injustice and prepare them for political action. While the Algebra Project concentrated on creating new possibilities for the individual students, Gutstein's project had the ambition of creating new possibilities for a community suffering from social and economic oppression. We are dealing with an explicit example of turning the learning of mathematics into a political act.

By making this observation, the conceptual circle that I set out to explore has been completed. I find that learning mathematics can come to be part of political actions that might change features of the students' life-worlds, and also the life-worlds of broader communities. Learning mathematics is not to be considered only as a classroom activity. It is a way of being, and also a way of being political.

References

Albertazzi, L. (2006). *Immanent realism: An introduction to Brentano.* Springer.
Baber, S. A. (2007). *Interplay of citizenship, education and mathematics: Formation of foregrounds of Pakistani immigrants in Denmark.* Doctoral dissertation, Aalborg University.

Baber, S. (2012). Learning of mathematics among Pakistani immigrant children in Barcelona: A socio-cultural perspective. In O. Skovsmose & B. Greer (Eds.), *Opening the cage: Critique and politics of mathematics education* (pp. 144–166). Sense Publishers.

Beth, E. W., & Piaget, J. (1966). *Mathematics, epistemology and psychology*. Reidel.

Biotto Filho, D. (2015). *Quem não sonhou em ser um jogador de futebol? Trabalho com projetos para reelaborar foregrounds (Who never dreamed of being a soccer player? Working with projects for elaborating foregrounds)*. Doctoral dissertation, Universidade Estadual Paulista (Unesp).

Bourdieu, P. (1996). *The state nobility: Elite schools in the field of power*. Polity Press.

Brentano, F. (1995a). *Psychology from an empirical standpoint* (Translated by A. C. Rancurello, D. B. Terrell, and L. L. McAlister; with an introduction by Peter Simons). Routledge.

Brentano, F. (1995b). *Descriptive psychology* (Translated and edited by Benito Müller). Routledge.

Carrijo, M. H. S. (2023). *Beyond borders and racism: Towards an inclusive mathematics education for immigrant students*. Doctoral Dissertation, Universidade Estadual Paulista (Unesp).

Frankenstein, M. (1983). Critical mathematics education: An application of Paulo Freire's epistemology. *Journal of Education, 165*(4), 315–339.

Frankenstein, M. (1989). *Relearning mathematics: A different third R – Radical maths*. Free Association Books.

Frankenstein, M. (1998). Reading the world with math: Goals for a critical mathematical literacy curriculum. In E. Lee, D. Menkart, & M. Okazawa-Rey (Eds.), *Beyond heroes and holidays: A practical guide to K-12 anti-racist, multicultural education and staff development*. Network of Educators on the Americas.

Frankenstein, M. (2012). Beyond math content and process: Proposals for underlying aspects of social justice education. In A. A. Wager & D. W. Stinson (Eds.), *Teaching mathematics for social justice: Conversations with mathematics educators* (pp. 49–62). NCTM, National Council of Mathematics Teachers.

Freire, P. (2014). *Pedagogy of hope: Reliving pedagogy of the oppressed* (With notes by Ana Maria Araújo Freire, and translated by Robert R. Barr). Bloomsbury.

Freud, S. (1997). *The interpretations of dreams* (Translation by A. A. Brill, Introduction by Stephen Wilson). Wordsworth Classics.

Gutstein, E. (2003). Teaching and learning mathematics for social justice in an urban, Latino school. *Journal for Research in Mathematics Education, 34*, 37–73.

Gutstein, E. (2006). *Reading and writing the world with mathematics: Toward a pedagogy for social justice*. Routledge.

Gutstein, E. (2009). Possibilities and challenges in teaching mathematics for social justice. In P. Ernest, B. Greer, & B. Sriraman (Eds.), *Critical issues in mathematics education* (pp. 351–373). Information Age Publishing.

Gutstein, E. (2012). Reflections on teaching and learning mathematics for social justice in urban schools. In A. A. Wager & D. W. Stinson (Eds.), *Teaching mathematics for social justice: Conversations with mathematics educators* (pp. 63–78). NCTM, National Council of Mathematics Teachers.

Gutstein, E. (2016). "Our issue, our people – Math as our weapon": Critical mathematics in a Chicago neighborhood high school. *Journal for Research in Mathematics Education, 47*(5), 454–504.

Gutstein, E. (2018). The struggle is pedagogical: Learning to teach critical mathematics. In P. Ernest (Ed.), *The philosophy of mathematics education today* (pp. 131–143). Springer.

Habermas, J. (1984, 1987). *The theory of communicative action* (Vol. I–II). Heinemann/Polity Press.

Hardy, G. H. (1967). *A mathematician's apology* (With a foreword by C. P. Snow). Cambridge University Press.

Hempel, C. G. (1965). *Aspects of scientific explanation*. The Free Press.

Husserl, E. (1970). *The crisis of European sciences and transcendental phenomenology* (Translated with an introduction by David Car). Northwestern University Press.

References

Husserl, E. (1998). *Ideas pertaining to a pure phenomenology and to a phenomenological philosophy: First book*. Kluwer Academic Publishers.

Husserl, E. (1999). *Cartesian meditations: An introduction to phenomenology* (Translated by Dorion Cairns). Kluwer Academic Publishers.

Husserl, E. (2001). *Logical investigations: Volume 1–2* (Translated by J. N. Findlay). Routledge.

Ince, J. (n.d.). *Belief in African Americans as learners in mathematics*. Doctoral dissertation, Universidade Estadual Paulista (Unesp) (in progress).

Jablonka, E., & Gellert, U. (2007). Mathematisation – Demathematisation. In U. Gellert & E. Jablonka (Eds.), *Mathematization and de-mathematization: Social, philosophical and educational ramifications* (pp. 1–18). Sense Publishers.

Jacquette, D. (Ed.). (2004). *The Cambridge companion to Brentano*. Cambridge University Press.

Jane, B., Fleer, M., & Gipps, J. F. (2007). Changing children's views of science and scientists through school-based teaching. *Asia Pacific Forum on Science Learning and Teaching, 8*(1), 1–21.

Kollosche, D. (2017). Auto-exclusion in mathematics education. *Revista Paranaense de Educação Matemática, 6*(12), 38–63.

Madsen, K. B. (1974). *Modern theories of motivation*. Munksgaard.

Mellin-Olsen, S. (1981). Instrumentalism as an educational concept. *Educational Studies in Mathematics, 12*, 351–367.

Moddleton, J. (2020). Motivation in mathematical learning. In S. Lerman (Ed.), *Encyclopaedia of mathematics education* (2nd ed., pp. 635–637). Springer.

Moran, D. (2005). *Husserl: Founder of phenomenology*. Polity Press.

Moses, R. B., & Cobb, C. E. (2001). *Radical equations: Civil rights from Mississippi to the Algebra Project*. Beacon Press.

Paul, S. (2021). *Philosophy of action: A contemporary introduction*. Routledge.

von Glasersfeld, E. (Ed.). (1991). *Radical constructivism in mathematics education*. Kluwer Academic Publishers.

von Glasersfeld, E. (1995). *Radical constructivism: A way of knowing and learning*. Falmer.

Penteado, M. G., & Skovsmose, O. (2015). Mathemacy in a mathematized and demathematized world. In B. S. D'Ambrosio & C. E. Lopes (Eds.), *Creative insubordination in Brazilian mathematics education research* (pp. 107–117). Lulu Press.

Piaget, J. (1970). *Genetic epistemology*. Columbia University Press.

Roncato, C. R. (2021). *Significado em educação matemática e estudantes com deficiências: Possibilidades de encontros de conceitos (Meaning in mathematics education and students with disabilities: Possibilities and meaning between concepts)*. Doctoral dissertation, Universidade Estadual Paulista (Unesp).

Schutz, A., & Luckmann, T. (1973). *The structures of the life-world* (Vol. I–II). Northwestern University Press.

Searle, J. (1983). *Intentionality*. Cambridge University Press.

Silva, G. H. G., & Skovsmose, O. (2019). Affirmative actions in terms of special rights: Confronting structural violence in Brazilian higher education. *Power and Education, 11*, 204–220.

Skinner, B. F. (1957). *Verbal behaviour*. Appleton-Century-Crofts.

Skovsmose, O. (1994). *Towards a philosophy of critical mathematics education*. Kluwer Academic Publishers.

Skovsmose, O. (2005a). *Travelling through education: Uncertainty, mathematics, responsibility*. Sense Publishers.

Skovsmose, O. (2005b). Meaning in mathematics education. In J. Kilpatrick, C. Hoyles, O. Skovsmose, & in collaboration with Valero, P (Eds.), *Meaning in mathematics education* (pp. 85–102). Springer.

Skovsmose, O. (2007). Foregrounds and politics of learning obstacles. In U. Gellert & E. Jablonka (Eds.), *Mathematisation – Demathematisation: Social, philosophical, sociological and educational ramifications* (pp. 81–94). Sense Publishers.

Skovsmose, O. (2014a). *Foregrounds: Opaque stories about learning*. Sense Publishers.

Skovsmose, O. (2014b). *Critique as uncertainty*. Information Age Publishing.
Skovsmose, O. (2016). An intentionality-interpretation of meaning in mathematics education. *Educational Studies in Mathematics, 92*(3), 411–424.
Skovsmose, O. (2018). Students' foregrounds and politics of meaning in mathematics education. In P. Ernest (Ed.), *The philosophy of mathematics education today* (pp. 115–130). Springer.
Skovsmose, O., Alrø, H., & Valero, P., in collaboration with Silvério, A. P. & Scandiuzzi, P. P. (2007). "Before you divide you have to add": Interviewing Indian students' foregrounds. In B. Sriraman (Ed.), *International perspectives on social justice in mathematics education. The Montana mathematics enthusiast*, Monograph 1 (pp. 151–167). The University of Montana Press.
Smith, D. W. (2007). *Husserl*. Routledge.
Smith, B. (1994). *Australian philosophy: The legacy of Franz Brentano*. Open Court.
Soares, D. A. (2022). *Sonhos de adolescentes em desvantagem social: Vida, escola e educação matemática* (Dreams of socially disadvantaged teenagers: Life, school and mathematics education). Doctoral dissertation, Universidade Estadual Paulista (Unesp).
Watson, J. B. (1913). Psychology as the behaviourists view is. *Psychological Bulletin, 31*(10), 755–776.
Wilson, G. (1989). *The intentionality of human action*. Stanford University Press.
Wittgenstein, L. (1953). *Philosophical investigations*. Blackwell.
Wundt, W. (1904). *Principles of physiological psychology*. Macmillan.

Part II
Crises and Formatting

Chapter 6
Crisis and Critique

Abstract The notion of crisis is related to the notion of critique. According to Marx and to orthodox Marxism, there exists a hierarchy of crises, where the principal crisis is formed by the antagonism between workers and capital owners. Contrary to this, I do not assume the existence of a pre-given structural relationship among crises, but see crises as interacting in many different ways. As there are different kind of crises, there are different forms of critique. Some critical approaches might be related, some might be independent of each other; some might find inspiration in other patterns of critique; some might be based on theoretical insight; and some might be formed by personal experiences. One should not assume the existence of a unifying and basic critical approach. As crises are manifold, so are critiques. It is this understanding of critique that I want to associate to critical mathematics education.

Keywords Crisis · Critique · Critical situation · Hierarchy of crises · Multiplicity of crises · Critical theory · Critical race theory

In 2017, the Ninth Mathematics Education and Society Conference (MES 9) took place in Volos in Greece. It had the title *Mathematics Education and Life at Times of Crisis*.[1] I joined the conference and got much inspiration for rethinking critical mathematics education from it.

Paying attention to the news, one learns about economic crises that affect families, companies, countries (such as Greece at that time); about security crises which may turn everyday life terrifying for many people; about energy crises that could turn into military conflicts; about environmental crises caused by ever-increasing pollution; and about health crises due to pandemic situations. Much evidence indicates that we live in times of crisis.

In the article "The Anthropocene", Paul J. Crutzen and Eugene F. Stoermer (2000) add to this evidence. The notion of anthropocene refers to the recent

---

[1] See Chronaki (2017).

historical period where human influence on the atmosphere is so significant that we might be witnessing a new geological epoch that needs a name: the anthropocene. A characteristic of this epoch is that we are moving through a multitude of crises, and that it is an illusion to assume that the occurrence of crises is temporary; rather, we have to deal with crises as ongoing phenomena.[2]

In the following, I relate the notions of crisis and critique. Then I present the assumption that there exists a hierarchy of crises, where the principal crisis is formed by the antagonism between workers and capital owners – the defining assumption for orthodox Marxists. I discuss an alternative idea, that crises might be interrelated in a multitude of ways. This idea allows for a broader conception of the notion of critique, which I want to associate to critical mathematics education.

6.1 Critical Situation

The Greek words *kritikós*, *kritikê*, and *kritikón* refer to the capacity of judging and deciding. These adjectives are derived from the verb *kríno*, which means to separate, to distinguish, or to decide. *Krísis* is a noun with meanings like "act of distinguishing", "act of choosing", and "act of deciding". Like *kritikós*, it also refers to a point of extreme illness. The words *kríno* and *krísis* have the same root, namely *kri*, which means to choose or to separate.[3]

Crisis calls for critique. This brings us to the notion of a critical situation. A patient at a hospital might be in such a situation. Things could go both ways: the patient might be cured, or the situation might end up being fatal. In the most direct way, a critical situation represents a crisis, which needs to be critically addressed. This not only applies to a person at the hospital, but to any critical situation. I use the notion of critical situations as referring to a crisis that is calling for a critique. I allow myself to be rather free in sometimes talking about a crisis and sometimes about a critical situation.

At MES 9, various sessions added to my conception of crisis. At the symposium *"Crisis" and the Interface With Mathematics Education Research and Practice: An Everyday Issue,* organised by Aldo Parra, Arindam Bose, Jehad Alshwaikh, Magda Gonzales, Renato Marcone, and Rossi D'Souza (2017), it was pointed out that discourses of crises might represent particular world-views.[4] When we hear about security crises, we might hear about acts of terror taking place in London, or Boston, or Paris; when we hear about economic crises, we might hear about difficulties in the European Union; and when we hear about refugee crises, we might hear about

---

[2] Coles (2016) opens a profound discussion of the anthropocene with respect to mathematics education.

[3] In this clarification of the Greek words, I got important help from Irineu Bicuco, who worked at Universidade Estadual Paulista (Unesp) at Rio Claro.

[4] See also Greer, Gutiérrez, Gutstein, Mukhopadhyay, and Rampal (2017), who also organised a symposium at the MES 9 that addressed the conception of crisis; and Chassapis (2017).

people from Syria and neighbouring countries crossing the border into Europe. All such references illustrate severe difficulties, but principally difficulties as experienced by the wealthiest countries in the world.

The organisers of the symposium pointed out that crises constitute everyday life for many people around the world, not least from the so-called Third World Countries. It was also pointed out that particular strategic priorities can be associated with crisis terminology, as crises call for attention and urgent action. References to security crises can be used for justifying implementations of new patterns of control and surveillance; references to economic crises can be used for justifying austerity measures; and references to refugee crises can be used for justifying the construction of new walls and legal barriers. Furthermore, we cannot ignore the possibility that some crises might be discursive fabrications. All this brings us to acknowledge both the relevance and the complexity of addressing crises.

I will use the notion of crises in a broad sense as referring to tensions, antagonisms, risks, conflicts, struggles, wars, disasters, genocides, massacres, and catastrophes. The notion of crises might indicate an event taking place during a condensed period of time. I will talk about the genocide of the Armenian people taking place during a 1-year period as being a crisis; about the Nazi genocide of Jewish people taking place within a 10-year period as being a crisis; of the North American genocide of the indigenous population taking place over more than 100 years as being a crisis; and about the brutal exclusion of the Dalit people due to the caste system – which took place over hundreds of years – as being a crisis. My use of the notion of crisis does not always indicate an event taking place during a concentrated period of time.

6.2 A Hierarchy of Crisis and Critique

Due to the works by Immanuel Kant, the notion of critique acquired a prominent position in philosophy, similar to that of "knowledge", "justice", and "truth". Critical investigations were conducted by Kant in three monumental works: *Critique of Pure Reason* (*Kritik der reinen Vernunft*) from 1781, *Critique of Practical Reason* (*Kritik der praktischen Vernunft*) from 1788, and *Critique of Judgment* (*Kritik der Urteilskraft*) from 1790.[5] Kant was rooted in the Enlightenment movement, according to which knowledge has a principal role to play in human development. As a consequence, it seemed obvious to ask: What is knowledge? Kant tried to clarify this question through profound analytical investigations without assuming anything about the nature of knowledge. To Kant, a critique was a philosophical activity addressing our basic epistemological conditions, and he wanted to provide a once-and-for-all critique of knowledge with permanent validity.

---

[5] Kant's works have been published in many editions; see, for instance, Kant (1929, 2007, 2010).

Kant inspired Georg Friedrich Hegel, who in turn inspired Karl Marx and many others. In the book *The Life of Jesus*, first published in German in 1835, David Strauss (2019) presents Jesus as a historical figure stripped of religious features. He discusses the nature of the empirical evidence one can collect for claiming the existence of a person called Jesus. At that time, such a presentation of Jesus was unheard of. *The Life of Jesus* symbolises the start of the left Hegelian movement. In *The Essence of Christianity*, first published in German in 1841, Ludwig Feuerbach (1989), also belonging to the left Hegelian movement, made a strong critique of Christianity. Inspired by Hegel, the left Hegelians questioned a range of religious dogma and taken-for-granted positions. The left Hegelians developed the critical enterprise profoundly, and far beyond what Kant had suggested. The so-called right Hegelians formed a different movement, also using Hegel's writing as a starting point. They found it possible to celebrate the political status quo as manifested by the Prussian state.

Initially, Karl Marx belonged to the left Hegelians, but soon he distanced himself from the movement, finding it important to develop a socio-political critique in a much more systematic way. In *The German Ideology*, Marx and Engels made a ferocious critique of some left-Hegelian positions, including the one adopted by Feuerbach. *The German Ideology* was drafted in 1846, but only published in 1932.[6] Marx agreed that it was important to critique religious dogmas, but he did not want such a critique to concentrate only on these sorts of ideological structures. A critique should also address the material living conditions of the working class.

Marx's principal work, *The Capital (Das Kapital)*, whose first volume appeared in 1867, has the subtitle *Critique of Political Economy (Kritik der politischen Ökonomie)*.[7] Like Kant, Marx made heavy use of the notion of critique. However, Marx's critique was of a different nature. His critique of the political economy addresses a range of economic and political theories, but simultaneously it addresses the very economic and political structures themselves. Marx's critique is also different from the one conducted by the left Hegelians. Marx found the work of the left Hegelians to be an unsystematic point-by-point critique; instead, he wanted critique to be based on a solid theoretical foundation, which he tried to establish through a meticulous investigation of the dynamics of a capitalist economy. In exploring these dynamics, he was inspired by Hegel's presentation of historical development as following a grandiose dialectic pattern. Simultaneously, Marx also distanced himself from Hegel by providing a strict materialistic interpretation of this dialectic as being driven by the antagonism between the ever-growing exploitation of workers and the accumulation of wealth by capital owners.

Marx did not use the notion of crisis, but one can talk about the antagonism between the working class and the capital owners as being a crisis. This crisis would become deeper and deeper, and Marx tried to capture this development in terms of the tendency of the rate of profit to fall, which he considered a crucial factor in the

---

[6] See Marx (1998).

[7] Marx's works have been published in many editions; see, for instance, Marx (1970, 1992\1993).

collapse of capitalism. When capitalist exploitation had been brought to its limit, the result would be an open crisis, a revolution, which clears the way for a new economic and political order.

Crises can be interrelated. Conflicts at a particular workplace can be interpreted as an instance of the overall worker-capitalist crisis. Crises in families due to poverty can be interpreted as instances of the same overall crisis. Marx – and, after him, orthodox Marxism – operated with a well-defined hierarchy of crises. The principal crisis of a capitalist society is the antagonism between workers and capital owners. All other crises are derivatives of this top-level crisis. The identification of this crisis-hierarchy is a main reason for Marx distancing himself from the left Hegelians, whom he found did not operate with any coherent conception of the political economy. The identification of a crisis-hierarchy reflects Marx's inspiration from Hegel; however, in order to use Hegel, Marx had to turn Hegel upside down and provide a material interpretation of dialectics. The idea of a hierarchical order of crises forms Marx's conception of critique. To be critical means, first of all, to address the principal crisis: that of the capitalist economy.

The Marxist conception of critique has implications for the formulation of Marxist pedagogy. The expression "Marxist pedagogy" refers to several quite different trends: some developed in the Soviet Union soon after the revolution and before Stalin came to power[8]; some developed in the Soviet Union during and after the Stalinist regime; some developed in Eastern Europe, and some in Western countries. A unifying idea was that education has to work for social change in alliance with the working class. Naturally, this idea appeared in one way from within the Soviet Union – where revolutionary changes had taken place – and from Western countries, where no revolution was in sight. Within the Soviet Union and countries dominated by the Soviet Union, Marxist pedagogy tended to accommodate students within the established social order and to celebrate this order, while in other countries Marxist pedagogy contributed to a critique of the capitalist economic order.[9]

6.3 A Multiplicity of Crises and Critique

Critical theory – as originally formulated by Max Horkheimer, Theodor Adorno, Herbert Marcuse, Erich Fromm, Walter Benjamin, and several others – provided critique with additional dimensions. Critical theory sought inspiration from Marx's writings, in particular the writings by the young Marx before he focused on elaborating a critique of the political economy. In his writings, Horkheimer addresses the worker-capitalist antagonism, but simultaneously he broadens the scope of critical activities. Adorno addresses music and art in general, and Marcuse criticised conceptions of science, in particular as formulated by logical positivism, and the

---

[8] As an example, I can refer to work by Makarenko (2001).
[9] As an example, see, for instance, McLaren (1997, 2016).

phenomenon of instrument reason. Erich Fromm creates new insights in psychological phenomena by interpreting them with reference to social phenomena. Walter Benjamin introduces a new methodological approach to critical investigations by collecting extensive data-material and presenting it in the form of a huge collage of quotations, expressions, and notices. In this way, he provides an interpretation of the construction of the Arcades in Paris as a manifestation of the capitalist order of things. In critical theory, we witness a fertile proliferation of critical approaches.[10]

After the conclusion of the Second World War, critical activities also addressed a new threatening phenomenon: the cold war. Atomic bombs were constructed and tested, representing a risk on a universal scale. New social movements were formed searching for global peace; the Pugwash movement was one of them. Environmental issues entered the agenda, and proliferated into different movements and non-governmental initiatives. Pollution and climate change were added to the critical agenda.

Colonialism has been a horrifying political phenomenon, although highly celebrated by imperial powers. Colonialism made the violent extracting of natural resources from the colonies possible, as well as using the indigenous population to do so. Furthermore, the colonies were resources for the trading of slaves. Now, the explicit and crude form of colonialism has disappeared, but the logic of colonialism is still in operation. Natural resources are no longer extracted from colonies, but from poorer countries around the world where underpaid workers are doing the job.[11] The possible relationships between colonialism and capitalism have been intensively discussed. Could capitalism be a kind of derivative of colonialism? Did capitalism come before colonialism? Are we dealing with two different, although resonating, phenomena? I have not seen clear answers to such questions, and I do not think it necessary to unify the critique of colonialism and the critique of capitalism. I find it more important to develop critique in multiple directions.

Critical race theory is an important manifestation of the struggle against racism. Here, one finds deep reflections on what racism might include and how it might manifest.[12] From functioning as an ideological cover for the brutality of slavery, racism has now taken different formats, serving as legitimisation for discriminating acts. I see critical race theory as bringing still more dimensions to the critical enterprise – and so do of the issues of sexism, homophobia, and xenophobia.

Marxism assumes the existence of a hierarchical order among crises, defining the tension between the working class and the dominant class as being the principal

---

[10] In the article "Tradtionelle und kritische Theorie", first published in 1937, Horkheimer presents in condensed form a description of the way critical theory distances itself from traditional theory. An English translation is found in Horkheimer (2002, pp. 180-243). See also Horkheimer and Adorno (2002), first circulated in a German version in 1944, and Marcuse (1964). For a presentation of the Arcade Project, see Benjamin (1999).

[11] Such phenomena have been addressed critically in post-colonial literature. See, for instance, Young (2001).

[12] See, for instance, Bridges (2019); Crenshaw, Gotanda, Peller and Thomas (1995); Delgado and Stefancic (2013); and Lynn and Dixon (2022).

one, and interpreting other crises as being "derivatives" of this "top-level crisis". I definitely share the concern for the exploitation of workers and their families. However, contrary to Marxism, I do not assume the existence of a pre-given structural relationship among crises. I see crises as interacting in many ways. We experience crises related to the exploitation involved with racism, sexism, homophobia, and xenophobia. We see crises due to lack of natural resources, due to climate change, and due to pollution. We see hunger crises, and huge problems related to lack of healthcare. We are surrounded by crises. I see the crises dynamics as being complex and unpredictable; I do not assume the existence of any hierarchical order of crises.

Assuming the existence of a multiplicity of crises – maybe interacting with each other, maybe intensifying each other, maybe operating in parallel, maybe operating independently of each other – brings about a concept of critique different from the concept assumed by Marx and by Marxism. By not assuming the existence of a hierarchical order of crises, the notion of critique becomes multi-directed. This is the concept of critique that I want to associate with critical mathematics education.

Marx included determinism in the conception of dialectics. The principal crisis in a capitalist society would develop according to an *a priori* logical pattern, leading to the classless society. Marx enveloped his conception of crisis and critique with a deterministic outlook. By not operating with the existence of a hierarchy of crises, but with a more or less intersecting multitude of crises, one leaves behind the assumption of the existence of a kind of crises determinism. This makes space for acknowledging the existence of a variety of critical approaches.

Critical activities can address different kinds of crises. Some critical approaches might be related, some might be independent of each other; some might find inspiration in other patterns of critique; some might be based on some theoretical insight; and some might be formed by personal experiences. One should not assume the existence of a unifying and basic critical approach. As crises are manifold, so are critiques. To provide a critique with a solid foundation is, to me, an illusion. I see any critical activity as being tentative and preliminary. Critique is an uncertain affair. This also applies to critical mathematics education.

Note

This chapter includes some pieces of texts already presented in "Crisis, Critique, and Mathematics" published in the *Philosophy of Mathematics Education Journal* (Skovsmose, 2019). I thank the publisher for kind permission to use this material.

References

Benjamin, W. (1999). *The arcades project*. The Belknap Press of Haward University Press.
Bridges, K. M. (2019). *Critical race theory: A primer*. Foundation Press.
Chassapis, D. (2017). "Numbers have the power" or the key role of numerical discourse in establishing a regime of truth about crisis in Greece. In A. Chronaki (Ed.), *Mathematics education*

and life at times of crisis: Proceedings of the ninth international mathematics education and society conference (pp. 45–55).

Chronaki, A. (Ed.). (2017). *Mathematics education and life at times of crisis: Proceedings of the ninth international mathematics education and society conference*. University of Thessaly.

Coles, A. (2016). *Mathematics education in the anthropocene*. Open access. University of Bristol.

Crenshaw, K., Gotanda, N., Peller, G., & Thomas, K. (Eds.). (1995). *Critical race theory: The key writing that formed the movement*. The New Press.

Crutzen, P. J., & Stoermer, E. F. (2000). The anthropocene. *Global Change Newsletter, 41*, 17–18.

Delgado, R., & Stefancic, J. (Eds.). (2013). *Critical race theory: The cutting edge* (3rd ed.). Tempe University Press.

Feuerbach, L. (1989). *The essence of Christianity*. Prometheus Books.

Greer, B., Gutiérrez, R., Gutstein, E., Mukhopadhyay, S., & Rampal, A. (2017). Majority counts: What mathematics for life, to deal with crises? In A. Chronaki (Ed.), *Mathematics education and life at times of crisis: Proceedings of the ninth international mathematics education and society conference* (pp. 159–163).

Horkheimer, M. (2002). *Critical theory: Selected essays*. Continuum.

Horkheimer, M., & Adorno, T. W. (2002). *Dialectic of enlightenment*. Continuum.

Kant, I. (1929). *Critique of pure reason* (Translated by Norman Kemp Smith). Macmillan.

Kant, I. (2007). *Critique of judgment*. Oxford University Press.

Kant, I. (2010). *Critique of practical reason*. Classic Books International.

Lynn, M., & Dixon, A. D. (Eds.). (2022). *Handbook of critical race theory in education* (2nd ed.). Routledge.

Makarenko, A. S. (2001). *The road to life: An epic of education. Volume 1*. University Press of the Pacific.

Marcuse, H. (1964). *One-dimensional man: Studies in the ideology of advanced industrial society*. Beacon Press.

Marx, K. (1970). *A contribution to the critique of political economy*. International Publishers.

Marx, K. (1992\1993). *Capital: A critique of political economy, I–III*. Penguin Classics.

Marx, K. (1998). *The German ideology*. Prometheus Books.

McLaren, P. (1997). *Revolutionary multiculturalism: Pedagogies of dissent for the new millennium*. Westview Press.

McLaren, P. (2016). *Pedagogy of insurrection: From resurrection to revolution*. Peter Lang.

Parra, A., Bose, A., Alshwaikh, J., Gonzáles, M., Marcone, R., & D'Souza, R. (2017). "Crisis" and the interface with mathematics education research and practice: An everyday issue. In A. Chronaki (Ed.), *Mathematics education and life at times of crisis: Proceedings of the ninth international mathematics education and society conference* (pp. 174–178).

Skovsmose, O. (2019). Crisis, critique and mathematics. *Philosophy of Mathematics Education Journal, 35*, 1–16.

Strauss, D. (2019). *The life of Jesus: Critically examined. Volume I*. HardPress.

Young, R. J. C. (2001). *Postcolonialism: A historical introduction*. Blackwell.

Chapter 7
Banality of Mathematical Expertise

Abstract Practices within research mathematics can and do serve as models for mathematics education. However, typically such inspirations impose a devastating narrowness in relation to reflections on mathematics. This narrowness I refer to as the "banality of mathematical expertise". Reflections on mathematics can be expressed through a philosophy of mathematics that goes beyond the traditional emphasis on ontological and epistemological dimensions, to become four-dimensional by also addressing social and ethical issues. Many working philosophies of mathematics are two-dimensional and operate within a narrow scope of reflections, seemingly located within an ethical vacuum. The consequence is a cultivation of banality, manifest in many university studies in mathematics as well as in dominant research paradigms in mathematics. This constitutes a serious limitation in providing models for mathematics education. By contrast, there exist examples of practices of mathematics education that demonstrate a richness of reflections on mathematics. Accordingly, I address the extent to which such practices of critical mathematics education could serve as models for research mathematics and mathematics education at the university level.

Keywords Role model · Philosophy of mathematics · Ethics · Ethical vacuum · Banality of mathematical expertise

No single well-defined practice of mathematics exists; rather, we are dealing with a multitude of practices in both research and application.[1] Neither does it make sense to talk about *the* practice of mathematics education; in this case, we are also dealing with a multitude of practices. Despite these ambiguities, I address the relationships between, on the one hand, the practices of mathematics as found at universities and research institutions, and on the other, the practices of mathematics education as found in school contexts.

---

[1] For a discussion of the diversity of mathematical practices, see Ferreirós (2016).

© The Author(s), under exclusive license to Springer Nature Switzerland AG 2023
O. Skovsmose, *Critical Mathematics Education*, Advances in Mathematics Education, https://doi.org/10.1007/978-3-031-26242-5_7

The relationships between the two sets of practices are opaque, which can be illustrated by an observation made by Leone Burton (2004). She completed an extensive study of how mathematicians were doing their research. She interviewed more than 70 research mathematicians working in different areas of mathematics, both pure and applied, and she identified a variety of approaches being used. Some mathematicians were thinking in terms of images and movements, while others engaged in extensive manipulations of symbols and formulas. Some created new ideas by writing out formulas, not on paper but with hand movements in the air. For some, the work could be undertaken after going to bed – and I wonder if such writing would work equally well in the dark. Some were grasping new connections while taking a shower. Through her study, Burton revealed the existence of a huge variety of creative practices in mathematics research.

Burton also asked the mathematicians what they found to be the best way of teaching mathematics, and here she observed an astonishing agreement: the appropriate way to make students learn mathematics was through lecturing. How the mathematicians got to this shared idea was not revealed through Burton's research, but the observation highlights that practices in mathematics and practices in mathematics education are not connected in straightforward ways.[2]

Mathematics could inspire many innovations and new approaches in mathematics education. In this chapter, however, I problematise the very idea that mathematics could serve as a role model for mathematics education in general. The reason for this is that practices of mathematics research often operate within a narrow scope of reflections on mathematics. These practices seem to be located in an ethical vacuum, and this phenomenon I am going to refer to as "the banality of mathematical expertise". In order to clarify what this banality involves, I will proceed along the following steps.

First, I refer to an example of the impact of mathematics on mathematics education, and in this way I try to clarify what it could mean to consider mathematics as a role model for mathematics education. Second, I outline a four-dimensional philosophy of mathematics, through which I point out what I consider important when reflecting on mathematics. Third, I pay particular attention to the ethical dimension of this philosophy and specify issues that it could include. Fourth, I highlight that ignoring ethical issues brings about a narrowness in reflections on mathematics. This narrowness I refer to as a banality of mathematical expertise. This banality makes me question the relevance of research in mathematics as a general role model for mathematics education. Finally, I present an example from the mathematics classroom that involves profound reflections on mathematics. These considerations lead me to observe that there exist practices in mathematics education that might serve as role models for the formation of mathematical expertise.

---

[2] In this short summary of Burton's research, I also draw on conversations I have had with her.

7.1 A Role Model

The notion "role model" was coined by the sociologist Robert Merton (1968), who in his work concerning reference groups pointed out that people often compare themselves with others. These could be people in positions one wants to obtain or people one wants to imitate. One finds role models among sports stars, celebrities, politicians, business people, etc. There might well be cases where mathematicians have served as role models for students. However, assuming some conceptual liberty, I will consider the possibility of mathematics serving as a role model for mathematics education.

Mathematics education might assume a certain terminology broadly applied in mathematics. Mathematics education might repeat topics that are considered important in mathematics. Mathematics education might imitate ways of reasoning through mathematical issues and problems. Mathematics can also serve as a role model for mathematics education by illustrating new ways of applying mathematics.

Let us look at an example. The Modern Mathematics Movement was initiated by the Royaumont Seminar that took place in France in 1959 (see OEEC, 1961). Until then, it had been generally assumed in the West that the West was well ahead of the East in terms of technological development. The launching of the Sputnik in 1957 by the Soviet Union shattered this assumption. Something had to be done, and at the Royaumont Seminar it was highlighted that the technological superstructures of the modern society should be ensured through education, and that mathematics education, in particular, would have to assume its role as the foundation of such structures.

But how was it to assume this role? The answer was clear: mathematics education had to be modernised. As pointed out by Jean Dieudonné in his lecture at the seminar (reprinted in OEEC, 1961), mathematics education was outdated. Apparently, mathematics education had ignored what had taken place in mathematics research during the last couple of hundred years. How could any modern technological superstructures be erected on top of such relics? If school mathematics were to be able to assume its glorious role, it needed a thorough modernisation.

Assuming a structuralism as manifested by the works of the Bourbaki collective, to which Dieudonné belonged, the modernisation of mathematics education took the form of a sweeping update of notions and foundational theories. The basic notions in Bourbaki's presentation of the architecture of mathematics were voted to be foundational to mathematics education as well.[3] As a consequence, set theory was stipulated as the starting point for learning mathematics, and was introduced as a central component of the curriculum of primary mathematics education in many countries. In the case of the Modern Mathematics Movement, mathematics served as a role model for mathematics education by propagating a conceptual update.

In the rest of this chapter, however, I want to concentrate on the observation that mathematics might serve as a questionable role model for mathematics education

---

[3] See Bourbaki (1950); Dieudonné (1970); and the discussion in Skovsmose (1990).

by limiting ways of reflecting on mathematics. In this case, the role modelling might impose a devastating narrowness on the scope of reflections.

7.2 A Four-Dimensional Philosophy of Mathematics

Deep reflections on mathematics are found within the philosophy of mathematics. In order to clarify what the content and scope of such reflections can be, let us take a careful look at dimensions of such a philosophy.

The philosophy of mathematics traces its ideas back to antiquity, where Platonism was originally formulated.[4] In the last century, we find formulations of logicism, formalism, and intuitionism. A recent trend in the philosophy of mathematics offers a critique of these classic positions by highlighting the importance of relating philosophical considerations to the practice of mathematics, and not primarily to already established philosophical discussions. Hersh (1979) was one of the first to suggest this redirection, also referred to as the "philosophy of mathematical practice".[5]

My colleague Ole Ravn and I have pointed out the importance of establishing a four-dimensional philosophy of mathematics (Ravn and Skovsmose, 2019). Here, I summarise the dimensions we have in mind, which make it possible to provide a clearer picture of what reflections on mathematics might themselves include.

The *ontological dimension* concerns what mathematics is about: that is, what entities we are addressing when we analyse numbers, points, functions, etc. Where do we locate such mathematical entities? Plato pointed towards the world of ideas; representatives of the scientific revolution – such as Nicolaus Copernicus, Galileo Galilei, and Isaac Newton – pointed towards nature; while Immanuel Kant pointed towards the mind, since he saw mathematics as a feature of human epistemic structures. Professional mathematicians might claim that there exists some kind of mathematical reality, which calls for further discoveries. These are examples of ontological claims.

The *epistemological dimension* addresses how we come to know mathematics. According to Platonism, we cannot make any empirical observations that reveal mathematical truths. The world of ideas is not accessible to us via our senses, but we can access it by our rationality and, in this way, grasp mathematical truths as being necessary truths. According to formalism, mathematical truths are also true with certainty, but for quite a different reason. Mathematics is a game with symbols. Whether or not a certain formula (a sequence of symbols) can be derived within a formal axiomatic system is a logical issue. If we identify a path of deduction, then

---

[4] Platonism exists in different versions. In Ravn and Skovsmose (2019), we address "Platonism Before Plato", "Platonisms After Plato", as well as "Plato's Platonism". Hacking (2014) carries out profound investigations of different versions of Platonism.

[5] More recommendations for such redirecting of the philosophy of mathematics can be found in publications by Ernest (1998); Hacking (2014); Mancosu (2008); and Tymoczko (1986). See Carter (2019) for an overview of the trend.

we know with certainty that the formula can be considered true within the system. Thus, mathematical truths are not absolute; here, we are dealing with truths relative to a well-defined formal system. Logicism and intuitionism also elaborate upon interpretations of the nature of mathematical truths.

The *social dimension* concerns the extent to which mathematical concepts and theories reflect social structures, and so addresses the social impact of mathematics. This dimension challenges the claim that mathematics deals with ahistorical entities, and highlights that the development of mathematics is sensitive to economic, political, religious, ideological, and cultural priorities. During the 1930s, Ludwig Wittgenstein (1978) formulated a radical social interpretation of mathematics. He compared mathematics with grammar, and mathematical theorems with grammatical rules. Naturally, one can make grammatical mistakes, but still grammatical rules are human constructions. They can be different (for example, in different geographical areas and in different cultural context), and they can change over time. In the same way, Wittgenstein found that mathematical rules are formed through long-term social processes.[6] This interpretation introduces an important dimension of a philosophy of mathematics, namely the social. Throughout his work, Sal Restivo has been dedicated to showing the social formation of mathematics, whether we are dealing with mathematical notions, ideas, theories, or applications. A most important contribution to the social turn was made by Paul Ernest (1998) when formulating social constructivism as a philosophy of mathematics.[7] The trend "philosophy of mathematical practice" has also contributed to the social dimension of the philosophy of mathematics.[8]

The *ethical dimension* of a philosophy of mathematics becomes visible when we consider the social impact of doing mathematics. Such impact is traditionally ignored when one concentrates on the significance of pure mathematics, as done by Godfrey Hardy (1967) in *A Mathematician's Apology*, first published in 1940. Hardy finds that mathematics can be useful, and adds that this is first of all true of elementary mathematics (in which he includes, for instance, calculus), and of what can be referred to as engineering mathematics. However, for what he considers "real mathematics", he does not see any applications. In this sense, mathematics is harmless and innocent, and no ethical controversies relate to it. A very different perspective has been elaborated through the notion of mathematics in action, which I return to in the following section. Mathematics – and also number theory, which Hardy considers the very epitome of purity – is brought into action in all kinds of contexts. Acting with mathematics can have all kinds of social impact and serve all kinds of interests. This calls for deep ethical considerations.[9]

---

[6] A detailed discussion of the historicity of mathematics is found in work by Grabiner (1974), raising the question: Is mathematical truth time-dependent?

[7] See also Restivo (1992); and Restivo et al. (1993).

[8] Many studies in ethnomathematics investigate the social formation of mathematics. A classic is Powell and Frankenstein (1997), while recent contributions are found in Rosa et al. (2016).

[9] Ernest (2016, 2018) has been focused on raising ethics as an issue in the philosophy of mathematics A systematic exploration of the ethical dimension of a philosophy of mathematics is found in

The philosophy of mathematics, as Ravn and I see it, has at least four dimensions: the ontological, the epistemic, the social, and the ethical. However, we do not claim that there are *only* these four dimensions – one can, for instance, consider the aesthetic dimension as well. One could also consider a political dimension of a philosophy of mathematics, but political issues I think of as being part of the ethical dimension.

The classic positions in the philosophy of mathematics – logicism, formalism, and intuitionism – concentrated on clarifying the nature of mathematical objects and how we came to know about them. They are paradigmatic examples of two-dimensional philosophies of mathematics. The original formulations of logicism were made in 1879 and 1884 by Frege (1967, 1978), and through their monumental work *Principia Mathematica*, Whitehead and Russell (1910–1913) elaborated in detail how mathematics could be developed from logic. By a strict axiomatisation of geometry published in 1899, Hilbert (1950) tried to eliminate any trace of intuition from geometric deductions, and in this way, he anticipated the formalist outlook. Later on, Hilbert (1922) turned this into a principal approach in addressing foundational issues. Contrasting with formalism, Brouwer (1913) presented the intuitionist programme that introduced a new perspective on mathematical intuitions by interpreting formal structures as imprecise and dead copies of genuine mathematical processes. Different as they are, these three schools of thought share the same quality, namely of being two-dimensional. They concentrate on the ontological and epistemological issues, and ignore the social and ethical. Many subsequent presentations of the philosophy of mathematics are trapped by the same two-dimensionality.[10]

Two-dimensionality in the philosophy of mathematics provides a problematic limitation of the scope of reflections on mathematics. Ravn's and my point in formulating a four-dimensional philosophy of mathematics is to highlight the importance of transcending two-dimensionality, and in particular to develop the ethical dimension. This vital in order to develop a broad scope of reflections on mathematics.

7.3 The Ethical Dimension

While the social dimension concerns the ways in which the social world impacts on mathematics, the ethical dimension concerns the ways in which mathematical activity impacts upon society.

One can make a distinction between a descriptive and a performative interpretation of science. Logical positivism presents the most elaborate descriptive interpretation, from which scientific theories should provide pictures of reality that are as

---

Part IV in Ravn and Skovsmose (2019).

[10] See, for instance, Bernacerraf and Putnam (1964); Körner (1968); and Shapiro (2000).

accurate as possible.[11] A scientific description should present the properties of what was observed, and not of who observed it, why it came to be observed, and why somebody took an interest in it. Any subjective features should be eliminated, and the best way to do so is to use mathematics as a means of description. According to logical positivism, mathematics can be considered the language of science.

Descriptive interpretations of science have been challenged by performative interpretations, which highlight that science has a range of social impacts: on production, management, communication, economy, security, and sustainability. The performativity highlighted with respect to science in general also applies to mathematics. Mathematics is also performative. *By doing mathematics, one is acting with a range of social implications*. Mathematics is not just a transparent descriptive device; it involves actions. No action takes place in an ethical vacuum, but calls for profound reflections. This also applies to mathematics brought into action. A performative interpretation of mathematics establishes ethics as an important dimension of a philosophy of mathematics.

One can identify different forms of mathematics-based actions, and I have previously referred to those such as technological imagination, hypothetical reasoning, justification, realisation, and dissolution of responsibility.[12] By investigating in detail such actions, one might get a better grasp of what ethical reflections on mathematics could include.

Technological imagination refers to the potential of identifying new technological possibilities. Imagination can be expressed through any type of language, and natural language is one possibility. However, the language of mathematics includes particular potentials for formulating possibilities. The very conception of digitalisation presupposes mathematics, and the computer can be seen as a mathematical construct. The conception of the internet is also formed through mathematics. No natural language would make it possible to detect such possibilities. Our life-worlds are not only formed by what is actually the case, but also by the possibilities that might be made available. This space of possibilities – including even the most questionable scenarios – is a mathematical construct. Simultaneously, it is an ethical challenge, as it is within this space that our future becomes directed.

Hypothetical reasoning refers to the possibility of operating in detail with counterfactual statements, which are statements that consider the implications of not-yet-implemented scenarios. Counterfactual statements occur in all kinds of contexts. When a company considers a new investment, it needs to explore the implications of making the investment before actually doing it. How may this be done? Through mathematical modelling! On this basis, the company makes decisions with respect to possible investments. This approach is not only applied with respect to financial decisions, it is a general pattern for decision-making. Any engineering project is

---

[11] See, for instance, Ayer (1959).

[12] See Skovsmose (2014). In Chap. 11, "Critique of Mathematics", I elaborate further on this presentation of mathematics brought into action. In Ravn and Skovsmose (2019), the notion of mathematics in action is elaborated in terms of fabrications of fiction, fabrications of facts, fabrications of risks, and the creation of illusions of objectivity. See also Yasukawa et al. (2012, 2016).

described and investigated mathematically, before any real construction takes place.[13] A constraint of mathematics-based hypothetical reasoning is that only features that can be expressed quantitatively are considered. Our whole method of reasoning about possibilities is profoundly formed, and simultaneously limited, by mathematics.[14]

Justification signifies forms of advocation for a decision to be made. One can imagine a medical company in a process of making a decision about whether or not to engage in a project developing a new kind of medicine. A decision will presuppose a range of calculations and estimations: How much investment is needed? How much time will it take before the medicine is developed? How expensive will the production become? What could be the price of the final product? How many people would need this medicine? How much advertising will be needed? No such questions can be answered without substantial use of mathematical modelling. If we are dealing with a private company, the question of profit becomes important, and one can come to meet a cynical pattern of reasoning: it may turn out to be non-profitable to produce the medicine. There might not be sufficiently many ill people at the moment who need the medicine. Better wait until the demand has grown "adequately". This is just one example of how mathematics can be brought into action to justify, or rather legitimise, certain decisions, sometimes of the most questionable nature. This calls for ethical considerations.

Realisation refers to the phenomenon that mathematics is not only used as a tool for decision-making, but that it may come to form part of reality as well. Implementations of automatic processes bring mathematical algorithms into operation. Such processes are found in all spheres of life. Davis and Hersh (1988) refer to the tax system as a prescriptive mathematical model. It not only describes what people are going to pay in tax, it also prescribes the taxation. Mathematics becomes real. According to Platonism, mathematical reality belongs to another world. Platonic ontology is explicit in locating mathematical entities in the world of ideas, which has nothing to do with the real world. What I highlight through the notion of "realisation" is also an ontological point, namely that mathematics becomes an integral part of our lived-through realities.

Dissolution of responsibility can take place when automatic decision-making is implemented. Crises can be caused by technological enterprises; they can be provoked and accelerated by bringing mathematics into action. Mathematics-based automatic decision making contributed to the economic crisis that exploded in 2008.[15] Decisions about buying and selling in the stock market were (and continue to be) based on mathematical risk calculations, and automatic algorithmic processes

---

[13] For a discussion of such phenomena, see Chapter 15 "Symbolic Power, Robotting, and Surveilling" and Chapter 17 "Mathematics as Part of Technology" in Skovsmose (2014); and Skovsmose and Yasukawa (2009).

[14] That hypothetical reasoning is an important feature of mathematics has been indicated by Peirce (1960), who states: "Mathematics is the study of what is true of hypothetical state of things." (p. 193). See also Carter (2014).

[15] See O'Neil (2016).

accelerated and got out of control. Who was responsible? The algorithms? The technicians who implemented the algorithms? The mathematicians who constructed the algorithms? The financial institutions that applied the algorithms? In fact, we are dealing with situations where the very operation of mathematics seems to dissolve responsibility. Responsibility is a crucial issue in an ethical discussion, concerning the persons (or group of persons, or institutions) who perform the actions. All actions have performers, who can be held responsible for the action. But with respect to mathematics-based actions, the identity of the performer is disguised. It appears that we encounter actions without performers.

By pointing out cases of technological imagination, hypothetical reasoning, justification, realisation, and dissolution of responsibility, I try to show the specificity that ethical reflections on mathematics in action might have. No actions automatically acquire praiseworthy qualities by being mathematics-based. Like all actions, so too can mathematics-based actions have any kind of qualities. They can be benevolent, dishonest, disastrous, innocent, pleasing, promising, mischievous, dangerous, generous, roguish, fatal, risky, etc. There is no particular quality to be expected from mathematics-based actions, due to the fact that they are *mathematics* based. Acknowledging this point, one can grasp the importance of operating with an ethical dimension of a philosophy of mathematics.

7.4 Banality

When interviewing mathematicians, as Burton did, one could imagine meeting a philosophy of mathematics that has been elaborated in detail, but this is seldom the case. Rather, one meets a mix of assumptions and presumptions about mathematics, which I refer to as a *working philosophy of mathematics*. This is a philosophy engraved in mathematical practice, although seldom explicit in the details. Hersh (1997) made the following observation:

> The working mathematician is a Platonist on weekdays, a formalist on weekends. On weekdays, when doing mathematics, he's a Platonist, convinced he's dealing with an objective reality whose properties he's trying to determine. On weekends, if challenged to give a philosophical account of the reality, it's easiest to pretend he doesn't believe it. He plays formalist, and pretends mathematics is a meaningless game. (p. 39)

In this elegant way, Hersh condenses the observation that, when asked, mathematicians tend to express themselves in formalist terms, but when doing research, they operate within a Platonist outlook. They feel that they are struggling with making mathematical discoveries.

I am not going to consider in detail how Platonism and formalism combine into a working philosophy of mathematics, or how they might, in fact, be expressed in particular practices. My point is that any such combination is two-dimensional. It is a working philosophy through which one may address ontological issues, such as (for instance) the nature of making discoveries, and epistemological issues – such as the degree of certainty that might be obtained by non-formal proof. As mentioned, the trend "philosophy of mathematical practice" highlights the importance of overcoming limitations set by the classics – logicism, formalism, and intuitionism – by relating the philosophical discussions to actual practices of doing research in mathematics. Considering, however, that many such practices neglect ethical issues, I wonder to what extent this philosophy might and help in bringing mathematical practices out of their ethical vacuum.[16]

A working philosophy of mathematics is expressed, for instance, through the defined curriculum. In many cases, university students meet an extensive list of mathematical topics that they have to come to master: Calculus 1, linear algebra, probability theory, topology, Calculus 2, etc. A working philosophy is also expressed through the nature of the exercises and tests that are presented to the students. It is displayed through the way in which students are supposed to learn mathematics; this could be by listening to lectures, as Burton found was a common priority among university mathematicians.

I see the common working philosophy of mathematics as demonstrating a highly problematic limitation integrated in the practice of mathematics as found in universities and research institutions. It limits ethical reflections on mathematics in action. Such a limitation, I refer to as the *banality of mathematical expertise*.

In order to understand my choice of terminology, I need to refer to the book *Eichmann in Jerusalem: A Report on the Banality of Evil* by Arendt (1977), first published in 1963. In 1960, Eichmann was captured in Argentina by the Israeli secret police and brought to Jerusalem. Here, Arendt followed the trial of Eichmann that took place in 1962. He was judged for his actions during the Second World War, and executed.

Arendt describes the process and her observations. Contrary to what one might have expected, she did not see an aggressive Eichmann roaring against the Jewish people. He did not behave as a representative of the Nazi terror. Eichmann did not appear as a demon, rather a narrow-minded clerk.

One gets the impression that Eichmann had operated as a technician, concentrating on keeping time schedules, finding resources for transportation, and against all odds maintaining efficiency in transporting Jews to the extinction camps. In such matters, he demonstrated remarkable administrative capacities. That he was transporting people to death camps, he seemed just to have disregarded. There is no indication that he, as part of his professional practice, reflected on what he was doing. Eichmann seemed only able to express himself in administrative terms. He claimed that he had followed instructions with great care, which he did; that he had operated with promptness and efficiency, which he did; and that he had observed a range of administrative quality priorities, which he also did. He operated within a

---

[16] Carter (2019), Mancosu (2008), and Ferreirós (2016), all contributing to a philosophy of mathematical practice, do not address ethical issues. However, in the book *Handbook of the History and Philosophy of Mathematical Practice*, edited by Sriraman (2020), Ernest (2020) contributes with "The Ethics of Mathematical Practice". By making this contribution, he makes a crucial addition to what, until then, has been addressed in the philosophy of mathematical practice.

particular and limited outlook focusing on what we can refer to as intrinsic administrative quality criteria. According to such criteria, he could claim that he had done a good job.

Arendt condensed all this into the expression: *banality of evil*. The evil, as represented by Eichmann, is not acted out through aggressive and dramatic statements and gestures. Rather, it is acted out through an administrative practice, which is not accompanied by reflections, just executed according to prescriptions.

I have been inspired by Arendt to use the expression *banality of mathematical expertise*. However, I am not claiming more than an inspirational relationship between the two conceptions of banality.

The banality of mathematical expertise refers to the high priority of being able to *do mathematics* compared to the capacity of being able to *reflect on* the ethical impact of what one is doing. This priority constitutes an integral part of university studies in mathematics, as these studies are most often structured. Everything therein concentrates on ensuring that students become able to master mathematical techniques and theories. Tests and examinations focus on intrinsic mathematical quality criteria. To develop the capacity to reflect on mathematics is not an integral part of the education of mathematicians. It is difficult to locate examples of university study programmes in mathematics that include the philosophy of mathematics as an integral part. It is even more difficult to locate study programmes where the ethical dimension of such a philosophy is addressed, and where one makes profound reflections on mathematics in action. The ethical dimension is missing, not only in the education of the mathematical expertise, but also in the very practice of this expertise. It appears that a quality of mathematics research is to be *context oblivious*. This is a consequence of operating within a narrow working philosophy of mathematics.[17]

The banality of mathematical expertise refers to a focus on doing mathematics, and doing so first of all by observing intrinsic quality criteria. It refers to being context oblivious. If mathematics serves as a role model for mathematics education, it might impose limitations on this education, making mathematics education concentrate on content-matter issues, and thus limiting accordingly the scope of reflections on mathematics.

7.5 From the Classroom

In mathematics education, one finds example of engaging students in reflections on mathematics, also of an ethical nature. Naturally, my point is not to make any claim about the existence in mathematics education of a general, critical reflective

---

[17] Naturally, one finds exceptions. Problem orientation and project-based education can be found in university mathematics education (see, for instance, Vithal et al. (1995). In such an educational format, one acquires possibilities for reflecting on mathematics in action.

approach to mathematics.[18] My point is only to highlight that, in mathematics education, one finds examples of broader socio-political and ethical reflections on bringing mathematics into action.

As part of the project "Math for Social Justice" organised by Eric Gutstein at the Social Justice High School in Chicago, the huge number of foreclosures that haunted the neighbourhood was examined. People had difficulties in paying their rent, and if they wanted to sell their houses, it turned out to be very difficult to do so. The neighbourhood was not considered attractive any longer, and house prices were falling. For many families, the situation turned precarious. Starting from this situation, Gutstein (2018) addressed the payment of mortgages:

> Our class was trying to determine whether a family earning the median income in the neighbourhood (around 32,000 US Dollars a year) could afford a home mortgage of 150,000 US Dollars, with an interest rate of 6% a year on a 30 year loan. According to the US Department of Housing and Urban Development, paying more that 30% on one's income for housing is considered a "hardship". (p. 131)

The students, about 17–18 years old, engaged in calculations and discussions, and it was clarified that a family taking a loan of 150,000 US Dollars could pay 808 Dollars a month without hardship. However, with such a monthly payment, it would not be possible to actually pay off the loan. It was calculated that, after paying 808 Dollars a month for 30 years, which would total 291,000 Dollars, the family would still owe about 92,000 Dollars. This observation was summarised in the equation:

$$150,000 - 291,000 = 92,000$$

What are we to think of this? It looks strange. It looks unfair. It could be read as a condensed expression of conditions for buying a house in this particular neighbourhood in Chicago. It could also be read as a condensed expression of the capitalist order of things. Certainly, it calls for a discussion of the roles of interest. Why do banks operate with interest, both when lending and borrowing? Why do interest rates change? Who makes the changes? It is impossible to understand the mathematical eq. 150,000–291,000 = 92,000 without understanding the mortgage system in which the equation is embedded.

In reflections on economic issues, hypothetical reasoning plays an important role. This also applies to a family trying to sell a house or to buy a house. Naturally, mathematics-based hypothetical reasoning can be limited, if not simply misapplied. Thus, there is no specific quality having to do with, for instance, accuracy, automatically accompanying mathematics-based investigations. Mathematics is a possible resource for making investigations, but not a reliable resource. The calculations that led to the formulation of the eq. 150,000–291,000 = 92,000 include different parameters, the value of which can be critically addressed. This opens up the whole space of hypothetical reasoning for profound reflections.

---

[18] What can be referred to as the school mathematics tradition is, in fact, characterised by a profound lack of critical reflections on mathematics. See Skovsmose and Penteado (2015).

Foreclosures are critical situations where mathematics becomes real. We are not only dealing with mathematical calculations that could be done correctly or not; we are dealing with the formation of real-life conditions experienced by many families. We are dealing with realisations. However, it is not easy to identify who, in fact, turned the results of the calculations into real-life conditions. Was it a mathematician? Or a bank assistant? Or a computer? Or a bank director? Or rather the capitalist organisation of society? Apparently, we are experiencing a case of dissolution of responsibility.

In mathematics education, one finds examples of engaging in profound reflections on mathematics. With these observations in mind, I feel tempted to recommend that research in mathematics should consider classroom practices – in particular, those inspired by critical mathematics education – as possible role models. Research in mathematics could try to bring its working philosophy out of its ethical vacuum. This is urgent today, where mathematics research brings mathematics into action in an ever-growing set of contexts. Mathematics is a powerful resource that can come to serve any kind of interests. The implications have to be addressed explicitly – otherwise we all too easily find ourselves surrounded by the banality of mathematical expertise.

Note
This chapter is a revised version of the paper "Banality of Mathematical Expertise" published in 2020 in the *ZDM Mathematics Education, 52*(6), 1187–1197. I thank the publisher for their kind permission to present it here.

References

Arendt, H. (1977). *Eichmann in Jerusalem: A report on the banality of evil*. Penguin Books.
Ayer, A. J. (Ed.). (1959). *Logical positivism*. The Free Press.
Bernacerraf, P., & Putnam, H. (Eds.). (1964). *Philosophy of mathematics*. Prentice-Hall.
Bourbaki, N. (1950). The architecture of mathematics. *The American Mathematical Monthly, 57*(4), 221–232.
Brouwer, L. E. J. (1913). Intuitionism and formalism. *Bulletin of the America Mathematical Society, 20*(2), 81–96.
Burton, L. (2004). *Mathematicians as enquirers: Learning about learning mathematics*. Kluwer Academic Publishers.
Carter, J. (2014). Mathematics dealing with 'hypothetical states of things'. *Philosophia Mathematica, 22*(3), 209–230.
Carter, J. (2019). Philosophy of mathematical practice: Motivations, themes and prospects. *Philosophia Mathematica, 27*(1), 1–32.
Davis, P. J., & Hersh, R. (1988). *Descartes' dream: The world according to mathematics*. Penguin Books.
Dieudonné, J. A. (1970). The work of Nicholas Bourbaki. *The American Mathematical Monthly, 77*, 134–145.
Ernest, P. (1998). *Social constructivism as a philosophy of mathematics*. State University of New York Press.
Ernest, P. (2016). A dialogue on the ethics of mathematics. *The Mathematical Intelligencer, 38*(3), 69–77.

Ernest, P. (2018). The ethics of mathematics: Is mathematics harmful? In P. Ernest (Ed.), *The philosophy of mathematics education today* (pp. 187–216). Springer.

Ernest, P. (2020). The ethics of mathematical practice. In B. Sriraman (Ed.), *Handbook of the history and philosophy of mathematical practice*. Springer.

Ferreirós, J. (2016). *Mathematical knowledge and the interplay of practices*. Princeton University Press.

Frege, G. (1967). Begriffsschrift: A formula language, modelled upon that of arithmetic, for pure thought. In J. van Hiejenoort (Ed.), *From Frege to Gödel: A source book in mathematical logic, 1879–1931* (pp. 1–82). Harvard University Press.

Frege, G. (1978). *Die Grundlagen der Arithmetik/The foundations of arithmetic*. Blackwell.

Grabiner, J. V. (1974). Is mathematical truth time-dependent? *The American Mathematical Monthly, 81*(4), 354–365.

Gutstein, E. (2018). The struggle is pedagogical: Learning to teach critical mathematics. In P. Ernest (Ed.), *The philosophy of mathematics education today* (pp. 131–143). Springer.

Hacking, I. (2014). *Why is there philosophy of mathematics at all?* Cambridge University Press.

Hardy, G. H. (1967). *A mathematician's apology* (With a foreword by C. P. Snow). Cambridge University Press.

Hersh, R. (1979). Some proposals for reviving the philosophy of mathematics. *Advances in Mathematics, 31*, 31–50.

Hersh, R. (1997). *What is mathematics, really?* Oxford University Press.

Hilbert, D. (1922). Neubegründung der Mathematik: Erste Mitteilung. *Abhandlungen aus dem Seminar der Hamburgischen Universität (Refoundation of Mathematics: First Communication), 1*, 157–177.

Hilbert, D. (1950). *Foundations of geometry*. The Open Court Publishing Company.

Körner, S. (1968). *The philosophy of mathematics*. Hutchinson University Library.

Mancosu, P. (Ed.). (2008). *The philosophy of mathematical practice*. Oxford University Press.

Merton, R. (1968). *Social theory and social structure*. Free Press.

O'Neil, C. (2016). *Weapons of math destruction: How big date increase inequality and threatens democracy*. Broadway Books.

OEEC. (1961). *New thinking in school mathematics*. Organisation for European Economic Co-operation.

Peirce, C. S. (1960). *The simplest mathematics: Collected papers* (Vol. IV). The Belknap Press of Harvard University Press.

Powell, A., & Frankenstein, M. (Eds.). (1997). *Ethnomathematics: Challenging eurocentrism in mathematics education*. State University of New York Press.

Ravn, O., & Skovsmose, O. (2019). *Connecting humans to equations: A reinterpretation of the philosophy of mathematics*. Springer.

Restivo, S. (1992). *Mathematics in society and history*. Kluwer Academic Publishers.

Restivo, S., van Bendegem, J. P., & Fisher, R. (Eds.). (1993). *Math worlds: Philosophical and social studies of mathematics and mathematics education*. State University of New York Press.

Rosa, M., D'Ambrosio, U., Orey, D. C., Shirley, L., Alangui, W. V., Palhares, P., & Gavarrete, M. E. (Eds.). (2016). *Current and future perspectives of ethnomathematics as a program* (CME-13 Topical Surveys). Springer.

Shapiro, S. (2000). *Thinking about mathematics: The philosophy of mathematics*. Oxford University Press.

Skovsmose, O. (1990). Perspectives in curriculum development in mathematics education. *Scandinavian Journal of Educational Research, 34*(2), 151–168.

Skovsmose, O. (2014). *Critique as uncertainty*. Information Age Publishing.

Skovsmose, O., & Penteado, M. G. (2015). Mathematics education and democracy: An open landscape of tensions, uncertainties, and challenges. In L. D. English & D. Kirshner (Eds.), *Handbook of international research in mathematics education* (3rd ed., pp. 359–373). Routledge.

References

Skovsmose, O., & Yasukawa, K. (2009). Formatting power of 'mathematics in a package': A challenge for social theorising? In P. Ernest, B. Greer, & B. Sriraman (Eds.), *Critical issues in mathematics education* (pp. 255–281). Information Age Publishing.

Sriraman, B. (Ed.). (2020). *Handbook of the history and philosophy of mathematical practice*. Springer.

Tymoczko, T. (Ed.). (1986). *New directions in the philosophy of mathematics*. Birkhäuser.

Vithal, R., Christiansen, I. M., & Skovsmose, O. (1995). Project work in university mathematics education: A Danish experience: Aalborg University. *Educational Studies in Mathematics, 29*(2), 199–223.

Whitehead, A., & Russell, B. (1910–1913). *Principia mathematica* (Vol. I–III). Cambridge University Press.

Wittgenstein, L. (1978). *Remarks on the foundations of mathematics*. Basil Blackwell.

Yasukawa, K., Skovsmose, O., & Ravn, O. (2012). Mathematics as a technology of rationality: Exploring the significance of mathematics for social theorising. In O. Skovsmose & B. Greer (Eds.), *Opening the cage: Critique and politics of mathematics education* (pp. 265–284). Sense Publishers.

Yasukawa, K., Skovsmose, O., & Ravn, O. (2016). Scripting the world in mathematics and its ethical implications. In P. Ernest, B. Sriraman, & N. Ernest (Eds.), *Critical mathematics education: Theory, praxis, and reality* (pp. 81–98). Information Age Publishing.

Chapter 8
Mathematics and Ethics

Abstract In the philosophy of mathematics, ontological and epistemological questions have been discussed for centuries. These two sets of questions span a two-dimensional philosophy of mathematics. I find it important to establish a four-dimensional philosophy of mathematics by adding two more dimensions, namely a sociological and an ethical. The sociological dimension addresses the social formation of mathematics, while the ethical dimension addresses the mathematical formation of the social. I concentrate on exploring the ethical dimension by showing the broad range of social implications set in motion through bringing mathematics into action. These implications I illustrate in terms of quantifying, digitalising, serialising, categorising, and imagining. With the term "banality of mathematical expertise", I refer to the phenomenon that the formation of this expertise takes place in an ethical vacuum. To me this is a devastating feature of mathematical research and application practices. It is important that a philosophy of mathematics brings mathematics out of this vacuum.

Keywords Two-dimensional philosophy of mathematics · Four-dimensional philosophy of mathematics · Ethics · Mathematics in action · Quantification · Digitalisation · Serialising · Categorisation · Imagination · Banality of mathematical expertise

Ontological and epistemological questions span a two-dimensional philosophy of mathematics, and many elaborations of the philosophy of mathematics can be located within these two dimensions.[1] In our book, *Connecting Humans to Equations*, Ole Ravn and I (Ravn and Skovsmose, 2019) acknowledge the importance of addressing ontological and epistemological issues, but we point out that

---

[1] As examples, I can refer to Bernacerraf and Putnam (1964), Bostock (2009), Brown (2008), George and Velleman (2002), Jacquette (2002), Körner (1968), Linnebo (2017), Mehlberg (1960), and Shapiro (2000).

two more dimensions need to be added to a philosophy of mathematics, namely a sociological and an ethical.[2]

The sociological dimension addresses the social formation of mathematics, exploring how mathematical notions and theories are formed through complex social processes. In this chapter, however, I concentrate on mapping out the ethical dimension of a philosophy of mathematics.[3]

In 1990, Ruben Hersh published a paper with the title "Mathematics and Ethics". This is the same title as the one I have given this chapter, so let us take a look at what Hersh had in mind. Hersh tried to relate mathematics and ethics by paying attention to the professional conduct of mathematicians. According to Hersh (1990), such conduct concerns the mathematicians' relationships to "staff, students, colleagues, administrators, and ourselves" (p. 21). The American Mathematical Society has published ethical guidelines for mathematical researchers, highlighting the importance of avoiding plagiarism, giving credit to colleagues, publishing without unreasonable delay, and correcting errors. The European Mathematical Society has published similar guidelines.[4] In line with such interpretations, Ruben Hersh (1990) highlights: "In pure mathematics, when restricted just to research and not considering the rest of our professional life, the ethical component is very small" (p. 22). I am not going to follow up on such profession-oriented approaches to ethical issues. I want to elaborate on the connections between mathematics and ethics with a much broader socio-political perspective.

One can make a distinction between a descriptive and a performative interpretation of mathematics.[5] According to the descriptive interpretation, mathematics is a language that can provide pictures of reality without interfering with reality. According to the performative interpretation, mathematics is a powerful tool for changing reality. Mathematics can constitute different forms of actions with a diverse range of qualities. I am going to explore features of bringing mathematics into action by addressing processes of quantifying, digitalising, serialising, categorising, and imagining.[6] As with any actions, mathematics-based actions call for profound ethical reflections.

---

[2] See also Hacking (2014) for a discussion of the possible scope of a philosophy of mathematics, and Bueno and Linnebo (2009) for an introduction of new issues.

[3] Ethical issues with respect to mathematics have been addressed by Ernest (2016, 2018, 2020), and his concerns resonate with Ole Ravn's and my own.

[4] See *Ethics in Mathematics* https://en.wikipedia.org/wiki/Ethics_in_mathematics

[5] See Chap. 10, "Picturing or Performing?", for a more careful discussion of this distinction.

[6] The notion of "mathematics" is a tricky one. It can refer to school mathematics, engineering mathematics, pure mathematics, applied mathematics, street mathematics, any kind of ethnomathematics. When, in the following, I elaborate on the ethical dimension of a philosophy of mathematics, I have first of all in mind what could be referred to as "academic mathematics". However, the ethical dimension of a philosophy of mathematics needs a number of further elaborations in order to address any of the many versions of mathematics we practise.

8.1 Quantifying

The quantification of nature was part of the so-called scientific revolution, and since then it has formed an integral part of the natural sciences. However, procedures of quantification have been utilised far beyond the limits of the natural sciences. One finds quantifications in the social sciences, psychology, medicine, economy, in any form of technical investigation.

The expansion of the scope of quantification brings about profound ethical issues. Here, let me just refer to one example, namely the calculation of the economic value of a human life. The identification of this value has been put on the research agenda for the last hundred years, even though such an identification appears to be a questionable example of quantifying.[7] Several philosophers, including Immanuel Kant, have pointed out that the value of a human life is unique and so cannot be evaluated, graduated, or subjected to any type of quantification.

Why try to calculate the value of a human being? One answer appears when one considers a health programme implemented by a government. One can think of any such programme as forming part of a humanitarian initiative. A health programme is an effort to save lives and prolong lives. Apparently, there is no need to provide any further justification for doing so. However, one can look at a health programme as an investment, and certainly there can be much money put into it. Considering health expenditure as an investment, it becomes consequential to ask about the economic output of this investment. What is the value of the saved lives and of the gained years of life? In order to answer such a question, one needs to identify the economic value of a human life.

One approach for doing this is to consider what a person would be able to produce during the rest of their lifetime. Such an approach can be referred to as the classic one, and it has been elaborated in the greatest detail.[8] A more recent approach is to consider life itself to be a commodity. When considered this way, the value of a life can be interpreted as the amount of money people are willing to pay for it. This value is normally referred to as the value of a statistical life. A procedure for identifying this value has been outlined in the following way:

> Suppose each person in a sample of 100,000 people were asked how much he or she would be willing to pay for a reduction in their individual risk of dying of 1 in 100,000, or 0.001%, over the next year. Since this reduction in risk would mean that we would expect one fewer death among the sample of 100,000 people over the next year on average, this is sometimes described as "one statistical life saved". Now, suppose that the average response to this hypothetical question was $100. Then the total dollar amount that the group would be willing to pay to save one statistical life in a year would be $100 per person × 100,000 people, or $10 million. This is what is meant by the "value of a statistical life".[9]

---

[7] See Hood (2017).

[8] For an elaboration of a particular human-life value model, see Hofflander (1966). Here, one gets an impression of the mathematical complexities such a model might include.

[9] The quotation is taken from the webpage of the United States Environmental Protection Agency: https://www.epa.gov/environmental-economics/mortality-risk-valuation#means. See also *Value of*

This approach is based on the quantification of people's willingness to pay for an expected extension of life.

One can immediately raise objections to such a methodology. A "willingness to pay" might reflect what people might be able to pay. People living in extremely poor conditions are not able to pay much, if anything. According to the outlined procedure for the quantification, the statistical value of their lives could turn out to become close to zero. The identification of the value of a statistical life might have a range of consequences, not least with respect to investments in healthcare programmes. Particular implications might be experienced when it comes to governments' decisions of saving (or not saving) lives during an epidemic. Business interests can easily be incorporated into such decisions, when the value of human lives has been quantified.

A recent approach to calculating the value of a human life has been to pay attention to the marginal costs of saving a life.[10] The marginal cost refers to the extra cost of saving one more life. The marginal cost of saving a life can be compared to the marginal revenue of saving a life. By considering marginal costs and marginal revenues, one can locate the discussion within an economic perspective. It is a classic economic insight that, when doing any kind of production where one wants to maximise profits, one needs to identify the level of production where the marginal cost equals the marginal revenue. Saving lives becomes conceptualised as a business that might be profitable, at least up to a certain level.

Whatever approach is used for defining the economic value of a human life, it paves the way for a variety of cost-benefits analyses, also of the most questionable nature. They can be applied by governments, but also by institutions and companies. As an example, I can refer to Ford Motor Company's way of dealing with the Ford Pinto case. In 1968, this car model was put into production, but soon it turned out that its fuel system was problematic. When the Pinto model was involved in car accidents, it tended to catch fire. Ford Motor Company needed to consider whether they should redesign the model such that the fuel tank was placed in a safer position, or whether they should continue production as it was.

The decision between these two options was made based on an explicit economic analysis. Ford calculated the costs of redesigning the model, and the costs of doing nothing. The cost of redesigning the model in production was estimated to be $137,000,000 US Dollars. The cost of doing nothing drew on the following estimations: 180 burn deaths, 180 serious burn injuries, 2100 burned vehicles. This estimation was naturally based on an extensive amount of already existing statistical information. The further calculation applied the following prices: 200,000 US Dollars per death, 67,000 US Dollars per injured person, and 700 US Dollars per vehicle. An important estimate was the cost of a death, which was what Ford had to pay according to US law, as a consequence for being responsible for the accident.

Life: https://en.wikipedia.org/wiki/Value_of_life

[10] For an introductory presentation of such an approach, see Thomas (2018). For an elaborated exposition rich of mathematical details, see Thomas and Vaughan (2015).

The total costs of doing nothing added up to 49,500,000 US Dollars. The calculations thus gave Ford a clear answer, and Ford continued production without changing anything.[11]

Here, we have reached a profound ethical issue. When addressing problems in natural sciences, quantifications might appear uncontroversial processes, but when quantifications are stretched out in all directions, they might turn problematic. Quantifications can be the carrier of questionable assumptions. They may direct us deep into brutal cynicism with respect both to decisions and to actions.

8.2 Digitalising

In the paper "On Computable Numbers, with an Application to the *Entscheidungsproblem*", Alan Turing (1937) provided an abstract presentation of the modern computer, although this was not the main purpose of his paper. In the title, the German word *Entscheidungsproblem* appears, it can be translated as "decision problem". The problem was formulated in 1928 by David Hilbert and Wilhelm Ackermann (1958) within a formalist outlook, which at that time was still operating with all its original ambitions.[12]

The *Entscheidungsproblem* in its general formulation assumes that we are dealing with a formalised mathematical theory. Considering an arbitrary formula from this formalism, one can ask whether this formula is a theorem or not – in other words, whether or not it can be derived from the axioms of the system. The *Entscheidungsproblem* raises the question whether an algorithmic procedure for deciding whether or not an arbitrary formula is a theorem exists.

Turing addressed this question, and in doing so he needed to establish a clear conception of what an algorithmic procedure is. He provided such a conception in terms of an abstract calculating machine that he referred to as an "*a*-machine". Later it was renamed a "Turing machine". By specifying the nature of possible calculations, Turing demonstrated that the *Entscheidungsproblem* "can have no solution" (Turing, 1937, p. 231). However, perhaps even more important is Turing's conception of the calculating machine, which turned out to be a precise conception of the electronic computer. Any such computer has an equivalent in a Turing machine. This mathematical representation of a computer makes it possible to study the potentials as well as the limits of computing, even before any real computer has been constructed. The *Entscheidungsproblem* represents an epistemic problem

---

[11] For more information about the Ford Pinto case, see Leggett (1999), and Birsch and Fielder (1994). The whole approach of putting value on human lives has also been critically addressed. One such attitude is established by the "vison zero". See Calibaba (2017); and https://en.wikipedia.org/wiki/Vision_Zero

[12] A heavy blow to the formalist ambitions only appeared in 1931, when Gödel (1962) showed that if a formal theory is consistent, it cannot be complete. A main ambition of formalism had been to create complete and consistent formal representations of mathematical theories.

formulated within a formalist outlook. By addressing this problem, Turing conceptualised the electronic computer and opened up a broad avenue from pure theoretical issues to real-life constructions.[13]

An electronic computer operates with numbers, and digitalisation is a crucial condition for any sort of computing. Within analogue computing, the computational representation of an object has apparent similarities with the original object. In digital computing, the representation is quite different: objects are represented by numbers. However, for the electronic computer to be able to operate technically with numbers, they need to be in a binary format. This possibility is also due to a mathematical insight, although an old one. In 1689, Gottfried Leibniz (1703) outlined the possibility of calculating within a binary number system. The avenue that leads to the construction of the digital computer is paved with mathematics.

It is common to refer to different industrial revolutions. The First Industrial Revolution is characterised by the invention of a range of new mechanical devices and machines. In 1764, the Spinning Jenny was constructed by James Hargreaves, and in the years that followed, many technical improvements of the spinning technology were introduced. In 1776, the first functioning steam engine was constructed. Before that, James Watt had been experimenting with the power of steam. The construction of the first steam locomotive was completed in 1813. The First Industrial Revolution is characterised by the further development of such technological constructions, and their implementation in the whole domain of production. The Second Industrial Revolution refers to a phase of standardisations and automatisations of production processes. Two important elements of this revolution were the decomposition of work processes into atomic units – known as Taylorism – and the invention of the conveyer-belt, which Henry Ford implemented in 1913 in one of his car production factories.

The Third Industrial Revolution has also been referred to as the Digital Revolution. This revolution is characterised by the construction of the electronic computer and its implementation in all spheres of life. This revolution brings about a range of social changes. Processes of automatisation become implemented in all kinds of production processes. The use of robots becomes common, not only in production, but also in, for instance, medicine and war machinery. Economic transactions become automatised and therefore accelerated. Computer-based patterns of surveillance become implemented in public as well as private spaces. The Digital Revolution has brought about dramatic changes of our life-conditions, and mathematics is a part of all of them.

In the first phase of the Digital Revolution, the expected role of the computer was described in optimistic terms. As an example, I can refer to the book *The Microelectronic Revolution*, by Tom Forrester (1981). However, it is important to recognise the complexities of the social impact accompanying the Digital Revolution. The First Industrial Revolution was accompanied by extreme oppression and exploitation of the workforce. Furthermore, this revolution assumed as a given that the

---

[13] For a further discussion of mechanical procedures and the Turing machine, see Sieg (1994).

colonial exploitation was maintained at its full scale. The second Industrial Revolution established a more systematic, if not science-based, exploitation of workers. As I will explicitly discuss in the following section, "Serialising", devastating patterns of oppression and exploitation also accompany the Digital Revolution.

The *Entscheidungsproblem* represented an intrinsic mathematical challenge, and it might be difficult to identify controversial ethical issues related to this problem. However, working with this problem opened up a route into further mathematical investigations, technological inventions, and industrial initiatives. This route brought us right into the centre of the Digital Revolution. It also brought us into an epicentre of ethical controversies. Digitalisation is a mathematics-based act with implications for our social and individual life. Digitalisations provide reconfigurations of our life-worlds.[14]

8.3 Serialising

As pointed out by Adam Smith (2009) in *The Wealth of Nations*, first published in 1776, the division of labour adds efficiency to the production process. Instead of being engaged in a complete process – say, of making a chair – each worker only completes a particular component of the process. Such a devision of labour, I will refer to as a serialisation. To Karl Marx, serialisations lead directly to the alienation of workers.[15]

A radical step in the further serialisation of work processes was proposed by Frederick W. Taylor (2006) in *The Principles of Scientific Management*, first published in 1911. After this publication, the Second Industrial Revolution got underway. In the book, Taylor presents the worker Schmidt, who is going to be exposed to the principles of scientific management. Taylor sees processes of production, or any work processes for that matter, as mechanical processes. The workers and the tools make up part of the same operational structure, which can be optimised through careful schemes of management. The workers thereby become part of the process, as objects, rather than as agents of their own lives. They become objectified and alienated. This is a fundamental part of Marx's analysis of the capitalist mode of production.

The particular work task that Taylor describes concerns how to load pig iron onto a railway wagon. Taylor and his staff had observed that an average worker, following the traditional work pattern, can load 12½ tons per day. However, Taylor estimated that the quantity could be much more, namely 47–48 tons per day. To reach such a goal presupposes, however, that a new form of organisation of the work

---

[14] As described in Skovsmose (2014b), my use of the notion of life-world is different from the use in phenomenology. With the term "life-worlds", I refer to the ways economic, political, and cultural format living conditions, and to how we experience such a formatting.

[15] Marx addressed the notion of alienation in his early writings (Marx, 1844), but elaborated upon it further in the *Capital* (Marx, 1992, 1993a, 1993b).

process is implemented. As an illustration of what that could mean, Taylor describes how the worker Schmidt came to work according to the new measures. The important thing is that Schmidt does not follow his own rhythm, but becomes totally subjected to the prescribed organisation of the work process. Schmidt has to be observed by a supervisor, and has to follow what the supervisor orders him to do. Taylor (2006) makes this clear to Schmidt in the following way: "When he tells you to pick up a pig and walk, you pick it up and walk, and when he tells you to sit down and rest, you sit down. You do that straight through the day" (p. 21).

The overall approach in Taylor's version of scientific management is to turn the worker into a robot. According to scientific management, complex work processes have to be broken down into their atomic units, and deciding which units to perform in what order has to be determined, not by the workers, but by scientific methods.[16] To me, Charlie Chaplin's film *Modern Times* creates an ironic illustration of what such serialising might imply, both with respect to the format of the work process and the condition of the workers.

Today, Taylor's version of scientific management has assumed a new format referred to as New Taylorism or as Digital Taylorism. For implementing Digital Taylorism, the computer plays a principal role. Digital Taylorism shares the basic ambition of classic Taylorism, namely to maximise the efficiency of any work process. However, by means of the computer, the serialising can be taken much further. The smallest atomic tasks can be identified, standardised, and turned into routines. Computer-based serialising also makes it possible to surveil and to control all details of the process, including the people involved in it. The atomic decomposition and the surveilling make it easy to substitute any worker that is not completing the prescribed task in an efficient way with another. Digital Taylorism drains the work processes for any specific person-related professionalism, as such professionalism becomes incorporated into the computer-structures.

In the article "Big Brother's Corporate Cousin: High-Tech Workplace Surveillance Is the Hallmark of an New Digital Taylorism", Christian Parenti (2001) makes the following observation: "According to the American Management Association, 80 percent of US corporations keep their employees under regular surveillance, and that percentage is growing all the time. From the low-tech body and bag searches at retail stores, to computerised ordering pads at restaurants and the silent monitoring of e-mail and phone traffic in offices, the American workplace is becoming ever more transparent to employers and oppressive for employees."[17] The new Digital Taylorism establishes an extensive and intensive Big-Brother control system.

Digital Taylorism has also been implemented in school systems. The processes of teaching and learning are broken down into specific units, and the students' learning is defined in terms of the mastering of well-defined tasks. It is part of Digital

---

[16] See also the presentation of Taylorism in Chap. 15, "Symbolic Power, Robotting, and Surveilling", in Skovsmose (2014a).

[17] See also Lanigan (2007).

Taylorism that the students' learning is checked along the route, making it possible to include loops and repetitions in the process. The role of the teacher is also specified and becomes easy to observe and to control. Learning and teaching become serialised tasks, and students and teachers come to operate in a digital panopticon.

With particular reference to the situation in the USA, Terry M. Moe (2003) makes the following observation with respect to the implementation of Digital Taylorism in school systems: "The movement for school accountability is essentially a movement for more effective top-down control of the schools. The idea is that, if public authorities want to promote student achievement, they need to adopt organizational control mechanisms – tests, school report cards, rewards and sanctions, and the like – designed to get district officials, principals, teachers, and students to change their behaviour in productive ways. [...] Virtually all organizations need to engage in top-down control, because the people at the top have goals they want the people at the bottom to pursue, and something has to be done to bring about the desired behaviours. [...] The public school system is just like other organizations in this respect ..." (p. 81).[18] Digital Taylorism captures the school system in the format of a production process that, like any such process, can be optimised and controlled.

The implementation of New Taylorism in workplaces and school systems is propelled forward by mathematical means. New Taylorism represents a digitalised serialising with an impact on all features of our life-worlds. This whole development calls for ethical considerations.

8.4 Categorising

Since the Scientific Revolution, it has been acknowledged that mathematics plays a fundamental role in the description of nature. The generally accepted explanation was that this was due to the properties of nature. The laws according to which nature was operating had a mathematical format; as a consequence, mathematics provides the best possible tool for describing nature. In *Critique of Pure Reason*, first published in 1781, Immanuel Kant (1929) created a radically different way of looking at mathematics. He agreed that mathematics is powerful in capturing our observations of nature. However, according to him, the reason is not the properties of nature, but properties of the epistemic structures that we, human beings, bring into operation when we observe natural phenomena. All our observations of such phenomena take place in space and time, and since mathematics captures the properties of space and time, mathematics will apply to any such observation. Mathematics is an integral component of our epistemic categories that become projected onto whatever we

---

[18] For a critical investigation of New Taylorism in the USA school system, see Au (2011). The quotation of Moe is also taken from there.

observe. According to Kant, this is the reason for the general applicability of mathematics.

At times, Michel Foucault has been referred to as the new Kant, and this is with good reason. Foucault was also deeply concerned about the formation of our knowledge, or rather with what we assume to be our knowledge. He also found that what we consider to be knowledge is formed through categories. However, quite different from Kant, Foucault considered these categories to be social fabrications. Such fabrications take place though complex historical processes that might incorporate all kinds of assumptions and preconceptions. Prejudices from different historical periods might be engraved in our categorical frameworks. Based on such categorises, we express what we think of as our knowledge.[19]

In our digitalised world, mathematics becomes part of the formatting of our knowledge. How could that be? When we search the net, information appears on the screen in front of us. Apparently, we are in charge of what we are doing. We type a searchword and see what appears. We might click our way onto some other webpages. We can change the searchword, if we want. We appear to be free in our walk on the net. We seem perfectly in charge of what we are doing. We might claim that we construct our knowledge based on what we locate on the net.

However, what appears on the screen in front of us is not simply determined by our choices and priorities. What appears is a pre-fabricated phenomenon. Our walk on the net is not any kind of free walk; it is a guided and pre-structured walk. Google's search engine is set into operation whenever we are searching. This engine is based on a mathematically elaborated PageRank model, which provides a ranking of web pages. As there are millions and millions of such pages, a random search would be without meaning. It would be like searching for information in a library, where all the leaves have been torn out of the books and put in a pile in the middle of the floor. Like the pages in a book, the webpages also need to be organised, and this is done by the PageRank model. This model establishes all kinds of priorities with respect to our search for information. One can think of it as a knowledge-structuring mathematical device.[20]

The PageRank model applies linear algebra and, in particular, matrix calculations. One parameter in these calculations is the number of links a webpage has got; this parameter is interpreted as an expression of the importance of the page. A collection of such priorities structures the page ranking. Huge business interests are associated with this ranking; it might be crucial to appear as one of the first pages in a search. Besides economic interests, any kind of stereotypes might also be incorporated in the ranking. As an example, previously a search for "lesbian" would bring the search onto pornographic pages. This stereotypical association has now been removed from the PageRank model, but many other stereotypes may be operating through the model, and still new stereotypes might become incorporated.

---

[19] See, for instance, Foucault (1994, 2000).

[20] For a further discussion of the PageRank model, see Langville and Meyer (2012), and Ravn and Skovsmose (2019).

That PageRank model provides a mathematical structuring of our walk on the net and therefore of the information we get from that. It is not any idealised structuring as suggested by Kant. It is an industrial structuring, which might include many kinds of preconceptions. It might include economic – as well as political – interest. It is a structuring with a tremendous impact on social life. It has an impact on what we might come to know about and what we come to ignore. It is not only Google that provides rankings of information. What we see and do not see on Facebook has also been subjected to algorithmic-based rankings. Any such structuring of information is in need of profound ethical reflections.

8.5 Imagining

The notion of sociological imagination was coined by Charles Wright-Mills (1959). By paying particular attention to this notion, Wright-Mills stepped away from the positivist position and claimed that social theorising also has to conceptualise alternatives to present states of affairs.

By *technological imagination*, I refer to the conception of technological possibilities that do not exist, but might come to exist. The First Industrial Revolution was initiated through several examples of technological imagination, such as the conception of the spinning machine and of the steam engine. Such imaginations anticipated mechanical constructions. In order to articulate such imaginations, no advanced mathematics was put in operation.[21] In arriving at the Digital Revolution, technological imagination has also been put into operation. However, contrary to what characterised the First Industrial Revolution and also the Second Industrial Revolution, mathematics plays the principal role in forming the technological imagination that bring us into the Third Industrial Revolution. The conception of the digital computer was impossible without mathematics.

Technological imagination paves the way for new possibilities. The connotations of the word "possibility" normally indicate that we are dealing with attractive potentials. However, we need to broaden this interpretation. I use the word "possibility" to include positive and attractive phenomena as well as risks, dangers, if not catastrophes. By using the word "possibility", I refer to what might come to take place in the future; this means that possibility is intimately connected to uncertainty.[22] Future technical possibilities will become formed through technological imaginations – and today, in particular, by mathematics-based technological imaginations.

---

[21] However, when the outputs of mechanical imaginations have to be constructed, mathematics becomes crucial. For the construction of, say, a locomotive, one needs to provide a geometric presentation in order to analyse its details.

[22] Technological determinism refers to the assumption that technological development is the principal force in social development. I do not assume any such determinism. Technological development operates on the global scene alongside political struggles, economic forces, and social movements. It is a dramatic and open-ended process.

Mathematics is a powerful creative force in opening up new technological possibilities.

We bring the pollution of today with us into the future. Will we be able to cope with this? What environmental policies might become implemented? Which dangers might be overlooked, or maybe deliberately ignored? Will new soluble forms of plastic be invented? Will we be able to clean up the oceans? What climate changes can we expect? Will we find alternative energy resources? How will the search for energy-alternatives becomes prioritised? Will hunger crises become global? Will new patterns of food production become developed? Will new patterns for redistributing wealth and resources become implemented? Will production processes get further accelerated? Will we come to experience unprecedented patterns of serialising and surveillance? Will Big Brothering further intensify? Will the distribution of information, as well as of misinformation, find new channels, with the implication that it becomes even more difficult to distinguish between information and misinformation?[23]

We do not know the responses to any of these potential uncertainties. Except for one thing: mathematics-based technological imagination will be brought into operation when dealing with issues about pollution, inventions of new material, search for alternative energy resources, further development of farming, gen-manipulation, redistribution of wealth and resources, acceleration of production, serialising, surveilling, and means of information.

Our life-worlds are formed through what is actually the case, as well as from what might become the case. What we experience as our life-world is not just composed of the actual space of facts. It is also composed of possibilities, including risks and dangers. It is composed of "what is" and of "what might become". In our present situation, very many "might-becomes" will be established through mathematics-based technological imaginations. This imagination is an important creative force in bringing about new technological possibilities, including possibilities that might have devastating implications. Technological imagination is a controversial and open-ended act.

8.6 Banality

By concentrating on ontological and epistemological issues, logicism, formalism, and intuitionism come to operate within an ethical vacuum. Let us now take a closer look at a more recent trend in the philosophy of mathematics, referred to as the "philosophy of mathematical practice".

One of the first to articulate this trend was Ruben Hersh (1979). He found that the philosophy of mathematics had developed with a pathological degree of

---

[23] In Chap. 9, "Mathematics and Crises", I have discussed how mathematics might contribute to the formation of critical situations.

8.6 Banality

inbreeding. The discussed problems had been expressed with reference to already formulated philosophical problems, not acknowledging the problems emerging from the actual mathematical practice. Hersh suggested making a new beginning of the philosophy of mathematics by paying attention to the mathematical practice. With such a practice, he first of all referred to what was taking place at universities and other research institutions, and not to practice as referred to by, say, ethnomathematical approaches.

The trend "philosophy of mathematical practice" follows Hersh's recommendation. The book *The Philosophy of Mathematical Practice*, edited by Paolo Mancosu (2008), illustrates what this could mean. In the preface, Mancosu highlights that he wants to "unify the efforts of many philosophers who were making contributions to a philosophy of mathematics informed by a desire to account for many central aspects of mathematical practice that, by and large, had been ignored by previous philosophers and logicians" (p. v).

The book addresses a range of issues related to ways of doing mathematics, and let me refer to three examples of visualisation in mathematics, both with respect to proving and discovering. According to a traditional formalist approach, visualisation might be relevant for making mathematical discoveries, but it has no role to play in mathematical proving. This dichotomy is challenged by Marcus Giaquinto (2008). Kenneth Manders (2008) investigates diagram-based geometric practices in order to provide a route towards "rigorous diagram-based reasoning practice" (p. 75). This is also a route that runs contrary to formalism.[24] Jeremy Avigad (2008) discusses the use of computers in mathematical inquiries. By doing so, he mixes empirical features with mathematical processes, while in classic approaches in the philosophy of mathematics this is considered problematic. Avidag makes the following observation which appears programmatic for a philosophy of mathematical practice: "The types of questions raised here are only meaningful in specific mathematical and scientific contexts, and a philosophical analysis is only useful in so far as it can further such inquiry. Ask not what the use of computers in mathematics can do for philosophy; ask what philosophy can do for the use of computers in mathematics" (p. 314).[25]

Most often, university studies in mathematics are defined through well-specified curricula outlining a set of mathematical topics. There is no mention of disciplines that might address ethical issues related to the production and application of mathematical knowledge. The social impact of bringing mathematics into action is ignored. This insertion of mathematics into an ethical vacuum is repeated by the philosophy of mathematical practice. In the book *The Philosophy of Mathematical Practice*, there is no discussion of any ethical issues. In the book, the very word "ethics" only appears once (and this is in a parenthesis).

---

[24] For a discussion of formal and informal provability, see Leitgeib (2009).
[25] More presentations of the philosophy of mathematical practice can be found in Carter (2019), Ferreirós (2016), and Tymoczko (1986).

The ignorance of the ethical dimension of a philosophy of mathematics, both in the actual practice of mathematics and in the philosophy of mathematical practice, I consider a huge problem. I consider it to be the most profound problem confronting a philosophy of mathematics today. The nature of this problem I have referred to as the "banality of mathematical expertise".[26] With this expression, I want to highlight an extremely problematic feature of mathematical expertise: it seems to operate in an ethical vacuum by first of all concentrating on technical features of doing mathematics without reflecting on the possible social impacts of doing mathematics.

Mathematics-based actions might have all kinds of impacts. I have tried to illustrate this by using the searchlights of quantifying, digitalising, serialising, categorising, and imagining along the routes connecting mathematics with contentious social issues. It is crucial that the impacts of bringing mathematics into action are critically addressed as part of both mathematical practice and a philosophy of mathematics. The ethical dimension of a philosophy of mathematics is an indispensable component of this philosophy.

Note

This chapter is a revised version of my paper "Mathematics and Ethics", published in 2020 in *Qualitative Research Journal, 8*(18), 479–503. It has been reprinted in the *Philosophy of Mathematics Education Journal*, (36). I thank the publishers for their kind permission to present it here.

References

Au, W. (2011). Teaching under the new Taylorism: High-stakes testing and the standardization of the 21st century curriculum. *Journal of Curriculum Studies, 43*(1), 25–45.
Avigad, J. (2008). Computers in mathematical inquiry. In P. Mancosu (Ed.), *The philosophy of mathematical practice* (pp. 302–316). Oxford University Press.
Bernacerraf, P., & Putnam, H. (Eds.). (1964). *Philosophy of mathematics*. Prentice-Hall.
Birsch, D., & Fielder, J. H. (Eds.). (1994). *The Ford Pinto case: A study in applied ethics, business, and technology*. State University of New York.
Bostock, D. (2009). *Philosophy of mathematics: An introduction*. Wiley Blackwell.
Brown, J. R. (2008). *Philosophy of mathematics: A contemporary introduction to the world of proofs and pictures* (2nd ed.). Routledge.
Bueno, O., & Linnebo, Ø. (Eds.). (2009). *New waves in philosophy of mathematics*. Palgrave Macmillan.
Calibaba, J. (2017). *The road to vision zero: Zero traffic fatalities and serious injuries* (2nd ed.). CreateSpace Independent Publishing Platform.
Carter, J. (2019). Philosophy of mathematical practice: Motivations, themes and prospects. *Philosophia Mathematica, 27*(1), 1–32.
Ernest, P. (2016). A dialogue on the ethics of mathematics. *The Mathematical Intelligencer, 38*(3), 69–77.

---

[26] See Chap. 7, "Banality of Mathematical Expertise".

References

Ernest, P. (2018). The ethics of mathematics: Is mathematics harmful? In P. Ernest (Ed.), *The philosophy of mathematics education today* (pp. 187–216). Springer.

Ernest, P. (2020). The ethics of mathematical practice. In B. Sriraman (Ed.), *Handbook of the history and philosophy of mathematical practice*. Springer. https://doi.org/10.1007/978-3-030-19071-2_9-1

Ferreirós, J. (2016). *Mathematical knowledge and the interplay of practices*. Princeton University Press.

Forrester, T. (1981). *The microelectronic revolution*. The MIT Press.

Foucault, M. (1994). *The order of things: An archaeology of the human sciences*. Vintage Books.

Foucault, M. (2000). *Power*. The New Press.

George, A., & Velleman, D. J. (2002). *Philosophies of mathematics*. Blackwell.

Giaquinto, M. (2008). Visualizing in mathematics. In P. Mancosu (Ed.), *The philosophy of mathematical practice* (pp. 22–42). Oxford University Press.

Gödel, K. (1962). *On formally undecidable propositions of Principia Mathematica and related systems*. Basic Book.

Hacking, I. (2014). *Why is there philosophy of mathematics at all?* Cambridge University Press.

Hersh, R. (1979). Some proposals for reviving the philosophy of mathematics. *Advances in Mathematics, 31*, 31–50.

Hersh, R. (1990). Mathematics and ethics. *Humanistic Mathematics Network Journal, 5*(9), 20–23.

Hilbert, D., & Ackermann, W. (1958). *Principles of mathematical logic*. Chelsea Publishing Company.

Hofflander, A. E. (1966). The human life value: A theoretical model. *The Journal of Risk and Insurance, 33*(4), 529–536.

Hood, K. (2017). The science of value: Economic expertise and the valuation of human life in US federal regulatory agencies. *Social Studies of Science, 47*(4), 441–456.

Jacquette, D. (Ed.). (2002). *Philosophy of mathematics: An anthology*. Blackwell.

Kant, I. (1929). *Critique of pure reason* (Translated by Norman Kemp Smith). Macmillan.

Körner, S. (1968). *The philosophy of mathematics*. Hutchinson University Library.

Langville, A. N., & Meyer, C. D. (2012). *Google's PageRank and beyond: The science of search engine rankings*. Princeton University Press.

Lanigan, C. (2007). From assembly line to just-in-time: Preparing a capable workforce for the knowledge economy. *CIO*. https://www.cio.com/article/2439128/from-assembly-line-to-just-in-time%2D%2Dpreparing-a-capable-workforce-for-the-knowledge-economy.html

Leggett, C. (1999). *The Ford Pinto case: The valuation of life as it applies to the negligence efficiency argument*. https://users.wfu.edu/palmitar/Law&Valuation/Papers/1999/Leggett-pinto.html

Leibniz, G. (1703). *Explanation of binary arithmetic*. http://www.leibniz-translations.com/binary.htm

Leitgeib, H. (2009). On formal and informal provability. In O. Bueno & Ø. Linnebo (Eds.), *New waves in philosophy of mathematics* (pp. 263–299). Palgrave Macmillan.

Linnebo, Ø. (2017). *Philosophy of mathematics*. Princeton University Press.

Mancosu, P. (Ed.). (2008). *The philosophy of mathematical practice*. Oxford University Press.

Manders, K. (2008). Diagram-based geometric practice. In P. Mancosu (Ed.), *The philosophy of mathematical practice* (pp. 65–79). Oxford University Press.

Marx, K. (1844). *Economic and philosophic manuscripts of 1844*. https://www.marxists.org/archive/marx/works/download/pdf/Economic-Philosophic-Manuscripts-1844.pdf

Marx, K. (1992, 1993a, 1993b). *Capital: A critique of political economy* (Vol. I–III). Penguin Classics.

Mehlberg, H. (1960). The present situation in the philosophy of mathematics. *Synthese, 12*(4), 380–414. Reprinted in D. Jacquette (Ed.) (2002), *Philosophy of mathematics: An anthology* (pp. 65–82). Oxford: Blackwell.

Moe, T. M. (2003). Politics, control, and the future of school accountability. In P. E. Peterson & M. R. West (Eds.), *No child left behind? The politics and practice of school accountability* (pp. 80–106). Brookings Institution Press.

Parenti, C. (2001). Big brother's corporate cousin: High-tech workplace surveillance is the hallmark of a new digital Taylorism. *The Nation, 273*(5), 26–30.

Ravn, O., & Skovsmose, O. (2019). *Connecting humans to equations: A reinterpretation of the philosophy of mathematics*. Springer.

Shapiro, S. (2000). *Thinking about mathematics: The philosophy of mathematics*. Oxford University Press.

Sieg, W. (1994). Mechanical procedures and mathematical experience. In A. George (Ed.), *Mathematics and mind* (pp. 71–117). Oxford University Press. Reprinted in D. Jacquette (Ed.) (2002), *Philosophy of mathematics: An anthology* (pp. 226–258). Oxford: Blackwell.

Skovsmose, O. (2014a). *Critique as uncertainty*. Information Age Publishing.

Skovsmose, O. (2014b). *Foregrounds: Opaque stories about learning*. Sense Publishers.

Smith, A. (2009). *The wealth of nations*. Thrifty Books.

Taylor, F. W. (2006). *The principles of scientific management*. Cosimo Classics.

Thomas, P. (2018). *Calculating the value of human life: Safety decisions that can be trusted* (Policy Report 25). University of Bristol. https://www.bristol.ac.uk/media-library/sites/policybristol/PolicyBristol-Report-April-2018-value-human-life.pdf

Thomas, J. P., & Vaughan, G. J. (2015). Testing the validity if the "value of a prevented fatality" (VPF) used to assess UK safety measures. *Process Safety and Environmental Protection, 93*, 293–298.

Turing, A. M. (1937). On computable numbers, with an application to the Entscheidungsproblem. *Proceedings of the London Mathematical Society, s2–42*(1), 230–265.

Tymoczko, T. (Ed.). (1986). *New directions in the philosophy of mathematics*. Birkhäuser.

Wright Mills, C. (1959). *The sociological imagination*. Oxford University Press.

Chapter 9
Mathematics and Crises

Abstract One can identify at least three different types of relationships between mathematics and crises. First, mathematics can picture a crisis. This is in accordance with the classic interpretation of mathematical modelling, which highlights that a mathematical model provides a representation of a piece of reality. Second, mathematics can constitute a crisis, meaning that mathematics can form an intrinsic part of the very dynamics of a crisis. This phenomenon can be illustrated by the economic crises that spread around the world in 2008. Third, mathematics can format a crisis. This formulation refers to a situation where a mathematical reading of a crisis brings about ways of acting in the critical situation that might be adequate, but also counterproductive, if not catastrophic. This is illustrated with reference to the potential crises due to climate change. As a conclusion, the paper addresses the politics of crises, which refers to the power that can be acted out through a crisis discourse, in which mathematics may come to play a deplorable role.

Keywords Mathematics picturing a crisis · Mathematics constituting a crisis · Mathematics formatting a crisis · Politics of crises

The human influence on nature, the environment, and basic living conditions appear many times to be devastating. We will come to face more and more crises rolling forward: crises due to drought as well as to floods; crises due to lack of clean drinking water or other natural resources; and crises due to pollution; and crises due to climate change. We will also see crises due to displacement of people, deportation, and immigration; crises due to overpopulation and hunger; and crises due to increasing inequalities with respect to basic welfare conditions. Political crises might take new formats, and so might crises caused by violence and war. We are entering a period where crises will be more and more frequent, if not a perpetual phenomenon. This is one possible reading of the future.[1]

---

[1] Due to the human influence on the atmosphere, we might witness a new geological epoch that needs a name, the anthropocene. For the formulation of the concept of anthropocene, see Crutzen and Stoermer (2000); for relating the discussion of the anthropocene to mathematics education, see

Instead of *crisis*, one can also talk about a *critical situation*. I am not going to discuss if it makes sense to try to differentiate between the two notions, but instead will use them as synonyms of one another.[2] Additionally, the notion of *risk* belongs to the same family of notions: a risk can be seen as a potential critical situation.

One can talk about crises in many different ways – a personal crisis, a psychological crisis, an identity crisis, a family crisis – but in this chapter I concentrate on large-scale crises by focusing on those that are epidemic, economic, and environmental. Naturally, there are strong connections between large-scale and personal crises. When a country enters a financial crisis, critical situations for families and individuals emerge; some might lose their jobs, some might need to sell their houses, some might need psychological support. However, the intensive dynamics between macro-crises and micro-crises are not the focus of this chapter.

I am concentrating here on possible relationships between mathematics and large-scale crises. I will refer to three examples, and in this way illustrate that *one can experience at least three different types of relationships between mathematics and crises*.

The first example illustrates that *mathematics can picture a crisis*. This is in accordance with the classic interpretation of mathematical modelling, which highlights that a mathematical model provides a representation of a piece of reality. This "piece of reality" can be a critical situation, and by investigating its mathematical representation, one might get a better understanding of the crisis itself. The second example shows that *mathematics can constitute a crisis*. A mathematical model need not operate as detached from a crisis, but can become an intrinsic part of the crisis. Mathematics-based algorithms may provide an underlying dynamic which brings about the crisis. In such a case, mathematics is not picturing a crisis; it becomes an essential ingredient in the crisis. The crisis might not exist without mathematics. The third example illustrates that *mathematics can format a crisis*. With this formulation, I have in mind a situation where a mathematical reading of a crisis brings about a way of acting in the critical situation. It might turn out that this form of action functions adequately, and that the crisis is brought under control. The opposite might also be the case: the reading might be a misreading, and the acting might be counterproductive. In both cases, mathematics formats both readings and actions, and in the end it formats the dynamics of the crisis itself. We have here three analytical types of mathematics-crisis relationships, but in real-life critical situations, the distinction between the three might be blurred.

I conclude by addressing the politics of the crisis. To call something a crisis focusses public attention onto the phenomenon. However, what are we to call a crisis, and what not a crisis? Who decides what to call a crisis? Many assumptions, presumptions, preconceptions, and particular interests are expressed through a

---

Coles (2016). See also Bostrom and Cirkovic (2008), who present a range of examples of global risks with catastrophic dimensions, and Klein (2014) and Latour (2018) who present broader political discussions of the issue.

[2] In the Chap. 6, "Crises and Critique", I point out some semantic connections between the notions of crisis and critique.

9.1 Mathematics as Picturing a Crisis

crisis discourse. As part of some such discourses, abominable and cynical ideas may be incorporated (and in some cases normalised), and we need to raise the question to what extent mathematics could come to constitute a part of such discourses.

9.1 Mathematics as Picturing a Crisis

In the *Tractatus*, Ludwig Wittgenstein (2002) interprets language as picturing reality. This is the essence of what has become referred to as Wittgenstein's picture theory of language. The language Wittgenstein has in mind is mathematics. The role of mathematics, according to the *Tractatus*, is to provide pictures of reality. This idea was further elaborated upon by logical positivism into the claim: mathematics is the language of science.

Wittgenstein's picture theory of language characterises a broad set of views about mathematical modelling. In order to specify a model, according to such a view, one needs to specify the part of reality that is being modelled, the mathematical entities that are going to be applied (parameters, functions, equations, etc.), and the way these entities are related to the features of reality.

Mathematical modelling of epidemics or pandemics is a developed discipline.[3] An important parameter in such a modelling is the number R, which refers to the number of infected persons generated by one infected person. As a further simplification, we can assume that this infection will take place within a certain time unit. If R is bigger than 1, the epidemic will develop and might get out of control. If t refers to the number of time units, then the number of newly infected people at time t, $N(t)$, will be $N(t) = R^t$. The total number of infected people at the time t, $S(t)$, will then be:

$$S(t) = N(0) + N(1) + N(2) + \ldots + N(t) = 1 + R + R^2 + \ldots + R^t = \left(1 + R^{t+1}\right)/(R-1)$$

The number $N(t)$ will follow an exponential growth pattern; the growth of total number of infected people, $S(t)$, will also have an exponential format. However, this is a simplified and abstract situation. As an epidemic takes place within a certain population, there is an upper limit to the number of people that can be infected. The graph of the function $N(t)$ will get a bell shape, while the graph of the function $S(t)$ will take an s-shape, like the format of a logistic growth. All of this is a gross simplification, but illustrates an important point: one way of trying to deal with an epidemic is to reduce the number R. This can, for instance, be done by establishing social distancing or social isolation.

---

[3] Capasso (2008) provides a careful presentation of the mathematics of epidemics, showing different mathematical approaches. An extensive presentation is also found in Brauer et al. (2019).

Such mathematical modelling fits Wittgenstein's picture metaphor. Mathematics might describe, more or less reliably, the dynamics of an epidemic.[4] That the mathematical picturing has no influence on the epidemics seems obvious. In this case, the language of mathematics seems detached from what it depicts.[5]

On 16 March 2020, the UK Imperial College COVID-19 Response Team (Imperial College, 2020) published Report 9: *Impact of Non-Pharmaceutical Interventions (NPIs) to Reduce COVID-19 Mortality and Healthcare Demand.* When the report was published, the coronavirus crisis was still in its initial phase. In many countries, the number of people catching the virus and the number of deaths did not appear alarming, and the governments in some countries – such as the USA and the UK – were not prepared to take any particular measures. Both the British Prime Minister, Boris Johnson, and the American President, Donald Trump, found the discourse of the press to be exaggerated, and some commentators and politicians even found it hysterical. In an interview, the 28th of February 2020, Boris Jonson indicated that washing hands would be an adequate way of dealing with the coronavirus problem.[6] Then the report emerged.

The report delineates the possible dynamics of the coronavirus epidemic. With particular reference to the USA and Great Britain, the report states:

> The epidemic is predicted to be broader in the US than in GB and to peak slightly later. This is due to the larger geographic scale of the US, resulting in more distinct localised epidemics across states [...] than seen across GB. The higher peak in mortality in GB is due to the smaller size of the country and its older population compared with the US. In total, in an unmitigated epidemic, we would predict approximately 510,000 deaths in GB and 2.2 million in the US, not accounting for the potential negative effects of health systems being overwhelmed on mortality (pp. 6–7).

This is shocking reading. These are the scenarios that were to be expected, had no measures been taken and the epidemic had developed according to its own dynamics, as pictured by the applied mathematical models.

The report, however, presents different future scenarios depending on what measures are taken. It shows both graphically and numerically the differences between the scenario with no precautions implemented, and scenarios where certain measures are applied. The report makes it evident what can be expected if one implements case isolation, or household isolation, or closing schools and universities, or if one establishes social distancing of more than 70% of the population. Even in the case that maximum precautions are applied, the prediction looks gloomy. The reports concludes:

---

[4] In order to describe the dynamic of an epidemic, system dynamics modelling becomes crucial; see, for instance, Duggan (2016).

[5] This appears to be common sense when we think of epidemics in biological terms. Kiss et al. (2017) address the mathematics of epidemics on networks. In this case, the situation is quite different, as we are dealing with a situation where mathematics constitutes the epidemic problem. It is through mathematics that one also needs to address the problem.

[6] See https://www.bbc.com/news/av/uk-politics-51682581/boris-johnson-advises-people-to-wash-hands-to-avoid-coronavirus

[…] that epidemic suppression is the only viable strategy at the current time. The social and economic effects of the measures which are needed to achieve this policy goal will be profound. Many countries have adopted such measures already, but even those countries at an earlier stage of their epidemic (such as the UK) will need to do so imminently. (p. 16).

Soon after the publication of the report, one can observe changes in the official policy in the USA and the UK. Measures became implemented, and social distancing became the official policy.

The change of strategies in the two countries is an example of the socio-political impact of a mathematical picturing of a crisis. Naturally, the mathematical modelling need not be precise. Any picturing depends on assumptions incorporated into the model. The model provided by the Imperial College COVID-19 Response Team might give a disproportionate picture of what might happen: the consequences could turn out to be less sinister; they could also be worse than predicted. A mathematical model provides an attempt at looking into the future, but it is an unreliable look.

The notion of "similarity gap" refers to the difference between the mathematical picture of reality and the reality itself. Such a gap might be minimal, but this need not be the case. The existence of similarity gaps is a direct implication of the fact that mathematical models always include simplifications and stipulations. In the case of a pandemic, the similarity gap refers to the difference between the dynamics described by a model, and the actual development of the pandemic. One thing, however, is to acknowledge the existence of a similarity gap; quite another thing is to estimate the size of such a gap. In cases where we are dealing with the picturing of an actual situation, a similarity gap can be estimated directly. A quite different situation occurs when we are dealing with predictions. In this case, the most reliable method for estimating the size of a similarity gap seems, unfortunately, to be to wait and see what comes to pass. Whenever we are addressing epidemics or pandemics through mathematical modelling, we have to acknowledge the possibility of even huge similarity gaps.[7]

9.2 Mathematics as Constituting a Crisis

When shopping in a supermarket, we pay with the credit card. This appears to be a trivial process. However, complex processes are running beneath the surface of this simple routine. My credit card is identified; the money I have to pay is subtracted from my bank account which is linked with the card; and the money enters the account of the supermarket. All the transactions are protected by certain security

---

[7] For a discussion of similarity gap, see Skovsmose (2005) and Chap. 4 in this book, "Sustainability and Risks".

measures. This entire process only operates due to an extremely complex configuration of mathematical algorithms.[8]

If we look, not at the supermarket, but at the stock market, we can observe a similar phenomenon. Transactions are again conducted by mathematics-based algorithms. A profound mathematical structuring of the stock market has taken place. There are no real bonds and no real money passed across the tables. Mathematical algorithms drive the processes. Mathematics is not doing any picturing of the stock-market transactions, but is conducting them. Mathematics constitutes the processes of transactions. One can consider to what extent the whole apparatus of a capitalist economy would function without mathematics constituting a range of its procedures.

In these processes, we are not dealing with a mathematical model that just pictures what takes place in the transactions. Mathematics performs the transactions. Mathematics constitutes the processes we are talking about.

When mathematics becomes part of the functioning of economic processes, it also becomes part of the possible dysfunctions of such processes. In the book *Weapons of Math Destruction*, Cathy O'Neil (2016) discusses the role of mathematics in handling huge data material. The subtitle of her book is: *How Big Data Increase Inequality and Threatens Democracy*. O'Neil was a research mathematician with a promising career in front of her, but she decided to quit her job at the university in order to work in a private hedge fund company. This job included the administration of huge investments, and for doing so, mathematical expertise is crucial.

In 2008, after working for about 1 year in the company, came the economic crisis. This crisis made O'Neil (2016) aware of some important features of mathematics:

> That crash made it all too clear that mathematics, once my refuge, was not only deeply entangled in the world's problems but also fuelling many of them (p. 2).

This is a principal observation: mathematics is not any detached academic discipline; it can be deeply entangled in real-life problems, and it may be fuelling such problems. O'Neil continues:

> The housing crisis, the collapse of major financial institutions, the rise of unemployment – all had been aided and abetted by mathematicians wielding magic formulas (p. 2).

This is a strong claim made with respect to mathematics that often has been considered "harmless and innocent". Contrary to such a position, O'Neil claims that by "wielding magic formulas" mathematics constitutes a range of socio-political and economic crises.[9] Mathematics might not only causes chaos, but also add to it:

> What's more, thanks to the extraordinary powers that I loved so much, math was able to combine with technology to multiply the chaos and misfortune, adding efficiency and scale to the systems that I now recognized as flawed (p. 2).

---

[8] See, for instance, Schneier (1996); Skovsmose and Yasukawa (2009); and Stallings (2017).

[9] It might be that O'Neil has something more particular in mind when talking about "magic formulas", thus the Black-Scholes formula plays a crucial role in algorithmic trading; see Chriss (1997); and Harford (2012).

9.2 Mathematics as Constituting a Crisis

How could mathematics come to assume such a role? How could mathematics come to add efficiency and scale to flawed systems? A particular feature of the modern stock market is algorithmic trading. Decisions about buying or selling depend on many factors, which can be systematised and condensed into algorithms, meaning that decisions about selling or buying can be executed automatically. These decisions can concern all the details of the process: the time to complete the business, the price levels that are considered acceptable, and the amount of money to be traded. In 2006, a third of all stock-market transactions in the European Union and the USA took the form of algorithmic trading.[10] Algorithmic trading makes it possible for economic transactions to run fast, and to be further accelerated. It can accelerate the whole situation out of balance, even with an economic crash as a consequence.

Naturally, an economic crisis can have different external causes, but as pointed out by O'Neil, we also have to look for the roots of the crisis in dysfunctions in the mathematical machinery that otherwise constitute the surface of economic normality. One can see the economic crisis of 2008 as being constituted by mathematics.

In her book, O'Neil describes other mathematical models brought into operation, not only with respect to economic transactions, but in all spheres of life: in administration, management, production, surveillance, military. O'Neil's point is that "many of these models encode human prejudice, misunderstandings, and bias into the software systems that increasingly manage our lives" (p. 3). According to her, such models are causing a range of critical situations for individuals and groups, as well as for whole sections in society. Mathematics in action can be a crisis-generating factor.

As an example, one can think of piloting an airplane.[11] The automatic pilot can take over, but even when the real pilot is in charge, many estimations and manoeuvres are made automatically. The steering by the real pilot is based on a huge number of mathematics-based automatic processes. The degree of automation gets more and more profound, and any such automatisation is constituted by mathematical algorithms. Like any such configuration, unexpected implications might occur. As with financial crashes, so also might airplane crashes have their explanations in automatised mathematical machinery going astray.[12]

Let me conclude this section by making a general remark about ontology. With respect to mathematics, ontology concerns the questions: What is a mathematical entity? In what sense does mathematics exist? Where is mathematics? A classic answer to such questions was provided by Plato, who claimed that mathematical entities – like points, lines, and triangles – have a real existence; however, this existence is not in any physical or empirical world, but in the world of ideas. No

---

[10] For a detailed presentation of algorithmic trading, see Johnson (2010). See also Miller (2014) for a discussion of mathematics-based risk management.

[11] In Skovsmose (2005), I have discussed how mathematics is formatting the airplane companies' approach to the booking of seats.

[12] See Hawkins (2019), who raises the question if the Boeing airplane crashes can be related to airplane automatisation.

empirical study of any physical triangle would be able to demonstrate the properties of real triangles. Access to the world of ideas can only be obtained by rationality. Contrary to any Platonic ontology, the discussion in the present section brings me to claim that mathematics constitutes parts of our reality as we experience it in our daily life, buying at the supermarket being just one example. Mathematics not only constitutes features of what we refer to as our normal, regular, routine-like normal life; it also constitutes many crises that we might experience. Mathematics makes up an integral part of our life-world.

9.3 Mathematics as Formatting a Crisis

Advanced mathematical modelling is applied in order to provide weather forecasts. This also applies to long-term forecasting with respect to the future climate of the planet. Such forecasting cannot be rooted in any laboratory experiments. Today, the principal way to try to look into the future is to make predictions through enacting some mathematical modelling. This has developed into an advanced discipline, where a range of mathematical topics and techniques are brought together.[13]

A climate model includes a multitude of parameters. In no case is it possible to include all possible parameters; a selection is necessary. This selection depends on what one considers to be important, and on the complexity of the model one is ready to try to build. Any model construction presupposes a number of pragmatic decisions to be taken, some of which might be rather arbitrary. The parameters need to be connected by equations, which also make up a process that includes decisions. Some connections might represent a profound theoretical insight, while others may have an *ad hoc* format. In the end, one ends up with a model through which one tries to make predictions with respect to the climate.

Nobody would think of climate modelling as doing anything to the climate. But now, look at the following article: "The Mathematical Formatting of Climate Change: Critical Mathematics Education and Post-Normal Science" by Richard Barwell (2013). First, the title: Barwell talks about the mathematical *formatting* of climate changes. But what could he have in mind? In what sense could mathematics format climate change?

Could it be that descriptive modelling is not just descriptive? Could it also do something? Let us listen to what Barwell (2013) says:

> The prediction of the likely future course of climate change is based on more advanced mathematics. Developing predictions about future global, regional or local effects of climate change draws on a range of advanced mathematical methods, including mathematical modelling, differential equations, non-linear systems and stochastic processes (p. 2).

---

[13] See, for instance, Coiffier (2011); and Warner (2011). For any such type of forecasting, system dynamic modelling is crucial.

9.3 Mathematics as Formatting a Crisis

Mathematics constitutes an integral part of any climate model. The whole conception of climate change is bound up with mathematical modelling, as mathematics provides a principal way of trying to look into the climatic future of the planet.

Furthermore, Barwell observes that any mathematics-based identification of climate change reflects not only some possible physical phenomena, but also the very nature of the mathematics being used. He highlights that mathematics provides a particular reading of climate change, making space for a selective set of possible actions. In Barwell's (2013) formulation:

> Mathematics also formats how we interact with the climate. Through the mathematised, model-based perspective prevalent in climate research, the climate is constructed in particular ways such as, for example, measurable, predictable, technical and controllable by humans, rather like the temperature in a high-tech sensor controlled greenhouse. (p. 2)

This is the crucial point: mathematics formats the way we interact with the climate. A mathematical model does not only describe a situation or provide a prediction, it forms our perception of the situation and, therefore, the way we tend to act in the situation. Mathematics formats actions and strategies with respect to climate change. As indicated by Barwell, they could repeat the assumption that we are looking at complications that can be regulated through technical interventions. As a consequence, climate change is constructed as a manageable issue. We can never expect mathematics to provide any neutral reading of a critical situation; rather, it associates with some specific perspectives that may serve certain political, economic, or industrial interests. In this sense, one can talk about mathematics as formatting climate changes.

That a mathematical reading of a critical situation formats how we may act in the situation, also applies to a pandemic. The mathematical model presented by the Imperial College COVID-19 Response Team (2020) appears to be a forthright descriptive model: it provides a picture of the situation and prospects for the future. At the same time, it formats the way we are acting in the situation. The model forms the way we try to handle the pandemic, and this means that it forms the very dynamics of the crises. In the same way as Barwell talks about the "mathematical formatting of climate change", we could talk about the "mathematical formatting of a pandemic".

Initially, we considered a mathematical model of an epidemic as a picture of the epidemic. However, we need to recognise that mathematics rather provides a formatting of the epidemic. This might be a more general observation, not only applying to an epidemic. We might be ready to talk about the mathematical formattings of any kind of critical situations. The picture metaphor refers to a particular analytical type of a mathematics-crises relationship. However, it might turn out that it is not a type, but rather a stereotype.

9.4 Politics of Crises

Huge political and economic interests are connected to calling something a crisis, while presenting other things in different terminologies.[14] Powers are exercised through discourses of crises. A discourse can be biased in favour of the world's richer countries. When they face serious problems, it can be referred to as a crisis, while similar problems for poorer countries can be referred to in other terms, if not simply considered normal. Eisensee and Strömberg (2007) point out that, for every person killed by a volcano, nearly 40,000 people have to die of hunger to get the same probability of coverage in the US television news.

How do we prioritise our actions when facing one or more crises? This can be a decision that reveals a range of economic interests, political priorities, and ideological preconceptions. An indication of what this could mean was experienced with respect to the coronavirus pandemic in 2020.[15]

One line of arguing, which we can refer to as the human-centred argument, maintains a clear priority: one first has to save lives. What might be needed in terms of financial resources for resolving the pandemic has to be spent, not only on ensuring direct medical care, but also on ensuring the livelihood of the very many people in precarious situations. They might have lost their jobs and possibilities for getting any form of income during the period where social distancing was maintained. The first priority is to save lives, then later on, the task becomes to take care of the economic crisis that has emerged.

Another line of arguing, which we can refer to as the economy-centred argument, claims that the medical precautions taken need to be measured with respect to their economic costs. This means, for instance, that the time period and the severity of social isolation or distancing need to be evaluated, not only with respect to their efficiency in reducing the pandemic, but also with respect to their economic consequences. Reference to economy can always mean different things. In the economy-centred argument, it might not be the workers' and poor people's daily economy that is in question, but the economy of bigger companies and bond owners.

The direct consequence of the economy-centred argument would be that some people are offered up in order to save the economy. The benefit is the economic gain, while the costs are "some people". Around March and April 2020, when the

---

[14] The political use of the notion of "crisis" was addressed several times during the Mathematics Education and Society Conference (MES 9) that took place in Volos in Greece. The title of the proceedings is *Mathematics Education and Life at Times of Crisis* (Chronaki, 2017). Important inputs to the discussion of crises came from Chassapis (2017); Greer et al. (2017); and Parra et al. (2017).

[15] In medicine, one finds different versions of cost-benefit analysis applied. While costs are measured in money, benefits can be measured in extra life years gained. This brings about an intensive use of mathematics in order to estimate the value of the amount of gained life years compared to the money invested. For a presentation of ways of completing economic evaluation of healthcare programmes, see Drummond et al. (2015).

9.4 Politics of Crises

two lines of argument were in circulation in Brazil, right-wing voices on social media stated explicitly that some increase in deaths might not be considered a problem. It was also mentioned that while previously, the young people had offered themselves in war, now the time had come for elderly people to do so.[16] One can think of such statements as cynical side-remarks, but there are economic interests and political powers behind such cynicism. Following the economic argument, some people becomes labelled as being disposable.[17] During the 2020 pandemic, we have heard the economy-centred argument, and I am afraid that we might come to hear it again and again in the future. If we are entering a period with more and more crises appearing, we can expect deep tensions between saving lives and saving money.

The preoccupation with defining the value of a human life has a long history. What it is all about is concisely formulated by Kathrin Hood (2017) in the following way:

> Every day, government analysts make calculations about how much human lives are worth compared to the cost of saving or prolonging them. (p. 442)

Hood adds that experts have spent "over a century trying to develop a scientifically sound way to measure the economic value of human life" (p. 442). One can easily see economic reasons for such efforts. Governments are saving or prolonging lives through the healthcare system. However, this is not only a humanitarian act; it is also a business investment. To what extent it is a good business depends on what is spent compared to the value of the lives saved. Therefore one needs to know the value of such savings. Whenever a production includes some risks for somebody – workers, neighbourhoods, consumers – cost-benefit analyses can come to guide the company's decision-making.

New crises will appear, new tensions between saving money and saving lives will appear, and mathematical tools for decision-making will be required. Let me perform a small *thought experiment*, indicating how a mathematical life-value model could be formulated: the guiding principle could be that the economic value of a person is the value of the production that the person would be able to make for the rest of their life. I imagine that the modelling includes careful classifications of people in terms of job, education, gender, age, and whatever else, in order to reach the average figure at the end: the economic value of a human life. However, due to the classifications, the model clears the way for much more specific estimations. What is the average value of a certain sub-group of people? People with further education, senior people, people with diabetes, etc. In a critical situation, the model might be applicable for pointing out which groups of people are worth saving, compared to the costs of saving them. Through such a modelling, younger people will

---

[16] See, for instance, https://istoe.com.br/dono-do-madero-diz-que-brasil-nao-pode-parar-por-5--ou-7-mil-mortes/; and https://economia.uol.com.br/noticias/redacao/2020/03/24/empresarios-coronavirus-o-que-dizem-criticas.htm

[17] See Bales (2012) for a discussion of disposable people and slavery in the modern globalised economy.

become more valuable than elderly people; people with more education will be allocated more value than people with less education; healthy people will get more value than people with serious disabilities; and people from certain nationalities will be more valuable than people from other nationalities. Whole populations might be modelled as if they have little value. The model might label groups of people as being disposable. As O'Neil also observed, all kinds of preconceptions might become integrated into such a mathematical model, and consequently all kinds of outputs can be expected. Still, such a model could become a tool for the government's decision-making in a critical situation.

I am deeply concerned that, when more crises roll across the world, arguments and actions might be based on such a kind of modelling. Mathematics might create pseudo-rationalities for addressing the dilemma between saving lives and saving money. Mathematics might help to fabricate cost-benefit analyses in a particularly cynical manner. Mathematics may format the way we are acting in critical situations in the most abominable and inhuman way.

Note

This chapter is a reworked version of the paper "Mathematics and Crises", published in 2021 in the *Educational Studies in Mathematics*, *108*(1–2), 369–383. I thank the publisher for their kind permission to present it here.

References

Bales, K. (2012). *Disposable people: New slavery in the global economy* (3rd ed.). University of California Press.
Barwell, R. (2013). The mathematical formatting of climate change: Critical mathematics education and post-normal science. *Research in Mathematics Education, 15*(1), 1–16.
Bostrom, N., & Cirkovic, M. M. (2008). *Global catastrophic risks*. Oxford University Press.
Brauer, F., Castillo-Chavez, C., & Feng, Z. (2019). Mathematical models in epidemiology. In *Texts in applied mathematics* (Vol. 69). Springer.
Capasso, V. (2008). Mathematical structures of epidemic systems. In *Lecture notes in biomathematica* (Vol. 97). Springer.
Chassapis, D. (2017). "Numbers have the power" or the key role of numerical discourse in establishing a regime of truth about crisis in Greece. In A. Chronaki (Ed.), *Mathematics education and life at times of crisis: Proceedings of the ninth international mathematics education and society conference* (pp. 45–55). Volos, Greece.
Chriss, N. A. (1997). *Black-Scholes and beyond: Option pricing models*. McGraw-Hill.
Chronaki, A. (Ed.). (2017). *Mathematics education and life at times of crisis: Proceedings of the ninth international mathematics education and society conference*. Volos, Greece.
Coiffier, J. (2011). *Fundamentals of numerical weather prediction*. Cambridge University Press.
Coles, A. (2016). *Mathematics education in the anthropocene* (Open Access). University of Bristol.
Crutzen, P. J., & Stoermer, E. F. (2000). The "anthropocene". *Global Change Newsletter, 41*(17), 17–18.
Drummond, M. F., Sculpher, M., Claxton, K., Stoddart, G., & Torrance, G. (2015). *Methods for the economic evaluation of health care programmes* (4th ed.). Oxford University Press.
Duggan, J. (2016). *System dynamic modelling with R*. Springer.

References

Eisensee, T., & Strömberg, D. (2007). News droughts, news floods, and US disaster relief. *The Quarterly Journal of Economics, 122*(2), 693–728.

Greer, B. Gutiérrez, R., Gutstein, E., Mukhopadhyay, S., & Rampal, A. (2017). Majority counts: What mathematics for life, to deal with crises? In A. Chronaki (Ed.), *Mathematics education and life at times of crisis: Proceedings of the ninth international mathematics education and society conference* (pp. 159–163). Volos, Greece.

Harford, T. (2012). *Black Scholes: The maths formula linked to the financial crash.* https://www.bbc.com/news/magazine-17866646

Hawkins, A. J. (2019). *Deadly Boeing crashes raise questions about airplane automatisation.* https://www.theverge.com/2019/3/15/18267365/boeing-737-max-8-crash-autopilot-automation

Hood, K. (2017). The science of value: Economic expertise and the valuation of human life in US federal regulatory agencies. *Social Studies of Science, 47*(4), 441–456.

Imperial College COVID-19 Response Team. (2020). *Impact of non-pharmaceutical interventions (NPIs) to reduce COVID19 mortality and healthcare demand.* Imperial College.

Johnson, B. (2010). *Algorithmic trading and DMA: An introduction to direct access trading strategies.* 4Myeloma Press.

Kiss, I. Z., Miller, J., & Simon, P. L. (2017). *Mathematics of epidemics on networks: From exact to approximate models.* Springer.

Klein, N. (2014). *This changes everything: Capitalism vs. the climate.* Simon and Schuster.

Latour, B. (2018). *Down to earth: Politics in the new climate regime.* Polity Press.

Miller, M. B. (2014). *Mathematics and statistics for financial risk management* (2nd ed.). Wiley.

O'Neil, C. (2016). *Weapons of math destruction: How big data increase inequality and threatens democracy.* Broadway books.

Parra, A., Bose, A. Alshwaikh, J., González, M., Marcone, R., & D'Souza, R. (2017). "Crisis" and the interface with mathematics education research and practice: An everyday issue. In A. Chronaki (Ed.), *Mathematics education and life at times of crisis: Proceedings of the ninth international mathematics education and society conference* (pp. 174–178). Volos, Greece.

Schneier, B. (1996). *Applied cryptography.* John Wiley and Sons.

Skovsmose, O. (2005). *Travelling through education: Uncertainty, mathematics, responsibility.* Sense Publishers.

Skovsmose, O., & Yasukawa, K. (2009). Formatting power of "mathematics in a package": A challenge for social theorising? In P. Ernest, B. Greer, & B. Sriraman (Eds.), *Critical issues in mathematics education* (pp. 255–281). Information Age Publishing.

Stallings, W. (2017). *Cryptography and network security: Principle and practice* (7th ed.). Pearson.

Warner, T. (2011). *Numerical weather and climate prediction.* Cambridge University Press.

Wittgenstein, L. (2002). *Tractatus logico-philosophicus.* Routledge.

Chapter 10
Picturing or Performing?

Abstract A picture theory of language, as elaborated by Wittgenstein in the *Tractatus*, is presented as a philosophical attempt to justify the general use of the picture metaphor. It is pointed out that a performative interpretation of language allows for a much deeper understanding of the possible social impact of language, captured by notions like discourse, symbolic power, and symbolic violence. A basic idea associated with a performative interpretation of mathematics is that mathematics forms not only our world views, but our worlds as well. This is illustrated through a discussion of Body Mass Index. Finally, it is observed that students might bring mathematics into action as part of a critical endeavour.

Keywords Picturing · Performing · Formatting power of mathematics · Quantification · Body Mass Index · Reading and writing the world with mathematics

The expressions "picturing" and "performing" refer to possible relationships between mathematics and reality. "Picturing" captures the idea that mathematics is a principal resource for describing reality. By means of mathematics, one can picture laws of nature and describe physical phenomena. "Performing" highlights a different feature of the mathematics-reality relationship, namely that, by means of mathematics, one performs interventions in real-life situations. Mathematics is not only a means for describing, but also a means for changing and acting.

In the following, I first show how a picture interpretation of mathematics has been elaborated upon as part of the formalist interpretation of mathematics. This interpretation, however, has been assumed far beyond the outlook of formalism. Then I present an interpretation of language as picturing, which has been formulated by Ludwig Wittgenstein. This interpretation has been one of the inspirations that tempted us to see the role of scientific theories as providing pictures of reality. Then I concentrate on the metaphor of "performing" and outline the emergence of a performative interpretation of language, confronting the picture theory of language. Finally, I indicate how the performative interpretation of language inspires a performative interpretation of mathematics.

10.1 Mathematics as Picturing

In 1951, Haskell Curry (1970) published the book *Outline of a Formalist Philosophy of Mathematics*. The draft of the book was, however, already written in 1939, and here Curry provides a concise presentation of the formalist conception of mathematics.

One can ask if a mathematical theorem is true or not, and also if a mathematical theory is true or not. Curry's point is that in order to answer them, we must engage in two different forms of investigations, as we are dealing with two different notions of truth. The distinction between the truth of a mathematical theorem and the truth of a mathematical theory is basic for the formalist understanding of mathematics as picturing.

Formalism is deeply inspired by David Hilbert's metamathematical programme that was formulated as a reaction to what has been referred to as the "foundation crisis of mathematics". Around 1900, a range of paradoxes related to basic mathematical concepts were identified, which until then had been taken as a solid given. Hilbert's approach for dealing with these paradoxes was to establish an analysis of mathematical theories themselves. He suggested that we should formalise mathematical theories, and then investigate these formalisations, in particular with respect to consistency and completeness.[1] By means of such metamathematical investigations, Hilbert wanted to make sure that paradoxes would not appear in mathematical theory building.

Formalism took a step beyond the metamathematical programme by asserting that formalisations of mathematical theories are not just copies of mathematics, but that they *are* the very mathematics. As a consequence, non-formal expressions of mathematical theories were interpreted as preliminary versions of mathematics. By identifying mathematics with formal systems, it became possible to provide a stringent definition of a mathematical theory.

According to formalism, a mathematical theory can be defined in in terms of its *alphabet, formulas, axioms, rules of inference, proofs,* and *theorems*. The *alphabet* refers to the list of symbols that can be applied. Such an alphabet can include symbols like, $p, q, r, (,), \vee, \Rightarrow, \neg, \forall, \exists$, and $\in$. For any formal system, the alphabet must be explicitly enumerated. The second step is to specify which sequences of symbols count as *formulas* in the system. This means specifying the grammatically correct sequences of symbols. The whole grammar has to be formulated in such a way that it is well-defined whether or not a sequence of symbols is a formula or not. The formulas can be defined in a recursive way using the patter: if A and B are formulas, then $(A \vee B)$ is also a formula. Some of the formulas have to be enumerated as

---

[1] That the formal system is consistent means that for any formula T, it is not possible to prove both T and $\neg T$. That the formal system is complete means that for any formula T, either T or $\neg T$ can be proved. In 1936, Gödel (1962) demonstrated that a formalisation of a mathematical theory (including the formalisation of the natural numbers) will be incomplete if it is consistent.

axioms. This set will serve as a starting point for the deductions to be made within the formal system. The *rules of inference* specify how one can, from one or more formulas, derive other formulas. When the rules of inference are clarified, a *proof* can be defined as a sequence of the formulas $F_1, \ldots, F_n$, where any formula F_i (where $1 \leq i \leq n$) is either an axiom or can be derived from one or more of the formulas in the sequence $F_1, \ldots, F_{i-1}$, in accordance with the rules of inference. A *theorem* can then be defined as a formula that occurs as the last formula F_n in a sequence of formulas $F_1, \ldots, F_n$ that composes a proof.

The formalist position is that only with reference to a mathematical theory, one can ask if a particular mathematical theorem T is true or not. This question is equivalent to asking if T is a theorem of the system or not. In the case where T is a theorem, it can be called true. It does not make sense to ask if a mathematical statement is true or not without referring to a specific, formalised mathematical theory. No absolute truths exist in mathematics, only system-related truths. A proposition might be a theorem in one system and not in another system. However, the theory-related truths are absolute in the sense that being a theorem in a mathematical theory (or not being one) is not a question of degree, but a question that can be answered definitively.[2]

It is Curry's point that, if we consider the truth of mathematical theory, the situation is quite different. The question of the possible truth of a mathematical theory concerns the applicability of a formal theory, taken as a whole. A theory can be called true in the case where it can be used for describing some real state of affairs. This conception of truth has nothing to do with provability within a formal system, but with the applicability of an entire formal system. Applicability is a question of degree, as a mathematical theory can "picture" reality in more or less accurate ways. While the possible truth of a mathematical theorem is a logical issue, the possible truth of a mathematical theory is an empirical question.

The formalist understating of mathematical truth is broadly assumed to be valid, even by many that do not explicitly assume formalism as their working philosophy of mathematics. One finds the understanding of mathematics as picturing expressed in many discussions of mathematical modelling, both with respect to advanced mathematics and with respect to school mathematics. A mathematical model has been described as a triple (R, M, f), where R refers to a set of empirical objects, M to a set of mathematical entities, and f to a function that relates R and M. With such a specification, the very process of picturing can be represented by the function f.[3]

---

[2] Although with limitations, as pointed out by Gödel.
[3] See Niss (1989); and Blum and Niss (1991).

10.2 Language as Picturing

Over a period of several years, Ludwig Wittgenstein elaborated on the ideas that were later presented in the *Tractatus Logico-Philosophicus*, published in a German-English parallel edition in 1922.[4] By the outbreak of the First World War, Wittgenstein had enrolled in the Austrian army. He was stationed at the Eastern front, and later at the Southern front towards Italy. In 1918, the Austrian army in Italy surrendered, and Wittgenstein was taken as a prisoner of war; he remained in a concentration camp until 1919. During the war, he worked on the manuscript of the *Tractatus*, and when the war finished, the manuscript had found its final form. The part of the *Tractatus* that I am going to concentrate on was added when Wittgenstein served at the Eastern front. It was the *general picture theory* of language.

An important component of this picture theory is Wittgenstein's conception of reality. The opening statement in the *Tractatus* claims that the world is composed of facts and not of things. This is an ontological claim, stating something about what exists. By making this claim, Wittgenstein distances himself from the classic position of empiricism, according to which we experience the world as a conglomerate of sense impressions. Consequently, the world appears as being composed of things, not of facts.

Tractatus is composed of numbered statements. It includes seven principal statements: §1, §2, ..., §7. The direct comments on §1 are numbered as §1.1 and §.1.2. By putting these three statements together, we get a condensed expression of Wittgenstein's ontological claim:

> The world is everything that is the case. The world is the totality of facts, not of things. The world is divided into facts.

Wittgenstein's epistemological claim is that any such fact can be pictured by means of language. Wittgenstein's initial inspiration for making this claim came when he read about a law case, where a traffic accident had to be clarified. In order to do so, the accident was illustrated in the court room by means of small figures of cars and of people. This way of picturing an accident inspired Wittgenstein for formulating a much more general idea: that by means of language we are able to picture reality.

However, according to Wittgenstein, it is not each and every kind of language that has this capacity: certainly not natural language, used in everyday communication. Wittgenstein shares the extreme suspicion of any form of natural language with analytic philosophers. In the article "On Denoting", Bertrand Russell (1905) illustrates how formulations created in natural language can cause misconceptions that, in turn, can lead to futile philosophical controversies. Such misconceptions will disappear when the original natural-language formulation is substituted by a formal-language formulation, which will uncover the proper logical structure of the

---

[4] In 1921, the book was published in a German journal, but this publication included so many typographical mistakes that Wittgenstein only considered the 1922 version as a publication of his work. See Wittgenstein (1992).

10.2 Language as Picturing

formulation. In line with this observation, Wittgenstein finds that the language that has the capacity to picture the world of facts is a formal language of the form presented by Whitehead and Russell (1910–1913) in *Principia Mathematica*.

When concentrating on §2 in the *Tractatus* and its two principal comments, §2.1 and §2.2, we get a condensed formulation of the picture theory.

> What is the case, the fact, is the existence of atomic facts. We make ourselves pictures of reality. The picture has the logical form of representation in common with what it pictures.

In this formulation, Wittgenstein refers to "fact" and to "atomic facts", which are the basis facts of which any fact is composed. One can make pictures of atomic facts, which can be referred to as atomic propositions. Any fact, being composed of atomic facts, can be pictured by a molecular proposition composed of atomic propositions.

An important feature of the formal language that Wittgenstein finds adequate for picturing reality is that it is truth-functional. A truth-function is a statement whose truth value is determined by the truth values of its parts. In a truth-functional language, the truth value of any molecular statement is completely determined by the truth value of the atomic propositions of which it is composed. While an atomic proposition can be referred to as $p, q, r, \ldots$, a molecular proposition is a composition of atomic propositions connected by logical connectives. An example of a molecular proposition could be $(p \Rightarrow q) \land ((r \lor q) \Rightarrow s)$, and its truth value will be determined by the truth value of its atomic parts p, q, r, and s.

When the truth value of a set of atomic propositions is determined, it is possible to calculate the truth value of any proposition that is a truth-function of some of these atomic propositions. If one could enumerate all possible atomic propositions and determine their truth values, then the truth value of any proposition could be determined. According to Wittgenstein, all scientific statements are truth-functions.

Wittgenstein finds that a language with the capacity to picture reality must have something in common with reality. The logical structure of the picturing language must be similar to the logical structure of the world of facts. The shared logical structure Wittgenstein refers to as the *logical form*. The language, and the only language, that shares its logical form with reality is the one presented in *Principia Mathematica* (or formal language equivalent to this).

As advocated in the *Principia Mathematica*, there is no difference between logic and mathematics. Logic provides the foundation of mathematics, as any mathematical notion can be defined by means of logical notions, and any mathematical theorem can be derived from logical theorems. In *Principia Mathematica*, logic and mathematics are brought together in one axiomatic system. To the extent that one assumes this unification of logic and mathematics, Wittgenstein's picture theory claims that, by means of mathematics, one can picture reality.

The claim that a formal truth-functional language is the unique all-embracing language is condensed in the proposition §6 in the *Tractatus*:

> The general form of a truth-function is: $\left[\overline{p}, \overline{\xi}, N(\overline{\xi})\right]$. This is the general form of a proposition.

Wittgenstein's condensed expression of a truth-function draws on the observation that the different logical connectives can be defined by means of one connective, Sheffer's stroke, which Wittgenstein refers to as N. Sheffer's stroke can be interpreted as "neither p nor q". Wittgenstein finds that any proposition of the language that has the capacity to picture reality is truth-functional. In the final §7 of the *Tractatus*, he exphasises that it is not relevant to consider any kind of language other than a formal truth-functional one:

> Whereof one cannot speak, thereof one must be silent.

Taken together, §6 and §7 claim that mathematics is the language, and the only language, by means of which the world of facts can be pictured.

10.3 Language as Performing

The picture theory of language portrays language and reality as running in parallel with each other, sharing a common logical form. We will now move beyond this metaphor, and see language and reality as interacting with each other. Language and reality are not two parallel universes, but are part of the same integrated and dynamic structure.

The picture theory from the *Tractatus* has been questioned, even by Wittgenstein himself. In *Philosophical Investigations*, Wittgenstein (1953) develops quite a different understanding of what language might be doing. Furthermore, he does not give any preference to formal languages, but sees a formal language as being only one among many other forms of language. Wittgenstein introduces the notion of "language game", which highlights that we are *doing* something by means of language (we are gaming), and also that we are dealing with a multiplicity of languages (there are many different games to play). In this way, Wittgenstein suggests a performative interpretation of language instead of a descriptive interpretation.

How to do things with words is the title of a book containing articles published by John L. Austin. The title is a concise expression of Austin's basic ambition: to show that language is performative. According to Austin, a statement has a locutionary content, an illocutionary force, and a perlocutionary effect. As an example, we can consider the statement: it is raining. The locutionary content refers to the fact that it is raining. The illocutionary force refers to the power of making the statement; it could be a warning to a person not to leave the house without an umbrella. The perlocutionary effect of the statement could be that the person changes their mind and stays at home. With a range of such examples – such as giving a warning, making a promise, cracking a joke, baptising a child, giving an order, and so on – it can be illustrated that any statement has a locutionary content, an illocutionary force, and a perlocutionary effect.[5]

---

[5] See Austin (1962).

10.3 Language as Performing

Austin operates within the tradition of analytical philosophy, paying particular attention to the role of language. He differs, however, from the approach advocated by those such as Russell, Wittgenstein in the *Tractatus*, and logical positivism. They look at natural language with great suspicion. They find that natural language has incorporated a range of ambiguities which make it inadequate for formulating scientific insight, as well as for addressing philosophical problems. According to this version of analytical philosophy, the proper answer to these difficulties is to critically investigate natural language by means of formal languages. The trend in analytical philosophy – advocated by Wittgenstein in *Philosophical Investigations* and by Austin – is different. They see natural language as a resource for philosophical investigation, meaning that it becomes important to explore natural language formulations, and not to substitute them with formalised structures.

When we leave the framework of analytical philosophy, we find much more radical interpretations of language as performing. Such interpretations have been expressed in terms of discursive theory.[6] By introducing the notion of discourse, the conception of speech acting acquires a new profoundness.[7] When considering a speech act like making a warning, the acting subject appears well-defined. Somebody can be considered responsible for the speech act. With respect to a discourse, the acting subject is not well-defined. A discourse might propagate a particular world view; it might include political priorities, doubtful ideologies, and preconceptions of any kind. But who can be held responsible for this? In order to address a discourse critically, one needs to draw on a much broader set of reflections, compared to an investigation of the impact of certain speech acts. The notion of discourse adds a new dimension to the performative interpretation of language by locating the possible impact of language in a broader socio-political framework. Pierre Bourdieu (1991) talks about symbolic power, which highlights the intimate relationship between discourse and power. Power can be acted out through different discourses, also with brutal consequences. Slavoj Žižek (2008) has talked about symbolic violence, which is a possible impact of symbolic power.

A basic idea associated with the performative interpretation of language is that language is not a transparent tool by means of which we describe things. We might think that we dominate language, but language also dominate us. It forms what we see, what we pay attention to, and what we might ignore. Language forms not only our world views; it forms our worlds as well.

---

[6] For presentations of discourse theory, see, for instance, Jørgensen and Phillips (2002) and Torfing (1999).

[7] For a discussion of speech act, see Searle (1969).

10.4 Mathematics as Performing

A basic idea associated with a performative interpretation of mathematics is that mathematics is not a transparent descriptive tool by means of which we may picture reality. Mathematics forms what we pay attention to, and what we might ignore. Mathematics forms our world views. It forms our worlds as well. In different ways – and by means of different terminologies – I have presented a performative interpretation of mathematics. In Skovsmose (1994), I outlined a performative interpretation by exploring the "formatting power of mathematics". In later publications, I have addressed mathematical performatives as features of "mathematics in action", and in Ravn and Skovsmose (2019), we have analysed mathematical performatives in terms of "mathematics-based fabrications".[8]

A characteristic feature of the so-called scientific revolution was that insight into nature needed to be based on quantifications. By means of quantifications, one can specify sizes, positions, movements, accelerations, and masses. Other features of the natural phenomena that one is observing – like smells, tastes, and colours – are ignored. However, by looking into the very processes of quantification, we come to recognise that quantification includes much more than picturing. Quantifying also includes powerful performatives.

In the article "The Mathematical Formatting of Obesity in Public Health Discourse", Jennifer Hall and Richard Barwell (2021) discuss how mathematical quantification might form what becomes considered normal and not normal with respect to health situations, and in particular with respect to conceptions of obesity. This is an example of how mathematics might do much more than "picturing" something; rather, it comes to be part of the formulation of a body ideal. One could consider many different parameters defining such an ideal, but quantification invites for a reduction of parameters to be considered. The Body Mass Index (BMI) is defined in terms of two parameters, namely the person's weight and height, and it is calculated in the following way:

$$\text{BMI} = w / h^2$$

where w refers to the weight of the person measured in kilos, and h to the height of the person measured in metres. For a person with a hight of 1.78 m and a weight of 68 kg we will get: BMI = $68/1.78^2$ = 21.5. The BMI is used for the following classification: a person is underweight if BMI < 18.5; the person has a normal weight if

---

[8] For further discussions of mathematical performatives, see also Yasukawa et al. (2012, 2016); and Part 4, "Mathematics and Power", in Skovsmose (2014, pp. 199–280). In Chap. 7, "Banality of Mathematical Expertise", I have expressed a performative interpretation of mathematics by looking at different forms of mathematics-based actions: technological imagination, hypotheticals reason, justification, realisation, and dissolution of responsibility. In Chap. 8, "Mathematics and Ethics", the performative interpretation of mathematics is explored in terms of quantifying, digitalising, serialising, categorising, and imagining.

10.4 Mathematics as Performing

$18.5 \leq \text{BMI} < 25$; the person is overweight if $25.0 \leq \text{BMI} < 30$; and the person is obese if $30.0 \leq \text{BMI}$.[9]

It all looks straightforward, but we are dealing with a tremendous simplification in relation to the number of parameters that could be considered. We are dealing with a mathematics-based stipulation of normality, which provides a demarcation between what to consider normal and what to consider as in need of treatment. Huge economic interests are associated with such demarcations. The medical industry will benefit from the identification of new needs for treatment. Mathematics becomes powerful by being used for formulating standards – in a seemly neutral and objective way – of great economic interest.

In terms of BMI, mathematics also becomes powerful by having an impact on the individual person's self-perception. Hall and Barwell (2021) highlight:

> Our point, then, is not that obesity does not exist or that obesity is not associated with health risks. Our point is that the certainty of science, through the use of mathematics, turns a fuzzy and complex phenomenon into a normative, prescriptive, and ideological abstraction, which in turn leads to concrete interventions (i.e., control), in the form of advice, medication, and penalties, such as, in some cases, exclusion from school. This normativity, in turn, is likely to feed into wider discourses relating to such topics as body image, femininity, masculinity, identity, and race. (p. 222)

Hall and Barwell highlight the importance of addressing the mathematical formatting of health discourses in mathematics education. They present some references to descriptions of how it is possible to work with BMI in the mathematics classroom by doing measurements, performing calculations, and making tables and graphs. They refer to cases where the calculations are made with reference to celebrities, fictitious individuals, family members, and neighbours. They also refer to cases where the calculations are based on measurements of the students themselves. Hall and Barwell find, and I agree with them, that this can be a highly problematic approach. Making differences among students public in terms of numbers, might provoke bullying and ruin the self-esteem of some students. We are dealing with a most important issue, which is simultaneously highly sensitive. As a symbolic system, mathematics can be violent.

Mathematical performatives can also create quite different possibilities. With inspiration from Paulo Freire, Eric Gutstein (2006) has elaborated on what reading and writing of the world with mathematics might include. Reading the world with mathematics refers to possible interpretations of socio-political phenomena, where mathematics contributes to the identification of cases of social oppression, exploitation, and marginalisation. Writing the world by means of mathematics refers to

---

[9] During different historic periods, the conception of "the beautiful body" has been expressed in various ways. As part of the renaissance, the conception was formulated in terms of a degree of similarity with some classic Greek sculptures. In a famous drawing, expressing the proportion of a human body, da Vinci provides a qualitative clarification of ideal human proportions, although with some quantitative indication. As an extreme simplification of this whole development, one gets to the formulation of the BMI. For a presentation of the historical development of the formulation of the BMI, see Eknoyan (2008).

making changes to social-political situations. Writing the world refers to the activist features of a critical activity. The expression "reading and writing the world with mathematics" captures a basic feature of a performative interpretation of mathematics: one can do things by means of mathematics.

References

Austin, J. L. (1962). *How to do things with words*. Oxford University Press.
Blum, W., & Niss, M. (1991). Applied mathematical problem solving, modelling applications, and link to other subjects: State, trends and issues in mathematics education. *Educational Studies in Mathematics, 22*(1), 37–68.
Bourdieu, P. (1991). *Language and symbolic power* (Edited with an introduction by John B. Thompson). Polity Press.
Curry, H. B. (1970). *Outlines of a formalist philosophy of mathematics*. North-Holland Publishing Company.
Eknoyan, G. (2008). Adolphe Quetelet (1796–1974): The average man and indices of obesity. *Nephrology Dialysis Transplantation, 23*(1), 47–51.
Gödel, K. (1962). *On formally undecidable propositions of Principia Mathematica and related systems*. Basic Book.
Gutstein, E. (2006). *Reading and writing the world with mathematics: Toward a pedagogy for social justice*. Routledge.
Hall, J., & Barwell, R. (2021). The mathematical formatting of obesity in public health discourse. In A. Andersson & R. Barwell (Eds.), *Applying critical mathematics education* (pp. 210–228). Brill and Sense Publishers.
Jørgensen, M. W., & Phillips, L. (2002). *Discourse analysis as theory and method*. Sage Publications.
Niss, M. (1989). Aims and scope of applications and modelling in mathematics curricula. In W. Blum et al. (Eds.), *Applications and modelling in learning and teaching mathematics* (pp. 22–31). Ellis Horwood.
Ravn, O., & Skovsmose, O. (2019). *Connecting humans to equations: A reinterpretation of the philosophy of mathematics*. Springer.
Russell, B. (1905). On denoting. *Mind, 14*(56), 479–493.
Searle, J. (1969). *Speech acts*. Cambridge University Press.
Skovsmose, O. (1994). *Towards a philosophy of critical mathematics education*. Kluwer Academic Publishers.
Skovsmose, O. (2014). *Critique as uncertainty*. Information Age Publishing.
Torfing, J. (1999). *New theories of discourse: Laclau, Mouffe and Žižek*. Wiley-Blackwell.
Whitehead, A., & Russell, B. (1910–1913). *Principia mathematica* (Vol. I–III). Cambridge University Press.
Wittgenstein, L. (1953). *Philosophical investigations*. Blackwell.
Wittgenstein, L. (1992). *Tractatus logico-philosophicus* (With an Introduction by Bertrand Russell). Routledge.
Yasukawa, K., Skovsmose, O., & Ravn, O. (2012). Mathematics as a technology of rationality: Exploring the significance of mathematics for social theorising. In O. Skovsmose & B. Greer (Eds.), *Opening the cage: Critique and politics of mathematics education* (pp. 265–284). Sense Publishers.
Yasukawa, K., Skovsmose, O., & Ravn, O. (2016). Scripting the world in mathematics and its ethical implications. In P. Ernest, B. Sriraman, & N. Ernest (Eds.), *Critical mathematics education: Theory, praxis, and reality* (pp. 81–98). Information Age Publishing.
Žižek, S. (2008). *Violence*. Picador.

Part III
Critique and Dialogue

Chapter 11
Critique of Mathematics

Abstract Several obstructions are in the way of us conducting a critique of mathematics. One is the general glorification of mathematics, which, for instance, was part of the so-called scientific revolution. Another obstruction is caused by the doctrine that mathematics is a harmless and innocent preoccupation without significant utilitarian values. I confront the assumption that mathematics can be glorified and that mathematics is harmless and innocent by showing that mathematics is indefinite. I argue that mathematics is indefinite with respect to concepts, proofs, culture, applications, and power. This observation makes a critique of mathematics not only possible, but also necessary.

Keywords Mathematics as glorious · Mathematics as harmless and innocent · Mathematics as indefinite.

A critique of mathematics is important, but such a critique meets obstacles. Mathematics has been glorified; it has been worshipped; and it has been ascribed unique epistemic qualities. Euclid's *Elements* has been assigned an exemplar role: knowledge in general has to be formulated according to a deductive pattern. In *Ethics*, first published in 1677, Baruch de Spinoza (2005) organises ethical considerations into an axiomatic system, starting with seven axioms. After this, theorem-proof-theorem-proof-theorem follow – including a proof of the existence of God. The *Elements* has been received as a magnificent epistemic model.

Another obstacle for conducting a critique of mathematics is the fact that mathematics has been considered a harmless and innocent occupation. Ways of working with mathematics have been described in artistic terms, and to ask about the utility of mathematics would be to ignore the possible beauty of logical reasoning and formal deduction. Mathematics has been associated with its intrinsic values, and not with any utilitarian values.

After discussing these two obstructions, I present mathematics as being indefinite, recognising that mathematical concepts, proofs, cultural embeddings, applications, and power relationships are ever-changing. This observation calls for a critique of mathematics. In launching such a critique, it is, however, important to remember that a critique not only means pointing out weaknesses and questionable applications, but also showing strengths and useful applications.

11.1 Mathematics as Glorious

In 1633, René Descartes had completed the manuscript of his book *The World*, where he presented a thoroughly mechanical worldview. He must have been extremely pleased with this work, but he also became afraid. In 1632, Galileo Galilei had to face the inquisition, accused of heresy for his claim that the earth was moving around the sun. Galilei was forced to withdraw his statement and to confirm that the earth occupies the centre of the universe, as stated in the Bible. The legend tells us that Galilei did, however – though only in a low voice – add that "still it is moving". Galilei was punished by being confined to his house for the rest of his life. This was, however, a mild punishment. In 1600, Giordano Bruno had faced the inquisition. His heresy was claiming that the universe was infinite. If infinite, the universe could not have a centre, and consequently the Earth could not be located in its centre. Bruno was burned alive at the Piazza de Fiori in Rome. Descartes knew about the fate of Bruno and he heard about the charge against Galilei. Consequently, he did not want to publish *The World*.

In *The World*, Descartes (1998) enumerates the laws according to which he claims nature is operating. The *First Law of Nature* states: "[E]ach particular part of matter always continues in the same state unless collision with others forces it to change its state" (p. 25). The *Second Law of Nature* states: "[W]hen one of these bodies pushes another it cannot give the other any motion except by losing as much of its own motion at the same time; nor can it take away any of the other's motion unless its own is increased by the same amount" (p. 27). The *Third Law of Nature* states: "[W]hen a body is moving, even if its motion [...] takes place along a curved line [...] each of its parts individually tends always to continue moving along a straight line" (p. 29). After presenting these three laws, Descartes adds, definitely with some proudness in his voice, that he tends to suppose that God did not impose any other laws on nature. The existence of God was assumed by everybody contributing to the scientific revolution: Nicolaus Copernicus, Galileo Galilei, Johannes Kepler, Isaac Newton, as well as Descartes.

Like a parliament can set the laws that regulate the movements of cars, God can set the laws that regulate the movements of physical bodies. The uniqueness of Descartes' presentation is that he enumerates exactly which laws God has implemented. The laws concern the mechanical movements and nothing else, and in this way nature is conceptualised as a mechanical clock. God could be considered a watchmaker, who has constructed a mechanical universe, and after this construction, the universe will move on forever, according to the imposed laws of nature.

There is no evidence that Descartes spent much time reading the Bible, but Isaac Newton did. Newton might even have spent more time doing so than studying physics and mathematics. In *The Principia: Mathematical Principles of Natural*

Philosophy, first published in 1687, Newton (1999) also presents three laws of nature. They are as mechanical as the laws formulated by Descartes and one sees a clear resonance between the two formulations. However, in *The Principia* Newton makes it as self-evident that God has created the laws of nature according to a mathematical format; consequently, by means of mathematics one can capture the master plan of God's creation. Mathematics reflects a divine rationality. In his studies of the Bible, Newton paid attention to information about the proportions of the Jewish temple. His idea was that it might be possible to identify similar proportions in nature, maybe in terms of relationships between certain distances, maybe in terms of relationships between some natural constants. However, Newton did not find any such indications of God's existence, and nor did the many subsequent studies that followed up on Newton's idea.

Even though conceptions of physics and mathematics were separated from religious outlooks, the glorification of mathematics continued, as mathematics seemed able to capture *eternal truths*. The three classic positions in the philosophy of mathematics – logicism, formalism, and intuitionism – also interpret mathematical truths as being eternal, although for different reasons. According to logicism – as launched by Gottlob Frege – mathematics rests upon a foundation of logic.[1] The logicist programme was reworked in all technical details in *Principia Mathematica* by Alfred Whitehead and Bertrand Russell (1910–1913). *Principia Mathematica* is the most ambitious attempt to ensure mathematical truths with certainty. *Principia Mathematica* follows the Euclidian paradigm, starting out with logical axioms that are so simple that their truths can be taken for granted – or, at least, so it was assumed by Whitehead and Russell. Then theorem-proof-theorem-proof-theorem follow. First the theorems belong to logic; later on they move into mathematics. Taken together, the whole exposition in *Principia Mathematica* was supposed to demonstrate that mathematical truths are eternal. The *formalist programme*, as launched by David Hilbert, also had the ambition of ensuring mathematical truths with certainty. Mathematical truths, however, are defined relative to formalised systems. Thus formalist absolutism has an if-then format: if the axioms of a formal system are true, then the theorem is true as well. While both the logicist and the formalist programmes were suspicious with respect to intuition and wanted to eliminate it from mathematical reasoning in order to ensure an absolutism, the *intuitionist programme*, as launched by L.E.V. Brouwer, positioned intuition in the centre of mathematical activities. However, according to Brouwer, mathematical intuition does not leave any space for relativism. He maintains an absolutism with respect to mathematics, claiming that "truth is always the same to those who understand" (Brouwer, 1996, p. 404).

---

[1] In the *Begriffsschrift*, first published in German in 1879, Frege (1967) indicates how logic itself can be organised as an axiomatic system. In *Die Grundlagen der Arithmetics*, published in 1884, Frege (1978) shows how the notion of number can be defined by means of logical notions. Finally, in *Grundgesetze der Arithmetik I–II*, Frege (1893/1903) shows in detail how logical mathematics can be presented as a deductive system.

The different versions of absolutism – as presented by logicism, formalism, and intuitionism – all contribute to the glorification of mathematics. Absolutism makes a critique of mathematics meaningless. It does not make sense to raise a critique of something that cannot be different.

11.2 Mathematics as Harmless and Innocent

Godfrey H. Hardy was deeply concerned about the role that science might be playing in warfare. In 1940, he published the book *A Mathematician's Apology,* where he addressed this concern. One can think of the publication as an immediate reaction to the outbreak of the Second World War, but it was, in fact, his delayed reaction to the First World War.

Hardy sees war as an abominable phenomenon, and it is terrible for him to think of science as a resource for war technology. But what about mathematics? Hardy (1967) acknowledges that sciences and branches of mathematics might be deeply involved in questionable war activeness, but states that real mathematics is not:

> There is one comforting conclusion that is easy for a real mathematician. Real mathematics has no effects on war. No one has yet discovered any warlike purposes to be served by the theory of numbers or relativity, and it seems very unlikely that anyone will do so for many years. It is true that there are branches of applied mathematics, such a ballistics and aerodynamics, which have been developed deliberately for war and demand a quite elaborate technique: it is perhaps hard to call them "trivial", but none of them has any claim to rank as "real" (p. 140).

Hardy finds that "real" mathematics has no effect on wars, and that no warlike purposes can be served by the theory of numbers or relativity. Consequently, he can state:

> [A] real mathematician has his conscience clear; there is nothing to be set against any value his work may have; mathematics is […] a "harmless and innocent" occupation (pp. 140–141).

This is the crucial claim in Hardy's conception of mathematics: we are dealing with a harmless and innocent occupation. However, Hardy is *not* talking about mathematics in general, but about what he refers to as "real" mathematics. Real mathematics is not driven by practical demands; it does not search for utilitarian justification. Real mathematics is driven by epistemic curiosity similar to aesthetic ambitions. A mathematical proof can be celebrated for its logical profoundness and its beauty.

Hardy acknowledges that sciences have abominable applications and "trivial" mathematics as well. The characteristics of being harmless and innocent only apply to the "real" mathematics. Hardy counted Maxwell and Einstein, Eddington and Dirac, among "real" mathematicians. He also counted Newton as "one of the world's three greatest mathematicians" (p. 71).

The crucial point in Hardy's argumentation is that "real" mathematics cannot be considered useful: "I have never done anything 'useful'. No discovery of mine has made, or is likely to make, directly or indirectly, for good or ill, the least difference to the amenity of the world" (p. 150). Real mathematics does not make a difference, neither for the good nor for the bad. Real mathematics has no utilitarian value whatsoever, whether in times of peace or in times of war. Hardy does not advocate any theory about the neutrality of science. In fact, he highlights that "science works for evil as well as for good", and that it does so in particular "in times of war". The only thing he insists upon is that "real" mathematics operates beyond any evil-good controversies. Hardy's thesis was dramatically falsified in 1945 by the dropping of the atomic bomb, and later number theory turned out to be crucial to coding theory and to cryptography, essential for modern war technology.

Even when being directly falsified, Hardy's thesis of "real" mathematics as being harmless and innocent has turned into a convenient *doctrine of neutrality*, broadly assumed as part of a working philosophy of mathematics. While Hardy's thesis is well-articulated, the doctrine of neutrality often operates as a discursive pattern that apparently refers to any kind of mathematics. The doctrine states that mathematics *as such* is harmless and innocent, and that one can conduct mathematics research without engaging in critical reflections about the possible social impact of mathematics. The doctrine is reflected in the organisation of university studies in mathematics, fully concentrated on mathematical topics. It is seldom that we see reflections of socio-political issues as an integral part of mathematical study programmes. Explicitly or implicitly, mathematics is considered an innocent and harmless preoccupation. Challenging this assumption is a principal step in addressing mathematics critically.

11.3 Mathematics as Indefinite

For elaborating a critique of mathematics, it is important to confront any automatic glorification of mathematics and to challenge the doctrine that mathematics is harmless and innocent. Mathematics is not. However, I do not claim that mathematics is intrinsically harmful, as it might also have extremely beneficial applications. I want to challenge the idea that mathematics is glorious, harmless and innocent, not by claiming that it is harmful, but by showing that it is indefinite.

I will specify what I mean by this expression by referring to how mathematical concepts and proofs are elaborated in different ways, by illustrating how mathematics becomes encultured in a variety of formats, and by highlighting that mathematics can be applied in a multitude of contexts and serve a multitude of interests in powerful ways. Mathematics is not an absolute monolithic structure; it is a human construction that, like any other human construction, can be reformed, remoulded, and reconstructed. Mathematics is a multitude of constructions that can serve a multitude of interests. This makes a critique of mathematics not only possible, but also necessary.

Mathematics can be *indefinite with respect to concepts*. An example of this is presented in *Proofs and Refutations*, where Imre Lakatos (1976) illustrates that the very notion of polyhedron is articulated, revised, reformulated, deconstructed, and reconstructed as part of a never-ending process. A polyhedron is not a stable mathematical entity, but is part of an ongoing conceptual dynamic. Lakatos shows how proof-generated concepts might facilitate the formation of proofs by allowing for alternative proof-strategies. Lakatos makes it clear that mathematical concepts are part of a dialectic process, propelled forward by proofs and refutations. This observation does not only apply to the concept of polyhedron, but to any mathematical concept. The conceptual landscape of mathematics, as we experience it today, is only a snapshot of an ongoing process.

Mathematics is *indefinite with respect to proofs*. In "Is Mathematical Truth Time-Dependent?" Judith V. Grabiner (1974) shows how paradigmatic forms of mathematical reasoning have changed over time, and consequently so have our conceptions of what count as mathematical truths. Karl Weirstrass was concerned about the way proofs were conducted in infinitesimal calculus. Previously, it had been common to operate with infinitesimals, at times considered very small units, at other times considered zeros. Weierstrass found that a complete revision of what to consider a mathematical proof in infinitesimal calculus was necessary. The result of his effort became the epsilon-delta terminology. However, later the very notion of infinitesimals was revived, and now it forms an integral part of non-standard analysis. Mathematical reasoning and deduction have been considered principally different from empirical deduction. Euclid's *Elements* has been celebrated for apparently not relying on any empirical observations, but only on logical deductions. This focus on logical deduction has been considered defining for mathematics. However, recently this focus has been questioned. The use of computers has opened up a new discussion of what to consider valid mathematical proof. Thomas Tymoczko (1979) discusses the computer-based proof of the four-colour theorem. Can a computer take over mathematical proofs? Computers have now also been used for performing experiments to try to reveal possible mathematical regularities. This also adds to the indefiniteness of mathematical proof.

Mathematics is *indefinite with respect to culture*. Ethnomathematical studies have shown that cultural diversities create profound differences with respect to mathematical conceptions, techniques, and applications. Looking at the history of mathematics, one can see different enculturations of mathematics.[2] During the period of Modernity, Eurocentrism became an integral part of the fabrication of the history for mathematics. Brutal processes of worldwide colonisation were an inherent part of the development of Modernity. The trading of slaves reached devastating dimensions, and simultaneously developed the idea that people from Africa and the East were inferior compared to people from Europe. This Modernist racism provoked questions like: How could it be that mathematics, with all its sublime epistemic qualities, could have its roots in Africa and the East? How could unique mathematical achievements be completed by inferior groups of people? How could the "white man's burden" have created such intellectual wonders? Answers to such

---

[2] For a discussion of mathematical enculturation, see Bishop (1988).

questions were formed through a reconfiguration of the history of mathematics. Egyptian, Mesopotamian, and Indian mathematics were relegated to being empirical-based and inferior, while Greek mathematics became glorified. Simultaneously, Greek culture was interpreted as a European phenomenon. A confrontation of the Eurocentric history of mathematics can be found in *The Crest of the Peacock: The non-European Roots of Mathematics*, where George Gheverghese Joseph (2000) outlines in detail the multicultural roots of mathematics. By paying attention to what has taken place in India, China, Japan, Africa, and South America, he demonstrates that the conception of mathematics as a European phenomenon is false. Mathematics is formed in particular ways in different cultural contexts. There is no singular principal enculturation of mathematics. I do not assume the existence of any genuine mathematics lingering behind the different forms of enculturation. Consequently, I do not find that any particular enculturation can claim to represent the genuine mathematics. These observations I condense in the claim that mathematics is indefinite with respect to culture.

Mathematics is *indefinite with respect to applications*. Mathematical theories can be developed due to an intrinsic mathematical curiosity, for later becoming applied in quite different areas. The two mathematical theories that Hardy referred to as representing real mathematics without any utilitarian value – the theories of numbers and the theory of relativity – have got huge and dramatic applications. When Albert Einstein reached the expression $E = mc^2$, it represented a deep theoretical insight, although without any possible application in sight. Only later it became recognised that it would be possible to create the huge amount of energy that the formula did promise, and the atomic bomb was constructed, and also thrown. For centuries, number theory was considered as being without practical application; however, during the 1970s it was recognised that number theory is of extreme importance for new approaches in cryptography.[3] Due to the application of computer technology, it became possible to encrypt and decrypt huge amounts of information in an efficient way. The possible applications of mathematics are indefinite: new application can grow out from old theories; whole clusters of applications can become obsolete and be substituted by other forms of applications.

Mathematics is *indefinite with respect to power*. It is broadly acknowledged that mathematics can be brought into action for many different purposes. In the book *Capitalism and Arithmetic: The New Math of the fifteenth Century*, Frank Swetz (1987) outlines how the first steps into a capitalist economy were accompanied by the construction of some new accounting strategies. Today, the intimate relationship between capitalism and mathematics is made up by a widespread fabric of connecting knots. Mathematical algorithms can be developed and implemented to make a production process more efficient. Such processes can be organised as a combination of manual work and automatic processes. Highly automatised machinery and robots have entered the production halls of cars, cell phones, and many other products. Specialised knowledge gets extracted from the workers and

---

[3] See Diffie and Hellman (1976); and Skovsmose and Yasukawa (2009).

integrated into the automatic processes. This means that the production process becomes independent of a specialised workforce, and it becomes unproblematic to substitute one worker with another. The capitalist demand for implementing further automatisation sets demands for mathematics. New mathematical techniques, mathematical algorithms, and more efficient computer programmes need to be constructed.

The title of Swetz' book was *Capitalism and Arithmetic*. I could be tempted to plan a book with the title for *Capitalism and Algorithms*. Let me indicate some of the themes that such a book could address.

Mathematics contributes to capitalist production machinery by creating new forms of management and control systems. Digital Taylorising is an expression of this. Taylorising in its original form, as presented at the beginning of the twentieth century, refers to the decomposition of complex work processes into their atomic units, which then can be assigned to the particular workers. In digital Taylorising, the atomising of work processes has reached a new degree of detail by means of the computer. Digital Taylorising can be found in all kinds of production processes, the modern slaughterhouse being just one example. Big chunks of meat roll forward, and a row of workers takes chunk after chunk and chops it into the proper sizes, which move on to other workers doing the wrapping of the parcels to be sent to the supermarket. Everything moves on in a continued process, which has been carefully optimised with respect to the speed and organisation. No time is lost on workers' unnecessary movements. Such kinds of optimisations have reached a new degree through digital Taylorising, implementing a range of algorithms for driving the work process forward.[4]

Production processes are carefully monitored. A classic panopticon is a construction where it is possible to observe everybody from a particular position, and in this way to keep things under control. Today, digital panopticons are installed in all kinds of production systems. While the classic panopticon is defined by the possibility of actually *seeing* everything, the digital panopticon is defined by the formation of a database where everything gets *registered*. This could be information about when the workers arrive at the workplace, how many units they produce during a day, and how often they take a break. The digital panopticon reaches its full potential at a computerised workplace. Information can be automatically collected about the time the workers log in, the number of clients that are attended to, the routes that are taken on the internet and the pages that are visited. The workers can be completely monitored. However, there need not be anybody that actually looks at the collected information. The database that defines the digital panopticon need not be studied by anybody. The database can be monitored automatically, meaning that by means of some pre-designed algorithms atypical behaviour is pointed out. Such cases then can be addressed by the managers of the whole working process. The

---

[4] See also the discussion of serialising as an feature of digital Taylorising in the Chap. 8, "Mathematics and Ethics".

formation of the database and the ways in which it is used is formed by mathematical algorithms. The digital panopticon is a mathematical construct.

Transfer of money is crucial for the capitalist economic order. Previously, the transfer of money had a physical appearance: payment was made in bank notes, and big scale payments maybe in gold. Today, economic transactions take place via the internet. At the stock market, selling and buying have turned into digital processes. Decisions about selling and buying have, to a great extent, even become automatised, thus algorithmic trading gets a bigger and bigger share of the stock market activities. Any electronic transfer of money needs to be carefully secured, and this makes the mathematical discipline of coding theory and cryptography extremely important.

In such different ways, mathematical algorithms operate as an integral part of the capitalist production machinery. Mathematics becomes a crucial component for generating profits. The relevance of mathematics for profit generation has strong impact on mathematics. Mathematical research priorities are far from an intrinsic mathematical affair; such priorities might be formed by opportunities for obtaining external funding and by business priorities in general. External funding can reflect economic priorities directly. The demand for profit-generating patterns of production sets the demand for what to research in mathematics. Much research is directed towards the investigation of algorithms; for instance, in order to increase the speed of an algorithm. This might be of high economic relevance, even though it might be of little mathematical significance.

But wait! Mathematics is indefinite with respect to power. The intimate relationship between capitalism and mathematics is not the only possible relationship. Mathematics can also be brought into action in order to identify and challenge particular manifestations of the capitalist production machinery. By means of mathematics one can point out degrees of exploitation. Students can become engaged in using mathematics for identifying cases of social injustice and for conducting a critique of economic, political, and social states of affairs. In this way, mathematics may also become powerful, and as a result so can the students who are engaged in these critiques in the classroom.

References

Bishop, A. (1988). *Mathematical enculturation*. Kluwer Academic Publishers.
Brouwer, L. E. J. (1996). Life, art, and mysticism. *Notre Dame Journal of Formal Logic, 37*(3), 389–429.
Diffie, W., & Hellman, M. E. (1976). New direction in cryptography. *IEEE Transactions on Information Theory, 22*, 644–654.
Descartes, R. (1998). *The world and other writings*. Cambridge University Press.
Frege, G. (1893/1903). *Grundgesetze der Arithmetik* (Vol. I–II). H. Poble.
Frege, G. (1967). Begriffsschrift: A formula language, modelled upon that of arithmetic, for pure thought. In J. van Hiejenoort (Ed.), *From Frege to Gödel: A source book in mathematical logic, 1879–1931* (pp. 1–82). Harvard University Press.
Frege, G. (1978). *Die Grundlagen der Arithmetik/the foundations of arithmetic*. Blackwell.

Grabiner, J. V. (1974). Is mathematical truth time-dependent? *The American Mathematical Monthly, 81*(4), 354–365.

Hardy, G. H. (1967). *A mathematician's apology* (With a foreword by C. P. Snow). Cambridge University Press.

Joseph, G. G. (2000). *The crest of the peacock: The non-European roots of mathematics*. Princeton University Press.

Lakatos, I. (1976). *Proofs and refutations*. Cambridge University Press.

Newton, I. (1999). *The principia: Mathematical principles of natural philosophy*. University of California Press.

Skovsmose, O., & Yasukawa, K. (2009). Formatting power of 'mathematics in a package': A challenge for social theorising? In P. Ernest, B. Greer, & B. Sriraman (Eds.), *Critical issues in mathematics education* (pp. 255–281). Information Age Publishing.

Spinoza, B. (2005). *Ethics*. Penguin Classics.

Swetz, F. J. (1987). *Capitalism and arithmetic: The new math of the 15th century*. Open Court Publishing Company.

Tymoczko, T. (1979). The four-color problem and its philosophical significance. *The Journal of Philosophy, 76*(2), 57–83.

Whitehead, A., & Russell, B. (1910–1913). *Principia mathematica* (Vol. I–III). Cambridge University Press.

Chapter 12
Critique of Critique

Abstract As a result of the work of philosophers such as Descartes, Kant, and Marx, critique became part of methodology. A different critical approach was launched by Nietzsche, who used critique as hammering. In contrast to these two approaches, I will present critique as dialogue. Inspiration for relating dialogue and critique comes from Freire. Critique as dialogue will be elaborated upon further by showing that critique, like dialogue, is temporal, partial, not dogmatic, argued, and inconclusive. This brings me to see critique as an expression of uncertainty. We cannot escape uncertainty; we have to live with it. Uncertainty constitutes part of our life-worlds. It forms an integral part of the formulation of opinions and knowledge, and it is embedded in any political positioning and actions. Uncertainty is a universal human condition and this applies to any critical enterprise, also to critical mathematics education.

Keywords Critique as methodology · Critique as hammering · Critique as dialogue · Dialogue · Critique · Uncertainty

The very notion of critique contains tension, if not contradictions. It has been used in many contexts and for different purposes. Ludwig Wittgenstein (1997) suggested that, in order to clarify the meaning of a notion, one should pay attention to how it is used. This sounds like a wise suggestion, and I will look more carefully at how the notion of critique has been used over different periods of time.

Different critical activities have developed throughout modernity, and this inspires me to talk about *critique as methodology*. I will try to clarify what this could mean by paying attention to works by René Descartes, Immanuel Kant, and Karl Marx. A different critical approach emerged from the writings of Friedrich Nietzsche who, in the middle of the golden age of modernity, launched an array of post-modern ideas. With this inspiration, I feel tempted to talk about *critique as hammering*.

Critical mathematics education operates explicitly with the notion of critique, but which particular approach? I will not see critique as methodology, nor as hammering. Rather, I am going to interpret *critique as dialogue*. This interpretation, however, is not independent of the two other interpretations, but draws on both of them. In order to better clarify what this does include, I present more carefully what I have in mind when talking about "critique as methodology" and "critique as hammering". Then, I consider "critique as dialogue", which I find crucial for understanding the outlook of critical mathematics education.

12.1 Critique as Methodology

Modernity is a broad notion which has been characterised in many ways. As a historical period, modernity is associated with extraordinary scientific developments, profound industrialisation, new patterns of exploitation of workers, brutal colonisation, the formulation of democratic ideals, slave trading, and the articulation of racist ideologies.

From a philosophical perspective, one can see modernity as a period preoccupied with ideas about truth, knowledge, science, progress, democracy, and social justice. The latter, "social justice", is a latecomer, as the term was first coined in 1848.[1] However, the critical endeavours that took place during modernity had a focus on such conceptions, and let me illustrate this by referring to the works of Descartes, Kant, and Marx.

René Descartes is often presented as the first modern philosopher. In his philosophy, he did not base his ideas on any previous philosophical system or set of assumptions, but tried to develop his thinking on its own terms. I am not aware of any text where Descartes explicitly uses the notion of critique. However, in his *Meditations on First Philosophy* (Descartes, 1993), first published in Latin in 1641, he presents the idea of a universal doubt, which embodies a radical critical attitude. Descartes' universal doubt has also been referred to as a *methodological doubt*. His aspiration was to display in a transparent way what could count as knowledge, and to Descartes, knowing means to know something with certainty. Looking at what was assumed to be knowledge at the time, he found that much had been included with dubious justification.

How might we operationalise this methodological doubt? Descartes' approach was to consider any statement that could possibly be doubted as being false; and consequently, it needed to be removed from the stock of assumed knowledge. This was a dangerous act at that time, when the inquisition was in full force. The methodological doubt implied that whatever the church was saying could be questioned and had to be removed from the assumed stock of knowledge. No religious dogma could count as knowledge.

Methodological doubt represented a radical version of scepticism. However, Descartes did not want to re-establish a form of scepticism, but to overcome it. After the drastic clearing-out of the stock of assumed knowledge, according to Descartes,

---

[1] See the discussion of social justice in Chap. 1, "Concerns and Hopes".

everything could be thrown out except from one single statement. He found it impossible to doubt the statement: *cogito, ergo sum*. This statement, and only this one, survived the application of methodological doubt, and so Descartes found that it was true, and true with certainty. From this truth, other truths could follow. To Descartes, as well as to many other philosophers at the time, knowledge had to be organised in a deductive system, in just the same way that Euclid's *Elements* organised geometric knowledge. By identifying a statement that could not be doubted, Descartes indicated an axiom from which all human knowledge could be deduced.[2] Although Descartes did not use the word critique, his methodological doubt is a paradigmatic example of a critical activity, which aims at distinguishing between on one hand knowledge, and on the other hand assumptions, beliefs, mistakes, and misunderstandings. In this way, Descartes contributed to a profound critique of knowledge.

As part of the Enlightenment, knowledge became a central issue, not only to philosophers, but to people in general. Knowledge became essential for both human and social progress, and therefore it became urgent to update and to make available all kinds of knowledge. In France, this project resulted in the publication of the *Encyclopédie* between 1751 and 1772, which later appeared with revisions and additions. During its first few years, Denis Diderot was the only editor; later, he collaborated with Jean le Rond d'Alembert. The aim was to make all kinds of knowledge public, and in this way to enlighten a whole population.

The attempt to publish all knowledge made it urgent to consider the fundamental question: What is knowledge? Immanuel Kant tried to answer this question and, contrary to Descartes, Kant explicitly referred to his analytical investigation of knowledge as critique. The title of his principal work is *Critique of Pure Reason* (*Kritik der reinen Vernunft*), which appeared in 1781. The title has a nice ambiguity. Kant (1929) tried to address pure reason through critique; at the same time, he tried to make a critique by means of pure reason. In his critique, however, he was not going to draw on any empirical observations; he wanted the critique to be purely analytical. Kant wanted his critique of knowledge to come before the formulation of knowledge. The critique should clarify the general conditions for obtaining knowledge and show the limits of knowledge. In this sense, Kant's critical approach was methodological.

Kant operated with the notion of *Ding an sich*, which translates as "thing in itself". According to Kant, we can never come to know anything about *Ding an sich*. Our knowledge is always based on our *experiences* of things. Such experiences are formed by certain universal categories, which structure the windows through which we look at "reality". There are no category-free windows through which we can look directly at the *Ding an sich*. That knowledge is formed through categories is a universal epistemic condition, according to Kant. Time and space make up two such categories, as everything we experience is located in space and time. We cannot

---

[2] Strictly speaking, *cogito ergo sum* is a deduction where the premise is *cogito*, the conclusion is *sum*, and the deduction is represented with *ergo*. However, Descartes makes use of the whole statement *cogito ergo sum*, as he needs a specification of what a deduction means in terms of clarity.

imagine an object not located in space, nor an event not taking place in time. According to Kant, mathematics captures the structures of time and space, and as a consequence he concludes that mathematics applies with necessity to any natural phenomena we might experience in space and time. In this way, Kant tried to provide an answer to the classic question: Why does mathematics apply so extraordinarily well to natural phenomena? According to Kant, the answer is not that nature is formed according to mathematical principles, as was normally assumed at the time, but that we human beings pre-organise or categorise our experiences according to mathematical principles.

By pointing out the inevitability of categorisations of human experiences, Kant tried to clarify the nature of not only mathematical knowledge, but of knowledge in general. To outline the nature and limits of knowledge is a paradigmatic and critical project, at least when critique is considered an epistemic enterprise.

Karl Marx brought new elements to the notion of critique. Like Kant, he was explicit about creating a critique, and the word is included in the title of his book *A Contribution to the Critique of Political Economy* (*Zur Kritik der politischen Ökonomie*), published in 1857. Marx's principal work, *Capital* (*Das Kapital*), the first volume of which appeared in 1867, has the subtitle *Critique of Political Economy* (*Kritik der politischen Ökonomie*).[3] Through his investigations, Marx provided a profound critique of a range of economic theories that were circulating at the time. He pointed out their questionable assumptions and apparent contradictions. However, Marx's critical enterprise was much broader. He not only wanted to address socio-political theories; he also wanted to address socio-political realities through his critique. To Marx, critique is not only an analytical but also a political activity.

Marx wanted his critique of the political economy to be based on a solid theoretical foundation, on an understanding of the economic dynamics of a capitalist economy. His key question was: What is the nature of the economic processes that systematically increase the exploitation of workers, and simultaneously create fortunes for the capital owners? Naturally, there are different interpretations of Marx's principal projects; according to one such interpretation, Marx finds inspiration for the natural-science paradigm. As Newton identified the laws that guide the movement of physical bodies, Marx wanted to identify the laws that guide the movement of economic structures. If properly identified, the political move towards a classless society would not be a utopian dream, but a socio-economic reality. Had Marx been successful in his critique, he would have established a proper foundation for formulating a political economy, as well as guidelines for engaging in political struggles. He would have put an end to the line of utopian socialism that speculated about what a better future might include.

Descartes, Kant, and Marx conducted methodological and systematic forms of critique; different as they were, they each contributed to the modern outlook.

---

[3] Marx's works have been published in many editions; see, for instance, Marx (1970, 1992/1993a/1993b).

Critique was aimed at securing knowledge with a solid foundation and of clarifying what "truth" could mean. Throughout modernity, critique also aimed at clarifying what could be meant by social progress, and in this respect, Marx provided one interpretation. During the age of modernity, critical endeavours addressed notions like *truth*, *knowledge*, *science*, *progress*, *democracy*, and *social justice*. The ambitions were that, by being systematic and methodological, one could establish a reliable foundation for both insight and political actions.

12.2 Critique as Hammering

A reaction to the methodological critique of modernity makes part of the post-modern outlook. However, in its most radical expression, the post-modern critique was formulated during the peak of modernity. Friedrich Nietzsche hammered down what the modern critique had been trying to construct. He formulated a critique of modernity long before the notion of post-modernity was coined. Many of his ideas have been incorporated into post-modern discourses, but here I will concentrate on Nietzsche's own formulations.[4]

Although modernity had taken steps away from orthodox Christianity, Nietzsche found that it was still lingering in a religious world-view. In *Ecce Homo,* first published in 1888, Nietzsche (2007) makes some comments about his own previous publications. About *Beyond Good and Evil*, first published in 1886, Nietzsche states that it is in essence a critique of modernity (see *Ecce Homo*, Beyond Good and Evil, §2).[5] It could naturally be added that Nietzsche confronts much more than the modern outlook; he disputes Christianity in general and the whole philosophical tradition since Plato. In *Beyond Good and Evil*, Nietzsche (1998a) highlights that, since Descartes, philosophers had tried "to assassinate the old concept of the soul", and in this way attempted to "assassinate the basic assumption of Christian doctrine". However, Nietzsche's point is that "overtly or covertly modern philosophy is anti-Christian, although it is by no means anti-religious" (*Beyond Good and Evil*, §54).

Nietzsche finds philosophy to be a deeply religious enterprise, as modern philosophy establishes a religious outlook by installing new idols. The modern idols are not the soul, or paradise, or eternity; they are instead the philosophical ideas through which reality is judged. Nietzsche hammers into pieces any such idols. By doing so, Nietzsche engages in a profound critical project quite different from the modern one. The entities that the modern critique tries to address in a systematic way are the objects that Nietzsche tries to destroy. Let us look at them one by one: truth, knowledge, science, progress, democracy, and social justice.

Truth. According to Nietzsche, there is no truth to be found anywhere. Modern philosophy talks as if it is able to discover truths. In order to maintain this way of

---

[4] See, for instance, DeLeuze (1986).

[5] As Nietzsche divided his texts into numerated paragraphs, I will use these numbers as references.

talking, the distinction between appearance and reality – such as the one formulated by Galileo Galilei and John Locke – plays a fundamental role.[6] The claim is that, behind appearance, one finds reality; knowledge concerns this reality and not appearances. But, according to Nietzsche (2003), there is no reality behind any appearance: "The 'apparent' world is the only one: the 'real' world has only been *lyingly added...*" (*Twilight of the Idols*, Reason in Philosophy, §2. Italics in the original.) The existence of reality behind appearance is a myth. It is a lie, and Kant was one of the liars by talking about a *Ding an sich*. There is no reality that truth can refer to. Truth has no "real" reference; truth is an illusion; it is an invention. Truth is not about anything; it is a way of talking; it is a "mobile army of metaphors" (Nietzsche, 2010, §1). The existence of eternal truths is just a modern idol that has substituted ideas about an eternal life after death.[7]

Knowledge. According to Nietzsche (1968), no position outside the stream of life exists from which one can look at things, judge things, and formulate true statements. One is submerged into the apparent world, behind which there is nothing else. This is the assertion of Nietzsche's radical perspectivism, which destroys any hope of establishing knowledge according to modern standards.[8] What is called knowledge does not represent any "insight" about certain states of affairs; "knowledge" is nothing but a strategy for survival:

> It is improbable that our "knowledge" should extend further than the strictly necessary for the preservation of life. Morphology shows us how the senses and the nerves, as well as the brain, develop in proportion to the difficulties in finding nourishment. (*The Will to Power*, §494)[9]

One can think of the human brain as an organ amongst other organs; and like any organ, it is important for our survival. Sweat is vital for our well-being, but there is no distinctive metaphysical quality related to sweat. In a similar way, the secretion produced by the brain, even though it is called knowledge, is no more unique than sweat or any other secretions produced by any of our organs. Like sweat, knowledge does not represent anything; it is nothing but a means for survival.[10] The problem is that the notion of knowledge, through immense philosophical misinterpretation, has become reified as an idol. Nietzsche hammers it into pieces.

Science. If knowledge is not about anything, and certainly not about truth, what then to think of science, the celebrated institution of modernity? Just a different idol. Nietzsche (1974) simply states:

---

[6] See Galilei (1957) and Locke (1997).

[7] See also Clark (1990).

[8] See also Hales and Welshon (2000).

[9] The original German version of *The Will to Power* was published in 1901, one year after Nietzsche died. It might contain edits not approved by Nietzsche.

[10] Nietzsche states that the "sense for truth" will have to legitimise itself as a "means for the preservation for man, as *will to power*" (*The Will to Power*, §495). The sense for truth is useful like any other sense. It helps to ensure our survival.

12.2 Critique as Hammering

A "scientific" interpretation of the world, as you understand it, might therefore be the most stupid of all possible interpretations of the world, meaning that it would be one of the poorest in meaning. (*The Gay Science*, §373)

Progress. It has been claimed that progress is a defining element in modernity. In the essay *A Discourse on Inequality* from 1755, Rousseau (1992) emphasises that, contrary to animals, human beings have a "faculty of self-improvement" (p. 88). This idea is axiomatic for the whole enlightenment movement, which claims that knowledge and education are crucial for the common welfare and for progress in general. The idea of progress penetrates the whole era of modernity. By developing knowledge and science, progress will take place in all spheres of life, both with respect to material welfare and with respect to political issues. Progress will ensure more social justice. However, Nietzsche just shrugs his shoulders: "The last thing I would promise would be to 'improve' humanity" (*Ecce Homo*, Foreword, §2). And even more explicitly that progress is "merely a modern idea, that is to say a false idea" (*The Anti-Christ*, §4). Progress is just an idol.

Democracy. As progress is an illusion, one should not expect Nietzsche to think any better of democracy, which has been celebrated as part of modern progress. Nietzsche has nothing positive to say about the French Revolution, and in general he finds nothing worthwhile captured by notions like liberty, equality, and fraternity. Instead, Nietzsche sees democracy as a miserable expression of a herd mentality:

> Indeed, with the help of a religion that played along with and flattered the most sublime desires of the herd animal, we have reached the point of finding an ever more visible expression of this morality even in the political and social structures: the *democratic* movement is Christianity's heir (*Beyond Good and Evil*, §202).[11]

Furthermore, he characterises modern democracy as "the decaying form of the state" (*Twilight of Idols*, §39). Nietzsche is not joining any democratic camp.

Social justice. According to Nietzsche (1998b), the slave morality turned values upside down by celebrating the poor, powerless, suffering, deprived, sick, and ugly, and by nominating the noble, powerful, and beautiful as evil, cruel, lustful, insatiable, and godless (see *On the Genealogy of Morality*, First Treatise, §7). Nietzsche finds that slave morality contradicts life as "life itself in its essence means appropriating, injuring, overpowering those who are foreign and weaker; oppression, harshness, forcing one's own form on others, incorporation, and at the very least, at the very mildest, exploitation" (*Beyond Good and Evil*, §259. Italics in original). Nietzsche also makes it clear that:

> [...] "exploitation" is not part of the decadent and imperfect, primitive society: it is part of the *fundamental nature* of living things, as its fundamental organic function; it is a consequence of the true will to power, which is simply a will to life" (*Beyond Good and Evil*, §259. Italics in original.)

---

[11] Nietzsche also observes: "The *overall degeneration of man*, right down to what social fools and flatheads call their 'man of the future' (their ideal!); this degeneration and diminution of man into a perfect heard animal; this bestialization of man into a dwarf animal with equal rights and claim is possible, no doubt about that!" (*Beyond Good and Evil*, §203. Italics in the original.)

The concept of social justice is far removed from Nietzsche's vocabulary, while "exploitation" is not, as it operates as part of the "nature of living things".

Nietzsche addressed critical issues that the modern conception of critique prepared for celebration. He condemned any festivity with respect to truth, knowledge, science, progress, democracy, and social justice. He considered any such notions as being idols that he simply wanted to hammer into pieces. We shall not forget that the subtitle of his book *Twilights of Idols* is: *Or How to Philosophize with a Hammer*.[12]

12.3 Critique as Dialogue

Critique as methodology and critique as hammering are extremely different approaches, but which approach could we adopt for critical mathematics education? My suggestion is to be ready to learn from both forms of critique, but not to adopt either of them. To me, critique need not try to reach out for certainty, nor hunt for foundations. Neither does critique need only to deconstruct and hammer into pieces. Throughout modernity, dialogue and critique were kept separate. Naturally, we need to be aware of the Platonic tradition of presenting discussion in the form of dialogue. This approach has been repeated many times; for instance, by Galilei. However, it is not this notion of dialogue that I have in mind.

I see *critique as dialogue*. An inspiring book for me has been *On Dialogue*, by David Bohm (1996). He contrasts dialogue with discussion, which might include forms of aggression and competition. One could try to win a discussion, but not a dialogue: "… a dialogue is something more than a common participation, in which we are not playing a game against each other, but *with* each other. In a dialogue everybody wins" (p. 7. Italics in the original). And further: "Dialogue is really aimed at going into the whole thought process and changing the way the thought process occurs collectively" (p. 9). Bohm relates dialogue to investigative processes, to creativity, and to collaboration. However, he does not try to relate dialogue and critique.

To me, the person who most clearly indicated connections between critique and dialogue is Paulo Freire. However, in *Pedagogy of the Oppressed*, Freire (2000) does not use the word "critique". He expresses his critical project through a different terminology. I am not aware of any place where Freire explicitly tries to define the notion of dialogue, but in many places he presents characteristics of dialogue; for instance:

> Dialogue is the encounter between men, mediated by the world, in order to name the world (Freire, 2000, p. 88).

Freire sees dialogue as a meeting between people. It is a meeting for "naming the world". This seems a simple activity, but to Freire "naming the world" means

---

[12] Foucault was inspired by Nietzsche, and he addressed issues that also preoccupied Nietzsche. See Foucault (1989, 1994, 2000).

12.3 Critique as Dialogue

interpreting the world, and also interpreting it as being open to changes. In the original Portuguese version, the quoted sentence runs: "*O diálogo é este encontro dos homens, mediatizados pelo mundo, para pronunciá-lo, não se esgotando, portanto, na relação eu-tu*" (Freire, 2021, p. 109). In this more extended formulation, dialogue is related to the I-you relationship, which is an implicit acknowledgement of Martin Buber's (1996) work *I and Thou*. Later, in *Pedagogy of the Oppressed*, Freire makes explicit reference to Buber.[13] While Buber engages with metaphysical and theological issues, Freire's project is political. As a consequence, Freire only uses Buber's conception of the I-you relationship as a general inspirational idea, and not as a theoretical footing.

Buber explores the fundamental human relationship in terms of a dialogic I-you encounters. I find that Freire operates with the same idea: dialogue represents a basic human relationship. It is not just a pattern that might appear in conversations. To Buber, as well as to Freire, dialogue is an existential category. Freire (2000) highlights the following:

> Dialogue is thus an existential necessity. And since dialogue is the encounter in which the united reflection and action of the dialoguers are addressed to the world which is to be transformed and humanized, this dialogue cannot be reduced to the act of one person's "depositing" ideas in another, nor can it become a simple exchange of ideas to be "consumed" by the discussants. Nor is it a hostile, polemical argument between those who are committed neither to the naming of the world, nor to the search for truth, but rather to the imposition of their own truth. Because dialogue is an encounter among women and men who name the world, it must not be a situation where some name on behalf of others. It is an act of creation; it must not serve as a crafty instrument for the domination of one person by another. The domination implicit in dialogue is that of the world by the dialogues; it is conquest of the world for the liberation of humankind (pp. 88–89).

A dialogue is more than an exchange of ideas; it is a creative process; it is a political process. It brings about reflections and actions; it brings about transformations; it brings about collective transformations; and it might bring about liberations. These are all profound critical activities.

Freire contrasts dialogical actions with anti-dialogical actions: "In the theory of anti-dialogical action, conquest [...] involves a Subject who conquers another person and transforms her or him into a 'thing'. In the dialogical theory of action, Subjects meet in cooperation in order to transform the world" (p. 167). The idea of dialogue as a collective action of transforming the world is further highlighted in the following:

> In the theory of anti-dialogical action, manipulation is indispensable to conquest and domination; in the dialogical theory of action the organization of the people presents the antagonistic opposite of this manipulation. Organization is not only directly linked to unity, but is a natural development of that unity. Accordingly, the leaders' pursuit of unity is necessarily also an attempt to organize the people, requiring witness to the fact that the struggle for liberation is a common task. This constant, humble, and courageous witness emerging from

---

[13] See Freire (2000, p. 167). See Chap. 14, "A dialogic Theory of Learning Mathematics" for more comments about Buber.

cooperation in a shared effort—the liberation of women and men—avoids the danger of anti-dialogical control. (pp. 175–176)

Freire introduces a perspective on dialogue that is much broader than the one presented by Bohm. While Bohm first of all sees dialogues as part of epistemic investigative processes, Freire sees dialogue as simultaneously part of transformative political actions. In this sense, Freire sees dialogic actions as critical actions.

I see critical processes as collective dialogic processes. In order to clarify this idea further, I will refer to some shared features of dialogue and critique.

To eliminate temporality from the domain of knowledge was a huge project of modernity. When seeing critique as emerging through dialogic interactions, we come to recognise that *critique is temporary*. A critique could come to take a new format, a new direction, a new focus, in just the same way a dialogue can do. A dialogue will always address some issues, leaving aside many others. The ambition of formulating an all-inclusive critique was part of the modern outlook, but contrary to any such ambitions, I see *critique as being partial*. It concerns some specific points of view, or some definite states of affairs, or some particular actions. *A critique is not conclusive*. A dialogue might have a beginning, but it does not start off from a platform of facts; rather, from questions, uncertainties, and doubts. Nor does a dialogue end up with definite conclusions. The "conclusions" of a dialogue might produce new questions, new ideas, and new doubts.[14] Like a dialogue, a critical investigation starts off from questions and uncertainties, and ends up with new questions and uncertainties. A dialogue, as well as a critique, is always tentative.

This brings me to see critique as an expression of uncertainty. We cannot escape uncertainty, but have to live with it. Uncertainty is part of our life-worlds. It forms an integral part of the formulation of opinions and knowledge, and it is part of any political positioning and action. *Uncertainty is a universal human condition*. This applies to any critical enterprise, and also to critical mathematics education.

References

Bohm, D. (1996). *On dialogue*. Routledge.
Buber, M. (1996). *I and thou*. A Touchstone Book, Simon and Schuster.
Clark, M. (1990). *Nietzsche on truth and philosophy*. Cambridge University Press.
Deleuze, G. (1986). *Nietzsche and philosophy*. Continuum.
Descartes, R. (1993). *Meditations on first philosophy*. Routledge.
Foucault, M. (1989). *The archaeology of knowledge*. Routledge.
Foucault, M. (1994). *The order of things: An archaeology of the human sciences*. Vintage Books.
Foucault, M. (2000). *Power*. The New Press.
Freire, P. (2000). *Pedagogy of the oppressed* (With an Introduction by Donaldo Macedo). Bloomsburry.

---

[14] This point is illustrated by Plato's first dialogues, often referred to as the aporetic dialogues. They begin with a question that might appear straightforward, but during the dialogue the question turns out more and more complex, in the end appearing to be impossible to answer.

References

Freire, P. (2021). *Pedagogia do Oprimodos, 80ª Edição (Pedagogy of the oppressed, 80th Edition)*. Paz e Terra.

Galilei, G. (1957). The assayer. In G. Galilei (Ed.), *Discoveries and opinions of Galileo* (pp. 229–281). Anchor Books.

Hales, S. D., & Welshon, R. (2000). *Nietzsche's perspectivism*. University of Illinois Press.

Kant, I. (1929). *Critique of pure reason* (Translated by Norman Kemp Smith). Macmillan.

Locke, J. (1997). *An essay concerning human understanding*. Penguin Books.

Marx, K. (1970). *A contribution to the critique of political economy*. International Publishers.

Marx, K. (1992/1993a/1993b). *Capital: A critique of political economy* (Vol. I–III). Penguin Classics.

Nietzsche, F. (1968). *The will to power*. Vintage Books.

Nietzsche, F. (1974). *The gay science*. Vintage Books.

Nietzsche, F. (1998a). *Beyond good and evil*. Oxford University Press.

Nietzsche, F. (1998b). *On the genealogy of morality*. Hackett Publishing Company.

Nietzsche, F. (2003). *Twilight of the idols and the anti-Christ*. Penguin Books.

Nietzsche, F. (2007). *Ecce homo: How to become what you are* (Translated with an Introduction and Notes by Duncan Large). Oxford University Press.

Nietzsche, F. (2010). On truth and lie in a nonmoral sense. In F. Nietzsche (Ed.), *On truth and untruth: Selected writings* (pp. 17–49).

Rousseau, J.-J. (1992). *Discourse on the origin and basis of inequality among men*. Indian Hackett Publishing.

Wittgenstein, L. (1997). *Philosophical investigations*. Blackwell.

Chapter 13
Initial Formulations of Critical Mathematic Education

Abstract Critical mathematics education arose from the idea that education has a political role to play. Education is part of the struggle against all forms of oppression and atrocity. The first formulations of critical education were made in the late 1960s, and from the beginning of the 1970s, one finds initial formulations of critical mathematics education, in particular in Germany and Scandinavia. A guiding idea was that any form of mathematics brought into action needed a profound critique, and furthermore that mathematics could be used for identifying and critically addressing any form of real-life problem. Now, critical mathematics education is in a powerful development stage, where young researchers from Brazil play an important role.

Keywords Critical education · Exemplary learning · Sociological imagination · Problem-orientation · Project organisation · Students' movement

When was critical education conceptualised? When was critical mathematics education conceptualised? There are no specific answers to these questions, but I want to refer to two publications that one can consider to have introduced the idea critical education.

In 1966, Theodor Adorno gave a presentation on the radio with the title "Erziehung nach Auschwitz" ("Education after Auschwitz"), and the following year the presentation was published in a written format. In 1971, "Erziehung nach Auschwitz" was included in the book *Erziehung zur Mündigkeit* (*Education for Autonomy*), where Adorno brings together more reflections important for formulating a critical education. The German word *Mündigkeit* has a legal interpretation, referring to the age when a person obtains rights as an adult, including the right to vote. Informally, it refers to a person claiming the power to speak for himself or herself. One can understand *Mündigkeit* as "legal maturity" but also as "political autonomy", which conveys the meaning that Adorno associated with the term when talking about "education for *Mündigkeit*". The first sentence in "Erziehung nach Auschwitz" states: "The very first demand for education is that Auschwitz should not happen again" (Adorno, 1971, p. 88, my translation). This statement can be taken literally, but also metaphorically, as claiming that education has to confront all forms of possible atrocities.

In 1968, Paulo Freire published the book *Pedagogia do Oprimido,* which in 1972 appeared in English as *Pedagogy of the Oppressed.*[1] The very title expresses a radical new way of thinking about education. By talking about a pedagogy of the oppressed, Freire makes it clear that education can be part of a political struggle for a particular group of people. One can engage in a struggle against oppression. Freire suggests that a pedagogy of the oppressed grows out of specific issues of injustice. This issue could, for instance, be the distribution of water in a particular neighbourhood. With respect to such an issue – which Freire refers to as a generative theme – one can raise questions such as: Who has easy access to water? Who has access to clean water? What health problems could be related to a lack of clean water? Through such questions, the "politics of water" can be explored. In a similar way, by engaging in other generative themes, a range of everyday issues might reveal profound oppressive structures engraved on the socio-economic order of society.

"Erziehung nach Auschwitz" and *Pedagogia do Oprimido* highlights that education has a political role to play. Education has to be a central part of a struggle against all forms of oppression and atrocity.[2]

13.1 Education Lost in a Dream World

In Germany during the Weimar Republic – initiated in 1919 when Germany adopted a new constitution, and destroyed in 1933 when Adolf Hitler came to power – one finds formulations of important educational ideas. Considering the many totalitarian ideas that were being articulated at that time, one could imagine such ideas being confronted with initial formulations of critical education. However, I have never come across any such formulations. I cannot even find preliminary formulations made during the Weimar Republic of the idea that education could play a political role by confronting cases of oppression or exploitation.

What educational ideas could one encounter during the Weimar Republic? Much progressive educational thinking was articulated as *Geisteswissenschaftliche Pädagogik.* The German word *Geisteswissenschaft* refers to disciplines belonging to the humanities, such as literature, history, language, and philosophy. While *Geist*

---

[1] See Freire (2021) for the Portuguese version and Freire (2000) for the English version. In 1973, the book was translated into Danish by one of my colleagues from the teacher training college where I was working during that time.

[2] One can find previous articulations of the idea that education has a political significance. In 1916, Dewey (1966) published *Democracy and Education*, where he points out intrinsic connections between educational activities – when formed according to the vision of progressive education – and a democratic way of life. Nevertheless, I do not consider Dewey as having a defining part in the critical enterprise, as his conception of society is formed by a general liberal outlook rather than by a recognition of particular patterns of oppression and exploitation.

means spirit, *Wissenschaft* means science. The use of the notion of *Geist* in this context dates back to the romantic period in Germany, and particularly relates to the concept of *Volkgeist*. *Volk* means "people", and preoccupations with *Volkgeist* reflect the assumption that the very spirituality of a people is an important entity. A principal idea for Wilhelm Dilthey was that the *Geisteswissenschaften* were different from the natural sciences. While natural sciences searched for explanations, human sciences should search for understanding.[3]

Geisteswissenschaftliche Pädagogik manifested a concern for developing education with emphasis on forming integral human beings. Through education, children should come to understand their historical roots, cultural context, and develop their full human potentialities. It was recommended that children be engaged in experimental and creative work, and that they were given the freedom to follow their own interests and priorities. A practical approach to such education was formulated by Rudolf Steiner through the Waldorf education system. The first Waldorf School opened in 1919 in Germany, the year the Weimar Republic was born.

Martin Wagenschein published both before and after the Second World War.[4] He was interested in natural sciences as well as in mathematics, but he did not assume any traditional educational interpretation of these subjects. He was deeply inspired by *Geisteswissenschaftliche Pädagogik*, and his ambition was to educate people to obtain a broad cultural outlook, including an understating of both natural science and mathematics.

Geisteswissenschaftliche Pädagogik, Waldorf education, and Wagenschein's initial ideas were all formulated during the Weimar Republic. These trends represent what, in general, has been referred to as progressive education. We are dealing with a strong contrast to the dominant authoritarian educational pattern of the time; however, progressive education, in any of its forms, did not position itself with respect to the dramatic political development of the time.

The Weimar Republic collapsed; the Nazis took control and drew the world into war. The Second World War caused tremendous suffering and destruction. Soldiers and civilians were killed in their millions, and the Holocaust killed millions more. Peace came with Germany laid to ruin. Despite all of this, in the years following the Second World War, progressive education (as well as educational theorising in general) moved on as if nothing had happened: during the 1950s, the pedagogical discussions from the time of the Weimar Republic were repeated and refined. *Geisteswissenschaftliche Pädagogik* continued to elaborate on what understanding could mean; the Waldorf education provided new examples of how fables and folklore could be applied in educational processes; and Wagenschein elaborated upon his conception of mathematics and science education in greater detail. When leafing through educational publications from Germany and the Scandinavian countries from the period 1920–1950, I have not been able to identify any indication that the

---

[3] Such ideas were elaborated upon carefully in 1910 when Dilthey (2002) published *Der Aufbau der geschichtlichen Welt in den Geisteswissenschaften* (*The Structure of the Historical World in the Humanities*).

[4] See, for instance, Wagenschein (1965/1970).

Second World War had taken place, except for an interruption in the dates of publication.

It was only in 1966 that the wake-up call for education came, when Adorno gave the talk "Erziehung nach Auschwitz". Did Adorno indicate that critical education could have been formulated during the Weimar republic? I think so. Did he imagine that it would be possible to establish critical education after the Nazis were in power? I do not think so. I imagine that Adorno fully recognised that confronting the Nazi terror was a military rather than an educational endeavour. However, the wake-up call from Adorno made it obvious that education, and also progressive education, had been lost in a dream world.

13.2 Education Turning Critical

During the spring of 1968, political protests took place in Paris, to a great extent initiated and organised by university students. More student-guided protests took place in other countries. On 11th April 1968, the German student leader Rudi Dutschke was shot in the head by a right-wing extremist. Dutschke survived and gradually recovered his partly lost abilities to speak and to remember.[5] In 1968 in Mexico, the student movement manifested a strong political opposition against the regime. On 2nd October, demonstrators were violently attacked by the armed forces, which resulted in a tragedy now known as the as Tlatelolco Massacre. During the 1960s in the USA, Martin Luther King participated in many protests and demonstrations. He was assassinated on 4th April 1968. Malcolm X had also been engaged in powerful anti-racist protests. He was assassinated on 21st February 1965.

From 1968 onwards, one also finds student-initiated protests in Copenhagen. In September 1968, I started my studies in mathematics at Copenhagen University, a year after I had concluded my teacher education. In mathematics and the natural sciences, there were not many signs of the student movement, which was mainly guided by students from psychology and the social sciences. In any case, to me the events were overwhelming, but it was only some years later that I grasped the full significance of these event. As in other countries, the protests targeted the USA – in particular, the Vietnam War, the capitalist exploitation of workers, and the use of atomic energy.

Protests also targeted the way university studies were organised. The student movements insisted that university courses should be profoundly reorganised by being focussed on actual socio-political problems, and the studies should be directed by the students themselves. The "dictatorship of the professors" should be broken and space made for dialogue between university teachers and students about both form and content of the study programmes.

---

[5] After his recovery, Dutschke obtained a temporary position at a university in Denmark.

13.2 Education Turning Critical

One ambition of the student movement was that existing university study-programmes should be subjected to a critique. The German word used for this critique was *Fachkritik*, which is composed of the word *Fach* meaning "discipline", and *Kritik* meaning "critique". Conducting a *Fachkritik* became one of the activities of the student movement. In Copenhagen one could see the word *fagkritik* (*Fachkritik*) painted on the walls of the old university buildings. I am not aware of any direct English translation of the term.

At that time the Danish government turned out to be relatively open to negotiations with the student movement. It was generally acknowledged that it was relevant to open two more universities in Denmark, and in 1972 Roskilde University Centre opened, followed in 1974 by Aalborg University Centre, later to be renamed Aalborg University. The organisation of these two universities was directly influenced by the student movement. In all disciplines, also mathematics, the study-programmes became problem-oriented and project-organised, and the students were supposed to have a strong influence on which problems to address.

In 1968, Habermas published *Erkenntnis und Interesse* (*Knowledge and Human Interest*), which provided inspiration for formulating a *Fachkritik*, in particular with respect to the social sciences. Habermas (1971) distinguished between different kinds of sciences. As mentioned, Dilthey had suggested that two kinds of science exist: the natural sciences and the humanities. By making this claim, Dilthey opposed the positivist conviction that there exists one – and only one – kind of science. In contrast to Dilthey and logical positivism, Habermas formulated the idea that there are three different types of sciences, and that these differences are rooted in different knowledge-constituting interests. The natural sciences, also referred to as the technical sciences, are grounded by a technical interest, which is an interest in doing things, fabricating things, and controlling things. More generally, it is an interest in manipulating things. The human sciences are guided by an interest in coming to understand, as can be closely associated with hermeneutics. Finally, Habermas claimed that the social sciences should be guided by emancipatory interest, confronting all forms of oppression.

Habermas' discussion of knowledge-constituting interests conveyed a critique of positivism that came to be an integral part of the 1968 student movement. This critique had a particular significance for the social sciences. With reference to Habermas, one could claim that the positivist position would establish technical interest as the knowledge-constituting interest also for the social sciences. This would imply that the social sciences came to serve an interest in manipulation. It became a principal aim of the student movement to establish social sciences as part of liberating social processes. Furthermore, it was claimed that education could not be guided by technical interest and associated with a behaviourist conception of learning, nor could it be guided simply by an interest in understanding, as assumed by *Geisteswissenschaftliche Pädagogik*. The point of establishing critical education was to make emancipation the guiding interest.

In *Soziologische Phantasie und exemplarisches Lernen: Zur Theorie der Arbeiterbildung* (*Sociological Imagination and Exemplary Learning: On the Theory of Workers' Education*), also published in 1968, Oska Negt (1968). indicates how such an idea could guide educational practice. The book concentrated on worker's education, and Negt suggested a politisation of this education. He wanted to address socio-political and economic issues in order to clarify the possible exploitation of workers. Negt's formulations had a strong impact on the initial formulation of critical education. Negt was well positioned in critical theory as a student of Adorno, and subsequently a research assistant for Habermas.[6] In 1971, Wolfgang Lempert (1971) published *Leistungsprinzip und Emanzipation (Performance and Emancipation)*, which addressed the relationship between education and emancipation; and in 1973, Klaus Mollenhauer (1973) addressed the same issue in *Erziehung und Emanzipation (Education and Emancipation)*.[7]

All these contributions make up the initial formulation of critical education. However, the very notion of critical education did not exist *a priori* to all these formulations; rather, it emerged as a more general label, and I use it to refer to educational approaches guided by the overall idea that education has a socio-political role to play in confronting oppression and atrocities.

13.3 Mathematics Education Turning Critical

Very soon after the initial formulations of critical education, proposals emerged for establishing critical mathematics education. However, these proposals were met with scepticism, even from many of those engaged in critical education.

Apparently, the very notion of critical mathematics education is contradictory, implied by Habermas' theory of the knowledge-constituting interest. If critical education should be guided by an emancipatory interest, and mathematics is guided by a technical interest, how could one possibly imagine critical mathematics education as being a possibility? Furthermore, the conception of critical education seems to assume an interdisciplinary approach. In what sense, then, could one talk about a critical *mathematics* education? Critical education expressed a deep suspicion towards any attempts to formulate critical mathematics education. In 1972, Negt founded the Glocksee School, located in Hanover. This school was organised according to visions of critical education, and it provided a space for students to engage in a range of free activities, as well as for addressing real-life social problems. Space was also made for students to develop what was referred to as "cultural techniques". This could, for instance, be typewriting. Within the Glocksee School,

---

[6] In 1975, Negt's book was translated into Danish, but I am not aware of any English translation of Negt's principal work.

[7] As far as I know, none of these works have been translated into English. However, Mollenhauer's (2013) last major work, *On Culture and Upbringing,* is available in English. The idea of critical education was presented in English by Young (1989). For more formulations in English, see, for instance, Apple (1982) and Giroux (1989).

mathematics was considered a cultural technique, and no critical potentials were associated with this discipline.[8]

Many times I have heard the comment that critical mathematics education is developed from critical theory, and I feel uncomfortable when hearing this claim. It seems clear to me that critical mathematics education is a part of the broader politisation of education and of the critical endeavour, as advocated by critical theory. However, when developing both theoretical and practical aspects of critical mathematics education, critical theory was of little significance. Critical mathematics education had to develop along its own pathway.

In 1974, Peter Damerow, Ulla Elwitz, Christine Keitel, and Jürgen Zimmer (1974) published the book *Elementarmatematik: Lernen für die Praxis* (*Elementary Mathematics: Learning for the Practice*) which discusses how school mathematics can become part of a critical enterprise. The subtitle of the book includes the word *Praxis*, which can be interpreted in two ways. On the one hand, *Praxis* can refer to all kind of work practices, and elementary mathematics can be relevant for such practices. We can think of farming, brick building, plumbing, and tailoring. Furthermore, elementary mathematics can be relevant with respect to a range of daily-life situations, such as doing shopping, controlling salaries, reading the newspapers, and making budgets. On the other hand, the notion of *Praxis* also refers to political practices, such as identifying cases of social and economic oppression, formulating political visions for the future, and engaging in strikes and protests. The notion of *Praxis* has these two different interpretations, and in the *Elementarmatematik* it is highlighted that elementary mathematics can be a learning experience for both. In this way, it becomes evident that mathematics education has a political role to play.

In the article "Plädoyer für einen Problemorientierten Mathematikunterrich in emanzipatorischer Absicht" ("Plea for Problem-Oriented Mathematics Education With an Emancipatory Aim") from 1975, Dieter Volk (1975) discusses how a problem-based mathematics education with a concern for emancipation can be established. He explores the debates over emancipation, and confronts the assumption that it is not possible to establish a form of mathematics education guided by an interest in emancipation.

The book *Kritische Stichwörter zum Mathematikunterricht* (*Critical Keywords for Mathematics Education*) from 1979, edited by Volk (1979a), provides a systematic presentation of topics relevant for critical mathematics education. The book is composed of short chapters where one finds discussions of topics such as women and mathematics, teacher education, set theory, Piaget, objectivity, project orientation, self-organisation, leaning aims, social experiences, and textbooks. These are all topics crucial for a fuller formulation of critical mathematics education. The discussion of set theory was important, as at the time, the Modern Mathematics Movement was in full swing. It was also important to address Jean Piaget's

---

[8] One can still encounter the sceptical opinions of critical education towards critical mathematics education. For instance, *The Routledge International Handbook for Critical Education*, edited by Apple et al. (2009), does not include any substantial discussion of critical mathematics education.

epistemology, as this represented a paradigmatic non-critical perspective on mathematics. Critical mathematics education distanced itself from both the Modern Mathematics Movement and Piaget's accompanying justification of this approach. In *Handlungorientierende Unterrichtslehre am Beispiel Mathematikunterricht* (*Action-oriented teaching using the example of mathematics education*) and *Zur Wissenschaftstheorie der Mathematik: Handlungorientierende Unterrichtslehre* (*On the philosophy mathematics: Action-oriented teaching*), Volk (1979b, 1980) provides a broader presentation of his conception of an action-oriented mathematics education. The action-orientation becomes related to both critical educational practices and to philosophical perspectives on mathematics.

In 1977, Stieg Mellin-Olsen (1977) published the book *Læring som Sosial Prosess* (*Learning as a Social Process*) based on his doctoral thesis, which was not approved as it was considered to be too political. This book had a tremendous impact on mathematics education in Scandinavia, particularly in Norway and Denmark. Mellin-Olsen established a new social perspective on mathematical learning processes, and on mathematics education in general. According to Piaget's conception of learning, processes of assimilation and accommodation are assigned particular roles, and consequently the learning of mathematics is conceptualised as an intrinsic personal activity. Through his book, Mellin-Olsen created quite a different social perspective: learning had to be understood as a set of social processes taking place within the context of political complexities. In *The Politics of Mathematics Education*, Mellin-Olsen (1987) reworked his previous studies and elaborated them further for the English-speaking mathematics education community.

When looking for initial formulations of critical mathematics education, it is important not only to look at what was published in articles and books, but also to look for what was actually taking place in educational settings. As part of their project work, mathematics students at Roskilde University Centre and at Aalborg University addressed a range of controversial issues related to mathematics and its applications. A particular concern was to elaborate a *fagkritik* with respect to mathematics. At times, this critique was interpreted as a *modelkritik* (modelling critique). Through studying particular mathematical models, it became evident that mathematics, when applied, does not automatically ensure objectivity and neutrality. A mathematical model might incorporate a range of assumptions, presumptions, and particular interests.[9] In Denmark at that time, the curriculum for primary and secondary education was not over-determined, and much space was left for teachers who wanted to engage in experimental work. The possibility of problem-based and project-organised learning inspired many mathematics teachers.[10]

---

[9] Students and researchers from IMFUFA (*Institut for studierne af Matematik og Fysik samt deres funktioner i Undervisning, Forskning og Anvendelser*) at Roskilde University Centre has provided a variety of critical studies of mathematics and its applications, published in the series *Tekster from IMFUFA*.

[10] For discussions of project-organised mathematics education, see, for instance, Münzinger (1977); and Vithal et al. (1995).

13.3 Mathematics Education Turning Critical

In 1975, I drafted my first proposal for a PhD study with the purpose of formulating critical mathematics education. My proposal was not approved. I was told that there was nothing called "critical mathematics education". I did a reworking of the research proposal, and in 1977 my redrafted project was approved. In 1979 and 1980 the first results of my PhD study appeared, first in terms of lectures for teachers and teacher educators, and then in written format. I did not perceive my PhD study as primarily scientific work, but rather as a piece of work on pedagogical reform. I was interested in presenting alternatives to the existing mathematics education in Denmark. I wanted to formulate t critical mathematics education as a possible pedagogical practice in primary and secondary school. Thus, I did not perceive the examples I described as empirical materials to be analysed, but rather as illustrations of pedagogical possibilities. I published three books: *Forandringer i Matematikundervisningen* (*Changes in Mathematics Education*) in 1980 (Skovsmose, 1980), *Matematikundervisning og Kritisk Pædagogik* (*Mathematics Education and Critical Pedagogy*), in 1981 (Skovsmose, 1981a), and *Alternativer i Matematikundervisningen* (*Alternatives in Mathematics Education*), also in 1981 (Skovsmose, 1981b). However, in order to acquire the PhD, I was asked to compile my work into a single dissertation. The result was *Kritik, Undervisning og Matematik* (*Critique, Education, and Mathematics*), which was defended in 1982 and published in 1984 (Skovsmose, 1984).[11]

In 1983, Marilyn Frankenstein (1983) published the article "Critical Mathematics Education: An Application of Paulo Freire's Epistemology". Through this publication, Frankenstein introduced the notion of "critical mathematics education" to the English-speaking educational community. She was specific in referring to Paulo Freire as her principal reference, and much of the further development of critical mathematics education in the USA has concentrated on this as a source of inspiration. Through the work of Frankenstein, an important new dimension was added to Freire's pedagogy of the oppressed.

In line with her emphasis, critical mathematics education should be an education for the oppressed, for the marginalised, and for students in vulnerable positions. However, remembering the direction indicated by Adorno, critical mathematics education concerns all groups of students, not just the disadvantaged, although it might manifest itself in quite different ways for students in different socioeconomic groups. I am concerned about developing critical mathematics education as an approach relevant for all different groups of students: students in vulnerable positions, students in comfortable socio-economic positions, and university students in mathematics. Naturally, though, the significance of critical mathematics education for these different groups might be rather diverse.[12]

---

[11] I was the first in Denmark to complete a PhD study in mathematics education. My supervisors were Bent Christiansen and Tage Werner. During the 1960s, Bent Christiansen had been the main person in Denmark promoting the Modern Mathematics Movement in primary and secondary schools. During the late 1970s, Christiansen was inspired by the work by Hans Freudenthal, and advocated that students got involved in investigative processes. During his whole career, Werner was dedicated to engaging students in mathematical investigations.

[12] In the Chap. 15, "All Students", I elaborate more on this point.

13.4 Sociological Imagination and Exemplary Learning

One overall connecting idea for critical education, and also for critical mathematics education, to assume a socio-political role, an idea which can be explored further in terms of sociological imagination and exemplary learning.

The notion of *sociological imagination* was coined by the American sociologist Charles Wright-Mills, who in 1959 published the book *The Sociological Imagination*. At that time, not least in the USA, sociology was dominated by the outlook of logical positivism. Consequently, the general ambition was to organise sociological studies according to the paradigm set by the natural sciences. The aim was to provide descriptions of social states of affairs, and to ensure that these descriptions were as accurate as possible. Such descriptions would, it was assumed, ensure scientific theories with objectivity and neutrality. During the 1950s and early 1960s, it was highly problematic to express leftist sympathies in the USA. The period was dominated by the Cold War, and McCarthyism was still very much alive. In this context, Wright-Mills expressed the leftist position that the role of sociology is *not only* to provide descriptions of states of affairs, but *also* to formulate alternatives to what can be observed. Sociology should not only describe, say, the poor living condition of workers and their families, but also conceptualise alternatives to such conditions. In this way, sociology could come to assume a critical role in society. Without using the notion of emancipation, Wright-Mills advocated a similar conception of the social sciences as Habermas when claiming that the social sciences should be guided by an emancipatory interest. In 1960, Wright-Mills (1960) published a short text with the title "Letter to the New Left". Here he coined the expression "new left", which was broadly applied in reference to the emerging reformulations of leftist positions.

Exemplary learning refers to an idea formulated by Wagenschein; namely, that one can gain profound and general insight by studying a particular case. He formulated this idea as a response to the contemporary problem of the apparently ever-growing curriculum. For every school subject – biology, physics, mathematics – it seemed relevant to include new topics and more information. Every school subject seemed to be stuffed with more and more facts and information. It seemed necessary to continuously add ever more chapters to the textbooks, and to put more and more emphasis on the teachers' expositions. Contrary to this trend, Wagenschein formulated the idea of exemplary learning. According to him, one need not present mathematics in a systematic way, as a deep understanding of mathematics can be obtained by studying a particular issue. As an illustration, he explained how working with the Pythagorean Theorem can bring about a mathematical understanding of a range of issues and of the very nature of mathematics.

As already referred to, Negt combines the two notions in the title of his book *Soziologische Phantasie und exemplarisches Lernen*. Concentrating on workers' education, Negt pointed out that this post-school education can take a starting point in a particular event at a particular workplace, and by exploring this event one can obtain a profound insight into the workers' conditions in the capitalist production machinery. The insight thus obtained can pave the way toward sociological imagination and, according to Negt, it is important as part of workers' education to conceptualise alternatives to the given social order. Negt's combination of a sociological imagination and exemplary learning can be applied far beyond workers' education. It can be applied to any form of education, and Negt's combination came to form an integral part of the general elaborations of problem–based and project-organised education. Freire does not explicitly use the notions of sociological imagination and exemplary learning; nevertheless, these conceptions are part of Freire's conception of *generative themes*. By focusing on a particular phenomenon, such as the water supply in a certain neighbourhood, one can obtain a more general insight into the living conditions in the neighbourhood, and also come to recognise that alternatives do exist.

The idea of engaging students in project work concentrated on particular problems is an expression of exemplarity. Furthermore, engaging students in critical activities assumes the importance of establishing a sociological imagination. A critique might include a description of a certain state of affairs, but it also means showing that the described facts are not essential facts, but contingent states of affairs. Things could be different. Sociological imagination and exemplary learning characterise many expressions and practices of critical mathematics education.

13.5 Brazil: Land of the Future

In 1934, after the rise of the Nazi movement in Austria and due to his Jewish origins, Stephan Zweig and his wife emigrated to England. In 1940 they moved to the USA and, from there, to Brazil. In *Brazil: A Land of the Future*, first published in German in 1941, Zweig (2000) expressed his fascination with the country and its many possibilities.[13] The land of the future also came to include dramatic retrogressions. In 1964, Brazil turned into a military dictatorship. People were prosecuted for assuming leftist positions and torture was frequent.[14]

Today, the future of Brazil appears to include contradictory tendencies. As in many other countries, extreme right-wing movements are in powerful positions, systemic poverty is increasing, totalitarian discourses are in circulation, racist manifestations are common, and fascist positions have been adopted. However, at the same time, strong movements are confronting these tendencies. Contrary to what

---

[13] Zweig and his wife also missed Austria, and it became too depressing for them to witness the apparent steady advancement of the German forces. They died together in February 1942 in what seemed to be a double suicide.

[14] Freire was accused of being a traitor, and was imprisoned for a period, after which he left Brazil and stayed for a short time in Bolivia before moving to Chile, where he worked for 5 years. During this time, he published *Pedagogia do Oprimido*. He stayed for a shorter period in the USA, before moving to Geneva. In 1979, he visited Brazil for the first time since his exile, and, feeling sufficiently safe, he returned to Brazil in 1980.

was taking place during the Weimar Republic in Germany, critical education is now a well-articulated movement, as is critical mathematics education.

This chapter is about initial formulations of critical mathematics education, but let me conclude by taking a brief look at present Brazilian contributions to critical mathematics education. The Épura research group was founded in 2008, associated with Universidade Estadual Paulista (Unesp) in Rio Claro.[15] The group provides a cooperative environment for a range of projects that are contributing to the further development of critical mathematics education, some of which are PhD studies, supervised by Miriam Godoy Penteado or by me.

Several of these studies address students at social risk, what kinds of future possibilities they might have in life, and how their foregrounds are socially structured. Denival Biotto Filho (2015) explored how children from an orphanage could become engaged in project work that gave them new experiences of learning mathematics and, in this way, opened up new educational possibilities for them. Daniela Alves Soares (2022) interviewed students in vulnerable positions and portrayed their dreams and hopes for their future. She demonstrated the complexities of such dreams and hopes, and consequently also the complexities of their possible motives for learning mathematics. Luana Oliveira (in progress) worked with students in vulnerable positions, and she concentrated on showing what "reading and writing the world with mathematics" could come to mean for this group of students.

Other groups of socially marginalised groups of students have also been addressed. For example, Guilherme da Silva (2016) studied the implementation of affirmative action in Brazil, and with reference to groups of indigenous students, he identified a range of exclusive patterns that also might be associated with well-intentioned affirmative actions. Denner Dias Barros (2021) engaged with the LGBT+ community and showed how mathematics can be a resource for analysing the preconceptions which this community often meets with. Manuella Carrijo (2023) studied possibilities and obstructions that students from immigrant families might meet when trying to engage with mathematics. In this way, she added new dimensions to the discussion of the social fabrication of learning obstacles. Jamaal Ince (n.d. in progress) worked with communities of African-American students from Newark in the USA. He showed how systemic racism prevents these students from engaging with mathematics, but also how these kinds of learning obstacles can be challenged.

Landscapes of investigation and dialogic relationships have been addressed with respect to all levels of mathematics education. Ana Carolina Faustino (2018) explored dialogic patterns of interaction and showed how dialogues among students – and between students and teachers – can turn into resources for critical activities related to environmental issues. João Luiz Muzinatti (2018, 2022) showed how to engage students from affluent neighbourhoods in critical investigations into the living conditions of people from poor neighbourhoods. In this way, he illustrated

---

[15] The Epura group was initiated by Miriam Godoy Penteado, and is now coordinated by her and me. More details about Épura can be found at https://igce.rc.unesp.br/#!/departamentos/educacao-matematica/grupos-de-pesquisa/epura/apresentacao/

the relevance of engaging all groups of students in addressing cases of social injustice. Raquel Milani (2015) showed how to "open a mathematics exercise" and, through this opening, to enter a landscape of investigation. To create such openings is important for changing the mathematics education that is captured by the exercise paradigm. Débora Vieira de Souza-Carneiro (2021) explored the possibility of addressing ways to bring mathematics into action as an integral part of project work in university studies with mathematical content. Through her study, she initiated a discussion on introducing problem-based learning in mathematics at universities in Brazil. With particular reference to teacher education, Priscila Coelho Lima (2022) further developed the notion of pedagogical imagination, and in this way, related more closely the discussion of critical mathematics education to the discussion of sociological imagination.

Further studies have related aspirations of inclusive mathematics education to the concerns of critical mathematics education. Renato Marcone (2015) coined the notion of deficiencialism, which refers to conceptions of people with disabilities as formed by discourses assuming the existence of "normality". Deficiencialism refers to a conception of what "we" find "them" not able to do. By means of the notion of deficiencialism, a range of preconceptions – for instance, related to racism – can be explored. Amanda Queiroz Moura (2020) studied mathematics classrooms where deaf and hearing students were engaged in the same mathematics activities. In this way, she provided a broader conception of inclusive education, acknowledging the importance of establishing dialogues across differences. Célia Roncato (2021) focussed on university students with different kinds of disabilities. She studied the difficulties that these students had to face and how they have struggled to overcome them.

Special attention has been paid to senior students. Luciano Lima (2015) worked with retired people who, after a lifetime of hard work, had some spare time. He engaged the seniors in different geometric topics – as for instance related to Euler's Polyhedron Theorem – as well as in the mathematics that appeared in daily newspapers. The seniors came to master a reasoning that they had not imagined to be within their reach. Matheus Pereira Scagion (in progress) addressed issues related to the life conditions of elderly people and focussed on how these conditions could become addressed in a school context.

If we look beyond what has been completed in terms of doctoral dissertations by members of the Épura group, we will find a richness of more Brazilian contributions to the further redevelopment of critical mathematics education.[16] If we look outside

---

[16] As an illustration of this richness, let me refer to: Araújo (2007, 2009); Azevedo (2019). Baroni et al. (2021); Campos (2013, 2018); Civiero (2016); Civiero et al. (2022a); Civiero et al. (2022b); Civiero and Oliveira (2020); Lima (2018); Lima (2015); Lima and Lima (2019); Milani et al. (2017); Miranda (2015); Oliveira (2015); Sampaio (2010); Silva (2005); Silva (2014); Silva et al. (2021); Soares (2008); and Souza (2021). In 2017, the journal *Revista Paranaense de Educação Matemática* launched a special issue dedicated to critical mathematics education (https://periodicos.unespar.edu.br/index.php/rpem/issue/view/311). Many studies in ethnomathematics share concerns with critical mathematics education, as, for instance, Knijnik (1996, 2007); Knijnik and Wanderer (2015, 2018), Silva (2013); and Silva and Severinho Filho (2021). The concerns of criti-

Brazil, one also finds many contributions to critical mathematics education, many of which refer to mathematics education for social justice.[17]

Considering all these studies, I feel very satisfied on behalf of critical mathematics education. The studies are driven by concerns about actual social, political, economic, and environmental problems, and by hope for ensuring both social and environment justice. However, this proliferation of critical mathematics education will meet many obstructions: some might be caused by traditional educational outlooks, but some might be caused by explicit totalitarian measures, neo-fascist ideologies, and new forms of economic and social oppression. These obstructions will attempt to make critical mathematics education impossible. However, simultaneously they make critical mathematics education necessary.

References

Adorno, T. W. (1971). *Erziehung zur Mündigkeit (Education for authority)*. Suhrkamp.
Andersson, A., & Barwell, B. (Eds.). (2021). *Applying critical mathematics education*. Brill and Sense Publishers.
Apple, M. W. (1982). *Education and power*. Routledge and Kegan Paul.
Apple, M., Au, W., & Gandin, L. A. (Eds.). (2009). *The Routledge international handbook of critical education*. Routledge.
Araújo, J. L. (2007). *Educação matemática crítica (Critical mathematics education)*. Argumentum Editora.
Araújo, J. L. (2009). Uma abordagem sócio-crítica da modelagem matemática: A perspectiva da educação matemática crítica (A socio-critical approach to mathematical modeling: A critical mathematics education perspective). *Alexandria, 2*, 55–68.
Avci, B. (2018). *Critical mathematics education: Can democratic mathematics education survive under neoliberal regime?* Brill and Sense Publishers.
Azevedo, S. S. (2019). *Educação financeira nos livros didáticos de matemática nos anos finais do ensino fundamental (Financial education in mathematics textbooks in the final years of elementary school)*. Master's thesis, Federal University of Pernambuco (UFPE).
Baroni, A. K. C., Hartmann, A. L. B, & Carvalho, C. C. S. (Eds.). (2021). *Uma abordagem crítica da educação financeira na formação do professor da matemática (A critical approach to financial education in mathematics teacher education)*. Editora Appris.
Barros, D. D. (2021). *Leitura e escrita de mundo com a matemática e a comunidade LGBT+: As lutas e a representatividade de um movimento social (Reading and writing the world with mathematics and LGBT+ commmunity: The struggles and representation of a social movement)*. Doctoral dissertation, Universidade Estadual Paulista (Unesp).

---

cal mathematics education are shared by many studies regarding mathematical modelling, as, for instance, Jacobini (2004); and Malheiros (2002). Soares (2013) presents an overview of studies (mostly by Brazilian authors),which are significant for critical mathematics education. Her list of references includes around 100 titles.

[17] Let me just mention some recent publications which explicitly refer to critical mathematics education in their titles: *Critical Mathematics Education: Can Democratic Mathematics Education Survive Under Neoliberal Regime?* (Avci, 2018); *Applying Critical Mathematics Education* (Andersson & Barwell, 2021); *Critical Mathematics Education and Climate Change* (Steffensen, 2021); and *Critical Mathematics Education: Theory, Praxis, and Reality* (Ernest et al., 2015).

References

Biotto Filho, D. (2015). *Quem não sonhou em ser um jogador de futebol? Trabalho com projetos para reelaborar foregrounds (Who never dreamed of being a soccer player? Working with projects for elaborating foregrounds)*. Doctoral dissertation, Universidade Estadual Paulista (Unesp).

Campos, I. S. (2013). *Alunos em ambientes de modelagem matemática: Aaracterização do envolvimento a partir da relação com o background e o foreground (Students in mathematical modeling environments: Characterisation of involvement from the relationship with background and foreground)*. Master's thesis, Universidade Federal de Minas Gerais (UFMG).

Campos, I. S. (2018). *A divisão do trabalho no ambiente de aprendizagem de modelagem matemática segundo a educação matemática crítica (The division of labor in the mathematical modeling learning environment according to critical mathematics education)*. Doctoral dissertation, Universidade Federal de Minas Gerais (UFMG).

Carrijo, M. (2023). *Beyond borders and racism: Towards an inclusive mathematics education for immigrant students*. Doctoral dissertation, Universidade Estadual Paulista (Unesp).

Civiero, P. A. G. (2016). *Educação matemática crítica e as implicações sociais da ciência e da tecnologia no processo civilizatório contemporâneo: Embates para formação de professores de matemática (Critical education and the social implications of science and technology in the contemporary civilizing process: Conflicts for mathematics teachers education)*. Doctoral dissertation, Universidade Federal de Santa Catarina.

Civiero, P. A. G., & Oliveira, F. P. Z. (2020). Landscapes of investigation and scientific initiation: Possibilities in civilizatory equation. *Revista Acta Scientiae, 22*, 165–185.

Civiero, P. A. G., Milani, R., Lima, A. S., & Lima, A. S. (Eds.). (2022a). *Educação matemática crítica: Múltiplas possibilidades na formação de professores que ensinam matemática (Critical mathematics education: Multiple possibilities in the education of teachers who teach mathematics)*. Editora da SBEM.

Civiero, P. A. G., Milani, R., Lima, A. S., & Miranda, F. O. (Eds.). (2022b). *Alçando voos com a educação matemática crítica: Discussões sobre a formação de professores que ensinam matemática (Get flying with critical mathematics education: Discussions about the education of teachers who teach mathematics)*. Editora IFC.

Damerow P., Elwitz U., Keitel C., & Zimmer J. (1974). *Elementarmatematik: Lernen für die praxis (Elementary mathematics: Learning for the practice)*. Ernst Klett.

Dewey, J. (1966). *Democracy and education*. The Free Press.

Dilthey, W. (2002). *The formation of the historical world in the human sciences. Selected works* (Vol. 3). Princeton University Press.

Ernest, P., Sriraman, B., Ernest, N., & N. (Eds.). (2015). *Critical mathematics education: Theory, praxis, and reality*. Information Age Publishing.

Faustino, A. C. (2018). *"Como você chegou a esse resultado?": O processo de dialogar nas aulas de matemática dos anos iniciais do Ensino Fundamental. ("How did you get to this result?": The dialogue in mathematics classes of the early years of Elementary School)*. Doctoral dissertation, Universidade Estadual Paulista (Unesp).

Frankenstein, M. (1983). Critical mathematics education: An application of Paulo Freire's epistemology. *Journal of Education, 165*(4), 315–339.

Freire, P. (2000). *Pedagogy of the oppressed* (With an Introduction by Donaldo Macedo). Bloomsbury.

Freire, P. (2021). *Pedagogia do Oprimodos, 80ª Edição (Pedagogy of the oppressed, 80th Edition)*. Paz e Terra.

Giroux, H. A. (1989). *Schooling for democracy: Critical pedagogy in the modern age*. Routledge.

Habermas, J. (1971). *Knowledge and human interests*. Beacon Press.

Ince, J. (n.d.). *Belief in African Americans as Learners in Mathematics (BAM)*. Doctoral dissertation, Universidade Estadual Paulista (Unesp) (in preparation).

Jacobini, O. R. A. (2004). *Modelagem matemática como instrumento de ação política na sala de aula (Mathematical modeling as an instrument of political action in the classroom)*. Doctoral dissertation, Universidade Estadual Paulista (Unesp).

Knijnik. G. (1996). *Exclução e resistência: Educação matemática e legitimidade cultural (Exclusion and resistance: Mathematics education and cultural legitimacy)*. Artes Médicas.

Knijnik, G. (2007). Mathematics education and the Brazilian landless movement: Three different mathematics in the context for the struggle for social justice. *Philosophy of Mathematics Education Journal, 21*, 1–18.

Knijnik, G., & Wanderer, F. (2015). Mathematics education in Brazilian rural areas: An analysis of the public policy and the Landless Movement Pedagogy. *Open Review of Educational Research, 2*, 143–154.

Knijnik, G., & Wanderer, F. (2018). Broadening school mathematics curriculum: The complexity of teaching mathematical language games of different forms of life. In K. Yasukawa, A. Rogers, A. K. Jackson, & B. Street (Eds.), *Numeracy as social practice: Global and local perspectives* (pp. 121–132). Routledge.

Lempert, W. (1971). *Leistungsprinzip und Emanzipation (Performance and emancipation)*. Suhrkamp.

Lima, L. F. (2015). *Conversas sobre matemática com pessoas idosas viabilizadas por uma ação de extensão universitária (Conversations about mathematics with elderly people made possible by a university extension action)*. Doctoral dissertation. Universidade Estadual Paulista (Unesp).

Lima, A. S. (2018). *A relação entre conteúdos matemáticos e o campesinato na formação de professores de matemática (The relationship between mathematical content and the peasantry in the training of mathematics teachers)*. Doctoral dissertation, Universidade Federal de Pernambuco (UFPE).

Lima, P. C. (2022). *Imaginação pedagógica e educação inclusiva: Possibilidades para a formação de professores de matemática (Pedagogical imaginations and inclusive education: Possibilities for mathematics teacher education)*. Doctoral dissertation, Universidade Estadual Paulista (Unesp).

Lima, A. S., & Lima, I. M. S. (2019). Diálogo, investigação e criticidade em um curso de licenciatura em educação do campo (Dialogue, investigation and critique in a degree course in rural education). *Revista de Matemática, Ensino e Cultura, 14*, 67–79.

Malheiros, A. P. (2002). *A produção matemática dos alunos em ambiente de modelagem (Students' mathematical production in a modeling environment)*. Master's thesis, Universidade Estadual Paulista (Unesp).

Marcone, R. (2015). *Deficiencialismo: A invenção da deficiência pela normalidade. (Deficiencialism: The invention of disability by normality)*. Doctoral dissertation, Universidade Estadual Paulista (Unesp).

Mellin-Olsen, S. (1977). *Læring som sosial prosess (Learning as a social process)*. Gyldendal.

Mellin-Olsen, S. (1987). *The politics of mathematics education*. Reidel Publishing Company.

Milani, R. (2015). *O processo de aprender a dialogar por futuros professores de matemática com seus alunos no estágio supervisionado (The process of prospective mathematics teachers learning to dialogue with their students during their supervised internship)*. Doctoral dissertation, Universidade Estadual Paulista (Unesp).

Milani, R., Civiero, P. A. G., Soares, D. A. & Lima, A. S. (2017). O diálogo nos ambientes de aprendizagem nas aulas de matemática (Dialogue in learning environments in mathematics classrooms). *Revista Paranaense de Educação Matemática, 6*(12), 221–245.

Miranda, F. O. (2015). *A inserção da educação matemática crítica na escola pública: Aberturas, tensões e potencialidades (The insertion of critical mathematics education in the public school: Openings, tensions and potentialities)*. Doctoral dissertation, Universidade Estadual (Unesp).

Mollenhauer, K. (1973). *Erziehung und Emanzipation (Education and emancipation)*. Juventa Verlag.

Mollenhauer, K. (2013). *Forgotten connections: On culture and upbringing*. Routledge.

Moura, A. C. (2020). *O Encontro entre surdos e ouvintes em cenários para investigação: Das incertezas às possibilidades nas aulas de matemática (Meeting amongst deaf and hearing student in landscapes of investigation: From uncertainties to possibilities in mathematics educations)*. Doctoral dissertation, Universidade Estadual Paulista (Unesp).

Münzinger, W. (Ed.). (1977). *Projektorientierter Mathematikunterricht (Project-oriented mathematics education)*. Urbau and Schwarzenberg.

Muzinatti, J. L. (2018). *A "verdade" apaziguadora na educação matemática: Como a argumentação de estudantes de classe média pode revelar sus visão acerca da injustiça social ("Truth" as a pacifier in mathematics education: How the argumentation of middle-class students can reveal their views on social injustice)*. Doctoral dissertation, Universidade Estadual Paulista (Unesp).

Muzinatti, J. L. (2022). *Matemática: Verdade apaziguadora (Mathematics: Soothing truth)*. Appris.

Negt, O. (1968). *Soziologische Phantasie und exemplarisches Lernen. Zur Theorie der Arbeiterbildung (Sociological imagination and exemplary learning: On the theory of workers' education)*. Europäische Verlagsanstalt.

Oliveira, M. (2015). *Laboratório de investigação matemática e educação financeira (Landscapes of investigaion and financial education)*. Master's Thesis, Universidade Federal de Juiz de Fora (UFJF).

Oliveira, L. P. F. (in progress). *Ler e escrever o mundo com matemática: Explorando possibilidades com estudantes dos anos finais do Ensino Fundamental (Reading and writing the world with mathematics: Exploring possibilities with middle school students)*. Doctoral dissertation, Universidade Estadual Paulista (Unesp).

Roncato, C. R. (2021). *Significado em educação matemática e estudantes com deficiências: Possibilidades de encontros de conceitos (Meaning in mathematics education and students with disabilities: Possibilities and meaning between concepts)*. Doctoral dissertation, Universidade Estadual Paulista (Unesp).

Sampaio, L. O. (2010) *Educação Estatística Crítica: Uma Possibilidade? (Critical Statistical Education: A Possibility?)*. Master's thesis, Universidade Estadual Paulista (Unesp).

Scargion, M. P. (in progress). *Ler e escrever o mundo com matemática: A condição de vida de pessoas idosas e a educação matemática (Reading and writing the world with mathematics: Elderly peoples' life conditions and mathematics education)*. Doctoral dissertation, Universidade Estadual Paulista (Unesp).

Silva, A. G. O. (2005). *Modelagem matemática: Uma perspectiva voltada para a educação matemática crítica (Mathematical modeling: A perspective of critical mathematics education)*. Master's thesis, Universidad Estadule de Londrina.

Silva, A. A. (2013). *Os artefatos e mentefatos nos ritos e ceremonias do danhono (The artifacts and mental facts in the rites and ceremonies of the danhono)*. Doctoral dissertation, Universidade Estadual Paulista (Unesp).

Silva, E. B. (2014). *O diálogo entre diferentes sujeitos que aprendem e ensinam matemática no contexto escolar dos anos finais do ensino fundamental (The dialogue between people who learn and teach mathematics in the final years of elementary school)*. Doctoral dissertation, Universidade Federal de Brasilia (UNB).

Silva, G. H. G. (2016). *Equidade no acesso e permanência no ensino superior: O papel da educação matemática frente ás políticas de ações afirmativas para grupos sub-representados (Equity in access and permanence in higher education: The role of mathematics education in relation to affirmative action policies for underrepresented groups)*. Doctoral dissertation, Universidade Estadual Paulista (Unesp).

Silva, A. A., & Severinho Filho, J. (Eds.). (2021). *Etnomatemática: Os múltiplos olhares para os saberes locais (Ethnomathematics: The multiple viewa at local knowledge)*. Editora Tangará.

Silva, G. H. G., Lima, I. M. S., & Rodríguez, F. A. G. (Eds.). (2021). *Educação matemática crítica e a (in)justiça social: Práticas pedagógicas e formação de professores (Critical mathematics education and social (in)justice: Pedagogical practices and teacher education)*. Mercado de Letras.

Skovsmose, O. (1980). *Forandringer i matematikundervisningen (Changes in mathematics education)*. Gyldendal.

Skovsmose, O. (1981a). *Matematikundervisning og kritisk pædagogik (Mathematics education and critical pedagogy)*. Gyldendal.

Skovsmose, O. (1981b). *Alternativer og matematikundervisningen (Alternaitves in mathematics education)*. Gyldendal.
Skovsmose, O. (1984). *Kritik, undervisning og matematik (Critique, education, and mathematics)*. Lærerforeningernes Materialeudvalg.
Soares, D. A. (2008). *Educação matemática crítica (Criticla mathematics educaion) XII Encontro Brasileiro de Estudantes de Pós-graduação em Matematica*. Universidade Estadual Paulista (Unesp).
Soares, D. A. (2013). *O Ensino de matemática em um perspective crítica: Dimensões teóricas e acadêmicas (Teaching mathematics in a critical perspective: Theoretical and academic dimensions)*. Novas Edições Acadêmicos.
Soares, D. A. (2022). *Sonhos de adolescentes em desvantagem social: Vida, escola e educação matemática (Dreams of socially disadvantaged teenagers: Life, school and mathematics education)*. Doctoral dissertation, Universidade Estadual Paulista (Unesp).
Souza, M. H. R. (2021). *A Matemática em ação no curso técnico de nível médio em agroecologia do SERTA (Mathematics in action in the high-level technical course in agroecology at SERTA)*. Master's thesis, Universidade Federal de Pernambuco (UFPE).
Souza-Carneiro, D. V. (2021). *A matemática em ação no ensino superior: Possibilidades por meio do problem-based learning (Mathematics in action in higher education: Possibilities through problem-based learning)*. Doctoral dissertation, Universidade Estadual Paulista (Unesp).
Steffensen, L. (2021). *Critical mathematics education and climate change: A teaching and research partnership in lower-secondary school*. Molvik AS, Western Norway University of Applied Sciences.
Vithal, R., Christiansen, I. M., & Skovsmose, O. (1995). Project work in university mathematics education: A Danish experience: Aalborg University. *Educational Studies in Mathematics, 29*(2), 199–223.
Volk, D. (1975). Plädoyer für einen problemorientierten Mathematikunterricht im emanzipatorischer Absicht (Plea for a problem-oriented mathematics education with an emancipatory interest). In M. Ewers (Ed.), *Naturwissenschaftliche Didaktik zwischen Kritik und Konstruction. (Natural science education between critique and construction)* (pp. 203–234). Belz Verlag.
Volk, D. (Ed.). (1979a). *Kritische Stichwörter zum Mathematikunterricht (Critical keywords for mathematics education)*. Wilhelm Fink Verlag.
Volk, D. (1979b). *Handlungorientierende Unterrichtslehre am Beispiel Mathematikunterricht. Band A (Action-oriented teaching using the example of mathematics education. Volume A)*. Päd Extra Buchverlag.
Volk, D. (1980). *Zur Wissenschaftstheorie der Mathematik: Handlungorientierende Unterrichtslehre. Band B (On the philosophy mathematics: Action-oriented teaching. Volume B)*. Päd Extra Buchverlag.
Wagenschein, M. (1965/1970). *Ursprüngliches Verstehen und exaktes Denken, I–II (Original understanding and exact thinking, I–II)*. Ernst Klett Verlag.
Wright Mills, C. (1960). Letter to the new left. *New Left Review, 5*.
Young, R. E. (1989). *A critical theory of education: Habermas and our children's future*. Harvester Wheatsheaf.
Zweig, S. (2000). *Brazil: A land of the future*. Ariadne Press.

Chapter 14
A Dialogic Theory of Learning Mathematics

Abstract As a preliminary for outlining a dialogic theory of learning mathematics, I make a few comments about the notions of theory, dialogue, and critique. I see learning as an interactive process, and in order to highlight this, I describe different learning interacts, namely getting in contact, exploring, positioning, foregrounding, externalising, and doubting. The clarification of these interacts constitutes my attempt to formulate a dialogic theory of learning mathematics. Predominantly, I do not see such a theory as being first of all a way of describing what actually might take place in a classroom setting, but rather as a way to conceptualise new educational possibilities.

Keywords Dialogic theory of learning mathematics · Dialogic act · Learning interact · Getting in contact · Exploring · Positioning · Foregrounding · Externalising · Doubting

Much has been written about "the social turn" in mathematics education research. This expression refers to a move away from looking at learning processes as intrinsic, epistemic processes towards looking at them as social interactions.[1] I want to suggest a "dialogical turn" in mathematics education. With this expression, I refer to a move towards paying particular attention to dialogic processes in the leaning of mathematics. As a contribution to this turn, I outline a dialogic theory of learning mathematics.[2] Such a theory is of particular interest for critical mathematics education.

The genetic epistemology, as presented by Jean Piaget, depicts the processes of assimilation and accommodation as definitive for learning processes, which thereby are seen as taking place in the individual. Lev Vygotsky introduces a social

---

[1] Lerman (2000) has pointed out the significance of this turn for mathematics education.
[2] In Chap. 5, "To Learn or Not to Learn", I have presented different conceptions relevant for interpreting learning phenomena, namely action, intention, foreground, motive, mathematics, and lifeworld. These notions all contribute to the formulation of a dialogic theory of learning mathematics.

perspective on learning by seeing processes as interactions. The social turn in mathematics education research is marked by a growing interest in Vygotsky's work among mathematics education researchers. I am paying particular attention to dialogic processes, which are particular forms of social processes. One can see the dialogic turn in mathematics education as inspired by the work of Paulo Freire.

As a preliminary to outlining a dialogic theory of learning mathematics, I make a few comments about my conceptions of theory, dialogue, and critique. Thereafter I concentrate on the notion of *learning interacts*, which are the foundations of my attempt to formulate a dialogic theory of learning mathematics.

14.1 Theory

In the article "Purposes and Methods of Research in Mathematics Education", Alan H. Schoenfeld (2000) points out some criteria that research in mathematics education should satisfy. These criteria, among other things, concern descriptive accuracy, explanatory power, predictive capacities, falsifiability, and multiple sources of evidence (triangulation). Such criteria worry me: they look threatening. Would my intended formulation of a dialogic theory of learning mathematics ever be able to meet such criteria?

From where might Schoenfeld have gathered such criteria? In *Aspect of Scientific Explanations*, Carl G. Hempel (1965) characterises the aim of a scientific theory assuming the outlook of logical positivism. According to this outlook, any kind of scientific theories addressing physical, chemical, biological, anthropological, sociological, pedagogical, or psychological phenomena can be brought together in a unified theoretical structure. Scientific explanations are of the same format for all disciplines. Scientific theories aim to provide descriptions of reality, and such theories should be possible – in principle – to verify or to falsify.[3] By means of scientific theories, one can provide explanations which, according to logical positivism, is the same as being able to make predictions.[4] In 1961, Ernest Nagel published *The Structure of Science: Problems in the Logic of Scientific Explanations*, which proliferates the outlook of logical positivism tremendously. It became common sense to claim that a scientific theory should be objective by describing reality as it is, and that it should be neutral by eliminating subjective factors from scientific theorising. The process of triangulation came to be recommended, referring to the process of looking at the same thing from different angles.

Schoenfeld's criteria are well-founded in the philosophy of science, but rooted in a particular view on science. An important move beyond the positivist outlook with respect to social theorising was taken by Charles Wright Mills (1959) when he published the book *Sociological Imagination*. He highlighted that the principal purpose

---

[3] See also Hempel (1959).

[4] In *The Logic of Scientific Discovery*, first published in German in 1934, *Karl* Popper (1965) suggests we should operate with a principle of falsification instead of a principle of verification.

of formulating a sociological theory cannot only be to provide descriptions; it is also to provide resources for formulating alternatives to actual states of affairs. By making this claim, Wright Mills hugely distanced himself from logical positivism and related conceptions of science.

A theory of learning mathematics need not only be preoccupied with capturing what is actually taking place in a classroom setting, but also with identifying new forms of educational settings. I am interested in trying to formulate a dialogic theory of learning mathematics, not only concentrating on what is taking place in the context of learning, but also showing what could come to take place. In this sense, I try to formulate a theory of learning that contributes to pedagogical imagination.[5]

14.2 Dialogue and Critique

Dialogue is an open concept, applied in different contexts, and at times with contradictory interpretations. In *Dialogue and Learning in Mathematics Education*, Alrø and Skovsmose (2004) have tried to characterise what we mean by dialogue. We do so by describing eight specific forms of interaction, which we refer to as *dialogic acts*.

The first dialogic act is *getting in contact*, meaning establishing emotional contact and showing interest in each other. Alrø and I highlight that getting in contact "means tuning in to each other in order to prepare for co-operation" (p. 62). The dialogic act of *locating* refers to attempts to grasp the overall concern of the other. It refers to the clarification of topics and concerns of the conversation. A teacher can try to locate the student's perspective by examining, for instance, how the students "understand a certain problem" (p. 62). *Identifying* is a process through which one tries to be more specific about the issues one is addressing. With reference to mathematics, one can try to identify, say, which equation one is going to investigate: "When the student is able to express his or her perspective, then it can possibly be identified in mathematical terms – identified not only by the teacher, but also by the student. Thus, the process of identification will provide a resource for further inquiry" (p. 61).

Advocating means "putting forward ideas or points of view not as absolute truth, but as something that can be examined" (p. 63). It means "proposing arguments for a certain position, but it does not mean sticking to a particular position" (p. 63). *Thinking aloud* occurs when one shows others how one is reasoning. We are dealing with a process of making thoughts not only audible, but also visible, maybe in the form of figures and sketches. *Reformulating* serves an important role in coming to understand each other better; thus, perspectives can be reformulated by the teacher in order to "understand what the student says" (p. 63). Furthermore, reformulations can be practised by the students "in order to check out his or her understating of the teacher perspective" (p. 63). *Challenging* means questioning a certain statement or

---

[5] For a careful discussion of pedagogical imagination, see Lima (2022).

perspective, but it does not mean attacking the other person. In challenging, the teacher could "play the role of opponent as well as the role for a partner" (p. 64). Challenging is important for experiencing the strength of advocating: "Furthermore, it is important that the teacher [...] is ready to be challenged. Making challenges can happen both ways" (p.64). *Evaluating* is an important dialogic act. It could address questions like: Did the teacher and student see the same problem? Did they try to solve it in the same way? The point of an evaluation is "not to obtain the 'right perspective', but to obtain a shared responsibility for the inquiry process" (p. 63). Having these dialogic acts in mind, Alrø and I interpret a dialogue as being a communicative process, rich in such acts.[6]

Critique is often taken as something negative. It could be, but it need not be so. A critique of literature, films, theatre, music, and art may highlight strong and exceptional features of what becomes evaluated. Naturally, it could also point out weaknesses and defects. When addressing mathematics and its applications critically, it is both positive and negative issues that I have in mind.

Dialogue and critique are closely related activities. I see dialogic interactions as being an important resource for producing a critique. This is the reason that I consider a dialogic theory of learning mathematics as being important for critical mathematics education.[7]

14.3 Learning Interacts

Learning is a complex and integral process of interaction. I will try to enumerate some particular features of such a process in terms of *learning interacts*. The enumeration is directly inspired by Alrø's and my identification of dialogic acts. The learning interacts I find relevant for formulating a dialogic theory of learning mathematics.

I use the expression "interact" in order to highlight that I see leaning as being first of all a social process, not an individual undertaking. When learning, students might interact with teachers, other students, or with other resources in order to gain knowledge. I do not try to contrast learning with teaching, as I see the two processes as integrated, and much of the time teaching simply means learning. In what follows, I will just talk about learning whenever referring to shared teaching-learning processes. Furthermore, my formulations reflect the idea that dialogic interaction is a resource for critical activities.

---

[6] Faustino and Skovsmose (2020) have characterised eight non-dialogic acts: ignoring, disqualifying, and lecturing being some of them. When analysing communicative processes, it is important to be aware of both dialogic and non-dialogic acts. In a mathematics classroom, operating according to the school mathematics tradition, we can identify many non-dialogic acts. See also the related discussion of inquiry gestures in Milani and Skovsmose (2014).

[7] For exploring the close relationship between dialogue and critique, see also the Chap. 12, "Critique of Critique".

14.3.1 Getting in Contact

Getting in contact was one of the dialogic acts that Helle Alrø and I paid attention to. This very same act I find to be an important learning interact as well. In order to establish a learning process, it is important to tune in to each other.

We are dealing with a rather intuitive notion; however, it is possible to relate "getting in contact" to more profound philosophical ideas. As an illustration, let me refer to Martin Buber's (1996) book *I and Thou*, first published in German in 1923. Buber finds that the interrelationship between an "I" and a "Thou" is crucial for establishing human relationships: "All actual life is encounter" (p. 62). Buber contrasts an I-Thou relationship with an I-It relationship. An I-It relationship is a relationship between a person and a thing; however, it could be between a person and another person who is considered to be a thing. Getting in contact is a way of moving away from an I-It relationship, and trying to establish an I-Thou relationship.

A learning theory might consider the students to be "things", or half-empty containers that need to be filled with knowledge. The students might also consider a teacher a thing that might be difficult to cope with. When creating dialogic relationships, it is important to establish I-Thou relationships, and one can think of getting in contact as an initial step in establishing such an relationship. While Buber highlights that "life in encounter", I want to highlight that "learning is encounter".

The act of getting in contact is an indication of being ready to listen to the other. Roger and Farson (1969) highlight that active listening is so-called because the listener has a responsibility: "He does not passively absorb the words which are spoken to him. He actively tries to grasp the fact and feeling in what he hears..." (p. 481). Listening does not only consist of registering what the other person is saying, it also means trying to grasp the perspective of the other. It means trying to assume, although temporary, the point of view of the other. Getting in contact is part of the preparation for active listening, and consequently or establishing learning encounters.

14.3.2 Exploring

John Dewey, one of the exponents of progressive education, paid particular attention to research processes that were associated with the natural sciences.[8] He found that learning could come to reflect basic features of such processes by engaging students in formulating questions and hypotheses, in collecting information and data, and in trying out the hypotheses.

Let me present an example of children, around 7 years old, doing explorations. The example is from a Danish context, where I was cooperating with teachers to implement project work in primary mathematics education. One project we referred

---

[8] See, for instance, Dewey (1966, 1974).

to as *The Playground*. The school was planning to construct a small playground right outside the children's classroom. The children had opinions, and naturally they knew about several different playgrounds. But what was a good playground? The mathematics teacher suggested that they could try out different playgrounds, and in this way find out which one they liked the best. The class went on several excursions around the city. They tried the swings; they tried the slides; they tried the carousels. But what then? How could they explain why the good swings are good? The children got the idea. We could measure! The teacher brought several measuring tapes and also a folding ruler, as used by a carpenter.

The children measured. But what should they measure, and how should they do it? For instance, it was not obvious to many of the children that the start for the tape measure should be put next to the start of what to measure. And there were numbers to read. It became obvious, also to the children themselves, that they needed to learn how to measure. Back home in the classroom they started to measure everything: the length of a table, the height of a chair, the length of a shoe, the height of the teacher, the length of the classroom. They went for a new excursion when they were ready to measure everything. Then came a new problem. How could they remember what had been measured? They brought some small notebooks where they wrote down many numbers. Back in the classroom, they realised that it was not so easy to remember what the different numbers referred to. However, in the end the playground was constructed, and all the children found that this was the best one they had ever tried.

During the project work, the children engaged in different explorations, all of them expressions of collaborative work. Trying out the different playgrounds was a task for the whole class. Some liked one kind of swings; some preferred a different kind. Preference was constantly communicated. When doing the measuring, the children often worked in pairs or in small groups. How to use the measure band was a collective discussion. What number should they make a note of? And who were going to make the notes? The students' explorations were driven forward by dialogic interactions.

Between the investigations of playgrounds, the students also did office work. They worked with the techniques of measuring, and they had practice in adding up numbers. However, such periods of consolidation and routinising need not degenerate into the traditional classroom pattern. They were thought of as periods of "office work". The classroom was imagined to be an office landscape: for instance, like one could see in a bank. The children were sitting at their desks occupying themselves with office tasks, in this case with solving exercises. Like in any office, the voices were kept low. Just as the bank assistant might go and get a coffee when needed, the children could go and get a soft drink. When working, they could also enjoy a biscuit. They could listen to soft music, and at times the teacher played the guitar. Such periods of consolidation and routinisation are also part of processes of exploration.

Exploring includes many different activities, such as *presenting*, *listening*, *questioning*, and *advocating*. In many situations, students are asked to *present* the result of their group work. This might be part of a strategy for controlling the students. However, first of all I see presenting as an important component of the very process of learning. *Listening* is important for both students and teachers. It is important that the teacher grasps how possible new knowledge relates to what the students already know. It is important that the teacher listens to what the students express in the way of ideas, ambitions, and hopes for the future. It is important that listening takes an active format, where listening comes to form part of a shared construction of new insights. A teacher might *question* the students, and such questioning might be part of the teacher's control over the students. However, this form of question might be pathological, in the sense that the teacher is asking the students about something that the teacher already knows in advance. By questioning, I have in mind a concern for acquiring the knowledge that one wants to know. *Advocating* is one of the dialogic acts that Alrø and I referred to. One can think of a mathematical proof as an example of advocating. However, advocating refers to more general processes of trying out possible lines of argumentation. Advocating is an integral part of establishing new knowledge.

Exploring – including presenting, listening, questioning, and advocating – is of dialogical nature. When one presents something, one presents it for somebody. When listening, one is listening to somebody. When questioning, one might question opinions formulated by others. And when advocating, one does so in conversations with others.

14.3.3 Positioning

One can see learning as adding something to already existing knowledge. However, I do not see learning as being a smooth accumulation. A person can have knowledge about boiling an egg, a physical theory, buying on the internet, figuring out price reductions, calculating the volume of a cylinder, where to find a cheap restaurant, reading a map, etc. I consider knowledge being a patchwork, composed of different pieces of knowledge. A person can come to learn new things by adding to the patchwork; maybe also by revising what is in the patchwork. Some pieces might fit together, others might contradict each other, and the patchwork will definitely include holes and gaps. An overall homogeneity of knowledge is not to be expected.

A student's mathematical knowledge can also be considered a patchwork. Studies completed in Brazil have shown that children spending much of their time as street sellers manage perfectly well dealing with money, adding up prices, returning change, and estimating possible bargains. However, these very same children might have difficulties solving mathematics problems with the same content, when formulated in the context of a mathematics classroom.

Knowledge can be local, it can be temporary, and it can be mistaken. It could be that I, instead of talking about knowledge, should talk about *assumed* knowledge. This is in fact what I have in mind when talking about knowledge. However, as I do not assume the existence of some kind of "genuine knowledge", it does not make much sense to me to make a distinction between "knowledge" and "assumed knowledge". So I will just talk about knowledge.

This understanding of knowledge brings me to the concept of *positioning*, which refers to the process of relating some pieces of knowledge to some other pieces of knowledge. Positioning plays an important role in a learning process. It is important to try to relate what is learnt with what has previously been learnt. New knowledge might appear unrelated to what is already known. It might also appear to contradict what is known. Through positioning, one establishes a perspective on what is being added to the patchwork of knowledge. Compared to what is already known, the new knowledge might appear relevant, or doubtful, or transparent, or well-grounded, or isolated. Whatever we think of what we are learning, we do so by relating it to the knowledge we are already operating with. Positioning is a crucial learning interact that can provide a critical perspective on what we already know, as well as on what we come to know.

Having the learning of mathematics in mind, Eric Gutstein (2006) talks about three kinds of knowledge: community knowledge, classic knowledge, and critical knowledge. By community knowledge, he refers to knowledge that might exist in a distributed format in a community. It could, for instance, be knowledge about how to solve practical problems, about how to save money, and about how to treat certain illnesses. By classic knowledge, having mathematics in mind, Gutstein refers to what students might be presented with in the regular mathematics classroom, such as the knowledge defined in terms of the curriculum. By critical knowledge, Gutstein refers to conceptions about what to consider injustice and exploitation, and about how to change such situations. Critical knowledge is directed towards political actions. Gutstein finds all three kinds of knowledge to be important for the learning of mathematics. Positioning concerns the relationship between community knowledge, classic knowledge, and critical knowledge; it concerns relationships between all kinds of knowledge.

14.3.4 *Foregrounding*

The word "foreground" is a noun, but let us also think of it as a verb. What could "to foreground" then mean? Foregrounding could refer to processes of recognising opportunities in life. Foregrounding could also mean to direct the students' attention to the existence of possibilities for further study, and to indicate job possibilities that the students might not have not been aware of. Foregrounding can refer to the construction of new and attractive possibilities.

Foregrounding might also include the deconstruction of perceived possibilities. A student might come to recognise that some attractive possibilities are out of reach. A child might dream of becoming a professional football player, but when playing football with other children, they might recognise that it is too difficult to become a good player. They might come to see that the dream was an illusion. A child from a favela in São Paul might dream of studying medicine and becoming a doctor, but they might recognise this as being impossible, as they cannot imagine any economic

14.3 Learning Interacts

resources for coming to do further studies. Foregrounding might include both constructions and deconstructions of possibilities.

Foregrounding might reveal risks and dangers. A student might be afraid of trying to do further studies, as it might become necessary to borrow money, since continuing to work, say as a street seller, would be impossible. The student might fear that they will not be able to pass an exam and not be able to conclude their studies successfully, and therefore not be able to repay the loan. The process of becoming aware of such risks and dangers is also part of a foregrounding.

Students influence each other, families might bring suggestions and warnings, and teachers might give them new ideas. A foreground is not a stable entity: new views on possibilities can be established and new dangers can be recognised. Foregrounding can reach far beyond individual students. It can concern the future of the family, the neighbourhood, or society. Students' hopes and fears become shaped through dialogic interaction, playing an important role in both constructions and deconstructions of foregrounds.

Foregrounding is important for establishing motives for learning.[9] Let me illustrate with an example. Years ago, I visited a school in a poor neighbourhood in Barcelona. Many of the students came from immigrant families. I was invited to do something with the students, who were around 16 years old.

I asked them all to close their eyes, and sit without doing anything else. – Just sit with closed eyes, and imagine yourself in 10 years' time. Imagine that everything that you could dream of will come true. Whatever you want to do, you will come to do. Whatever you want to become, you become. You will succeed! All the students, around 25 in total, were sitting with closed eyes. They were concentrating. Sometimes, a smile flickered around their lips. – Just take your time and imagine. And they took their time. After a while I clapped my hands and asked them to open their eyes. Many looked a bit troubled with all the light. They looked around as if they had become shy. Some smiled a bit.

I asked if somebody would like to tell us about what they had been imagining. Nobody said anything. I was thinking about doing the exercise again, although this time with open eyes. I would ask them to imagine themselves in 10 years' time, but this time to consider what they most likely would come to do, realistically speaking. My further idea was to talk about how mathematics could relate to their future, the dreamy future as well as the realistic one.

After a while a student raised her hand; she was ready to tell me what she had been dreaming about: she wanted to become a hairdresser. She looked a bit ashamed when she told about it. Then a boy raised his hand: he was dreaming about becoming an electrician. He would like to go to people's houses to set up lamps and lights and such things. I was about to interrupt and tell them that they had misunderstood the task. The realistic dreams would come in the second round, first I wanted to listen to their wildest dreams. I had imagined them talking about becoming film

---

[9] She the Chap. 5, "To Learn or Not to Learn?" for further discussion of relationships between motives and foregrounds.

stars, famous football players, rich business people, maybe university professors. However, before I opened my mouth, I realised that in fact I had been listening to their wildest dreams. Water came to my eyes. Given their situation, becoming a hairdresser or an electrician was out of reach. Such possibilities could only be part of dreams with their eyes closed.

We started talking about mathematics. The boy who wanted to become an electrician asked me if I knew how an electrician finds out how much wire there is in a reel without having to unroll everything and measure it. I did not know, so he explained that the electrician would count the number of rounds, measure the diameter of the reel, then multiply the number of rounds with the diameter of the reel, and finally multiply the result by three. Then he knows how much wire there is in the whole reel roll. I was happy to get this clarification, and other students became interested in explaining what mathematics had to do with their dreams for the future.

This episode in Barcelona taught me several things. First it made me aware that dreams for the future might be put into cages. Dreams could be too hurtful to consider, as they could be out of reach. Dreams could tell us about all the limitations that the students experience. The episode also taught me the importance of relating what takes place in the classroom with the students' possible aspirations for the future. The student who had a dream of becoming an electrician told me that it was his mathematics teacher that had spoken with him about the electrician's technique of estimating the length of wire in a reel, and that this made him interested in becoming an electrician.

Foregrounding – whatever we are thinking of the construction of attractive possibilities, the deconstruction of possibilities, or the identification of risks and dangers – forms an integral part of a learning process. It might establish new motives for learning. Foregrounding is a social process, manifested through dialogues.

14.3.5 *Externalising*

Many times, it has been highlighted that it is important for students to internalise what they have learnt. I want to highlight that *externalising* is important as well. By externalising, I mean activities that reach beyond the personal learning processes; they can be activities that reach beyond the regular classroom activities.

After being engaged in project work, the students can organise a presentation for another class in the school, maybe for students younger than themselves. They can prepare a presentation for adults; it can be for their parents and grandparents. The students can prepare presentations for students from another school. A presentation can take the form of an exhibition of posters, ot the students can produce a video where they speak about their observations. All such activities are examples of externalising, seen as a learning interact.

Let me refer to a particular example where externalising played an important role. Years ago, Miriam Godoy Penteado – my wife – as well as some of her graduate students and I were invited to do some teaching at a school in Rio Claro. The group of students, 12 in total (between 13 and 16 years old), were put together from different classes in the school. They were considered to be the most problematic students in the school. The teachers could not cope with them; they ruined lessons; they were violent; some of them were about to be expelled from the school.

The first time the 12 students entered the classroom we were going to use, they passed through the door, all of them, at the same time with much screaming and pushing. We had chosen to use the computer room as we knew that working with computers might fascinate them. The majority of the students – in fact, perhaps all of them – had no access to computers at home. Seeing all the computers made them curious. Yes, they were in fact going to work with the computers, and there was a computer for each of them. We had prepared some activities for them, and we had concentrated on dynamic geometry. We did not choose a topic that could be related to their actual socio-economic reality. The reason for our choice of topic I will return to.

During one of the lessons, I was sitting next to Daniel (pseudonym). He showed me how a dynamic system could work. He had drawn a triangle and grasped one of its corners with the cursor. He moved the corner round, and round, and round again. The movements got faster and faster, and the sound by which he accompanied the movements got higher and higher. It was the hissing and whistling of a good old black locomotive. I had never seen something so dynamic on a screen before.

After we had enjoyed the sound and the dynamics of the locomotive for a while, I asked him if he had noticed the small numbers shown on the screen. He had not. I asked him to try to stop the whole machinery. It was not easy, the locomotive was up to full speed, but in the end Daniel managed. We could now clearly see the numbers. – What about adding up these three numbers? I suggested. – Adding up the numbers? Daniel looked horrified. – Do I have to write them down? I tried to calm him down by explaining that there was a calculator installed on the computer. He added up the numbers:180, so what? The locomotive started again. After a while he managed to bring it to a full stop again. I suggested that he should add up the numbers again. – 190, so what? – Try to do the addition again. Denial found that he had made a mistake, now he got 180. He looked at me, he looked at the 180. He started the machinery again, not so fast. He brought it to a stop, this time with less hissing and whistling. He added up the three numbers. He got 180. He looked at me.

What about adding up the other three numbers? he suggested. – Good idea, try this. The three other numbers showed the lengths of the sides in the triangle. He added up these numbers. He looked preoccupied; the result did not look so nice. He started to drag a corner of the triangle, but this time very slowly, and without sounds. He looked at the rolling numbers, and tried to follow how they changed. He concentrated. Concentrated until the bell rang. Then he and the other students scrambled out of the door, all at the same time.

Some lessons later, we suggested inviting another class with younger children; the 12 students could teach them some dynamic geometry. The students liked the idea, and during the next lesson they started preparing themselves. Then came the day, and the younger children entered the computer room. They were impressed by everything, particularly by being received by the 12 students known by all teachers and children at the school – and not for anything good. But here they appeared nice; they welcomed the younger children into the computer room. Also the younger children came to enjoy working with dynamic geometry.

This is an example of externalising. The 12 students did not only work with dynamic geometry focusing on their own learning; they were also learning in order to present things to others. The externalising was a collective and dialogical process.

The different learning interacts are connected, and let me make an additional remark about foregrounding. By working with dynamic geometry, we were not relating the content to the students' actual situations, nor trying to relate it to their own experienced problems. However, we were trying to relate the learning activities to features of their foregrounds. The students were not doing well in school. They did not know any mathematics; they had no idea either if they would be able to come to master some mathematics. This has implications for how they imagine what they could eventually do in the future. Their foregrounds might be severely curtailed. By working with a topic like dynamic geometry, the students might come to learn some mathematics, but maybe something even more important: they might come to recognise new things about their own capacities. These might well be capacities that have been hidden for their teacher, as well as for themselves, due to their problematic behaviour. By engaging in externalising, the foregrounding took a new turn.

14.3.6 Doubting

In classic epistemologies, knowledge has been associated with truth and certainty, and contrasted with doubting. Doubting has been put under suspicion. It has been assumed that knowledge can be developed as a homogeneous and coherent structure. Every bit and piece of new knowledge will somehow fit into the patterns for already obtained knowledge. Building up knowledge is like adding new pieces to an enormous jigsaw puzzle showing a panoramic picture.

I am questioning any such conception of knowledge, and I do not see doubt as something that needs to be eliminated, but rather as something that is important to cultivate. Doubting can be articulated through positioning. When the homogeneity of knowledge is questioned, the act of positioning becomes important. When knowledge is seen as a patchwork, one needs to consider how pieces of knowledge might relate to each other. Relating does not mean integrating, but considering to what extent some pieces might fit together, some might appear unrelated, and some might contradict each other. Positioning might imply that one comes to doubt what one already assumes as knowledge; it might also imply that new things one is learning appear doubtful. New constellations of positioning bring about new constellations of doubt.

I see doubting as an important learning interact. Doubting is part of the formation of knowledge. Doubting can be carried forward by questioning. Naturally one can ask oneself questions, but in general questioning is part of dialogic interactions.

With respect to mathematics, there are many questions to ask. One question often asked by students is: *Did we get the right result?* Related but broader questions are: *Did we use the right way of solving the problem? Did we use the right algorithm? Are there other ways of doing the calculations? Does the result look reasonable?* Such questions relate to processes for solving pre-defined exercises, but more questions can be raised with respect to broader problem-based investigations: *Can we address the problem by means of mathematics? Do we need to simplify the problem in order to use mathematics? Could the use of mathematics create new problems? Could it be that mathematics is not relevant for addressing the problem?* One can continue with more general reflections: *What types of problems might be relevant to address by means of mathematics? What risks might be associate to the use of mathematics? Can mathematics create dangers?* One can question the often assumed neutrality of mathematics: *Can mathematics be used to serve particular interests? What kind of power can be exercised through mathematics? Can mathematics be the carrier of particulate world-views?*[10]

Raising such questions brings about a constructive doubt with respect to mathematics. Doubting does not signify a demolishing of knowledge; it is part of a critical construction of knowledge.

14.4 Was That a Theory?

Learning is an interactive process. In order to highlight this, I have presented the following learning interacts: getting in contact, exploring, positioning, foregrounding, externalising, and doubting. But does such enumeration add up to a dialogic theory of learning? Definitely not if we consider how Schoenfeld characterises a theory.

However, if we consider that a theory of learning mathematics need not only be preoccupied with capturing what is actually taking place in a classroom setting, but also with conceptualising new possibilities, the situation might look different. I try to formulate a dialogic theory of learning that also might indicate educational possibilities. This is an ambition of a dialogic theory of learning mathematics.

References

Alrø, H., & Skovsmose, O. (2004). *Dialogue and learning in mathematics education: Intention, reflection, critique.* Kluwer Academic Publishers.
Buber, M. (1996). *I and thou.* Touchstone by Simon and Schuster.
Dewey, J. (1966). *Democracy and education.* The Free Press.

---

[10] See the Chap. 4, "Sustainability and Risks" for a presentation of these questions as related to a mathematical modelling process.

Dewey, J. (1974). *On education* (Selected writings, edited and with an introduction by Reginald D. Archambault). The University of Chicago Press.

Faustino, A. C., & Skovsmose, O. (2020). Dialogic and non-dialogic acts in learning mathematics. *For the Learning of Mathematics, 40*(1), 9–14.

Gutstein, E. (2006). *Reading and writing the world with mathematics: Toward a pedagogy for social justice.* Routledge.

Hempel, C. G. (1959). The empiricist criterion of meaning. In A. Ayer (Ed.), *Logical positivism* (pp. 108–129). The Free Press.

Hempel, C. G. (1965). *Aspects of scientific explanation.* The Free Press.

Lerman, S. (2000). The social turn in mathematics education research. In J. Boaler (Ed.), *Multiple perspectives on mathematics teaching and learning* (pp. 19–44). Ablex.

Lima, P. C. (2022). *Imaginação pedagógica e educação inclusiva: Possibilidades Para a formação de professores de matemática (pedagogical imaginations and inclusive education: Possibilities for mathematics teacher education).* Doctoral dissertation. Universidade Estadual Paulista (Unesp).

Milani, R., & Skovsmose, O. (2014). Inquiry gestures. In O. Skovsmose (Ed.), *Critique as uncertainty* (pp. 45–56). Information Age Publishing.

Nagel, E. (1961). *The structure of science: Problems in the logic of scientific explanations.* Hardcourt, Brace and World.

Popper, K. R. (1965). *The logic of scientific discovery.* Harper and Row.

Roger, C. R., & Farson, R. E. (1969). Active listening. In R. C. Huseman, C. M. Logue, & D. L. Freshley (Eds.), *Readings in interpersonal and organizational communication* (pp. 480–496). Holbrook.

Schoenfeld, A. H. (2000). *Purposes and methods of research in mathematics education.* Notices of the American Mathematical Society.

Wright Mills, C. (1959). *The sociological imagination.* Oxford University Press.

Part IV
Width and Depth

Chapter 15
All Students

Abstract Critical mathematics education is concerned about all groups of students. The students could be those socially at risk who experience oppression, exploitation, and social injustice. They could also be students in conformable positions, who live in affluent environments and do not experience any lack of economic resources. It is important that this group of students also addresses cases of social injustice and comes to participate in a broader discussion of what social justice could mean. Critical mathematics education is also concerned about university students in mathematics. They might become part of the expertise that will bring mathematics into action and contribute to the power that can be exercised through such actions. It is important that university students in mathematics do not get inveigled into a banality of mathematical expertise, but become prepared for addressing different features of the mathematics-power relationship.

Keywords Students at social risks · Students in conformable positions · University students in mathematics · Mathematics and power · Banality of mathematical expertise

Critical mathematics education is concerned about students facing social risks. They could be students living in poor neighbourhoods, maybe in one of the huge ghettos that are part of metropolises the world around, maybe in remote rural areas with little access to healthcare systems and other public services. They could be students subjected to racial discrimination, sexism, violence, or caste systems. They could be students suffering social exclusion due to homophobia or other forms of preconceptions. They could be refugee students sheltering in temporary camps far away from their homes. They could be street children trying to get through to the next day. What mathematics education could come to mean for any of these groups of students is as open question. I do not assume it to be possible to identify any general educational approach for addressing students facing social risks.

© The Author(s), under exclusive license to Springer Nature Switzerland AG 2023
O. Skovsmose, *Critical Mathematics Education*, Advances in Mathematics Education, https://doi.org/10.1007/978-3-031-26242-5_15

However, I see critical mathematics education as being concerned not only about students at social risks. It is concerned about all groups of students.[1] In order to illustrate this point, I will consider three groups of students: *Students at social risks* who most directly experience one or another kind of social injustice. *Students in conformable positions* who live in affluent environments and do not experience any lack of economic resources. And *university students in mathematics* that will come to bring mathematics into action and deal with the power that can be acted out through mathematics.

15.1 Students at Social Risks

Much literature in critical mathematics education, many times referred to as mathematics education for social justice, focusses entirely on students at social risks. A principal inspiration for making this focus comes from Paulo Freire (2000), whose whole educational enterprise concentrates on people in oppressed situations. Marilyn Frankenstein (1983) brings the inspiration from Paulo Freire into mathematics education, and in her formulation, critical mathematics education first of all engages with students at social risks. Eric Gutstein (2006), also directly inspired by Freire, engages students at social risks in reading and writing the world with mathematics.

As part of his PhD study, Denival Biotto Filho (2015) worked with children living in an orphanage, located in a city in the interior of the São Paulo State. The children's parents may have been involved in crime, drug abuse, or prostitution. Some might still be in jail, some might have disappeared. It was a mixed age-group Biotto Filho was working with, between the ages of 9 and 13. He engaged these students in different landscapes of investigation, one of which was called *Football*.

Biotto Filho and I, being the supervisor of his PhD study, initially felt little enthusiasm about this theme. Over time, football has been the subject of innumerable projects, and we did not see the possibility of much originality emerging from this theme. We were uncertain about what kinds of investigations could become interesting for the students to engage in. Would the project end up in the students just playing football? We speculated about what could be the critical potential of the project. What kind of socio-political insight might be shaped by this project work? How could we address issues about social justice? Anyway, the students, both girls and boys, voted for *Football*.[2]

In Brazil, many dreams among children and young people are associated with football, and several times Marta Vieira da Silva, widely known in Brazil, has been nominated by FIFA as the best female football player in the world. The famous

---

[1] In Skovsmose (2016), I have explored what critical mathematics education could mean for students at social risks, students in comfortable positions, disabled students, senior students, displaced students, university students in mathematics, and engineering students.

[2] In the following presentation on the project, I draw on Skovsmose (2016).

clubs in Brazil are always in search of new talent, and talented children can be spotted. There are many well-known stories about famous players that were originally spotted in poor neighbourhoods. A project like *Football* can be accompanied by many hopes and dreams, as well as by illusions.

Engaging with *Football* not only touched the students' imaginations, it was deeply rooted in their everyday knowledge. Like many people in Brazil, the students knew the names of a huge number of players, their positions in the field, when they made a spectacular goal, if they were about to move to a new club, and so on. Many people have their favourite team and like to wear a T-shirt of the team with the number of their favourite player. The project *Football* was grounded in a solid amount of common knowledge. This turned out to be an important energy resource for the further investigations.

The landscape *Football* broadened out as it was being explored and came to include a range of issues; one concerned the working conditions of professional players. A professional player, although not any famous one, from the local club visited the orphanage. He explained about the working day of a professional, and most of this was new to the students. He explained about doing exercises and physical training. He told about the importance of eating properly and sleeping adequately. It became clear that the life of a professional football player includes very many things other than playing football.

The professional player also explained about salaries, and it was revealed that more than half of the professional football players in Brazil earn less than the legal minimum salary. This came as a huge surprise. We all know about the glamorous life of certain professional players, but the reality for the majority of players is generally overlooked. Not only the small salaries were pointed out, but also the fact that the daily life of a professional can be tough, tiresome, and stressful. Players who are not considered good enough are not selected for the team, but they might become better and maybe become selected in the future. They have to wait and hope, and to do whatever possible in order to become better. We are dealing with the survival of the fittest, acted out in the most explicit way.

What the students might have associated with becoming a football player was modified and elaborated upon during the project. Some illusions were discarded. The students came to acknowledge the differences between what appears on the TV when big teams are playing, and what is taking place during the weekdays.

Results of the project *Football* were presented by the students to a mixed group, including some of Biotto Filho's colleagues, teachers, and people from the university. At the presentation, the students managed the microphone with the most natural competence. Not that they ever had tried to use such a thing before, but as part of the project work, where a tendency had been that too many talked at the same time, Biotto Filho introduced the use of a microphone. It was an empty plastic bottle from a soft drink, and if one wanted to say something, one needed to hold the microphone in their hand. In this way, the discussion became well structured, and this experience was useful for the students when doing the real presentation. Here the students managed perfectly well with the real microphone.

The presentation grabbed people by the heartstrings. The students were very well prepared. They welcomed everybody and introduced the project, they presented topics, and they answered questions from the audience. They spoke about different football teams, about different football stars, about conditions for becoming professional, and about the deplorable economic conditions of many players. The students' competent way of doing the presentation surprised those who attended. The presentation was conducted by a group of students from whom nobody had expected anything.

What might the students have got from working with *Football*? Did they learn some mathematics? The project did not conclude with any sort of systematic evaluation, so we can only guess about that. Working with *Football* might lead to a range of challenging mathematical questions. Considering the results of the football matches during the weekend brings about questions like: Now with only five rounds left, will it be possible for team A, almost at the bottom at the table, to escape relegation? How many points is it possible for the team A to get? Who will their opponents become? How many of the five matches will A be playing at its home ground? How big is the risk that A will be relegated? How great is the chance that team B, now at the top of the table, in fact will become the champion? In Brazil, when tournaments are about to finish, sports channels on the TV give estimations of such likelihoods, in terms of exact numbers. But where do such exact numbers come from? What kind of calculations? Any such questions can also be addressed in a mathematics classroom.

Learning mathematics concerns much more than acquiring mathematical insight; it also concerns attitudes towards mathematics. A devastating situation for many students at social risks is the difficulty in finding something to hope for. When the social, political, and economic conditions can produce nothing but hopelessness, learning gets obstructed, including the learning of mathematics. To many students at social risks, mathematics appears hostile, incomprehensible, and therefore threatening. The project *Football* might have contributed to a different image of mathematics. Mathematics might have come to appear as something that can be mastered. This is an important experience for students that find themselves to be marginalised, both in a social and a school context.

One can interpret the project *Football* as giving new perspectives on the future to the students coming into the orphanage. They have experienced being able to master mathematical ways of thinking. They have experienced mathematics being not just an obstacle, but also a learning possibility. The students might have experienced: "Yes, we can!"

15.2 Students in Comfortable Positions

Addressing cases of social injustice is not only important for students suffering such injustice. Addressing the consequences of poverty is not only important for students living in poor neighbourhoods. Addressing cases of racism is not only important for students subjected to racism. Addressing consequences of pollution is not only important for students living in polluted neighbourhoods. Such issues are important for all groups of students.

In order to illustrate that critical mathematics education needs to look broadly, I will show how students in comfortable positions can come to work with problems associated with poverty. The expression "students in comfortable positions" could refer to students whose parents are rich, or at least economically well off. These students belong to the well-protected layers of society. In Brazil, these students most likely go to private schools, which gives them an advantage in getting access to further education. They can expect to come to enjoy the social benefits that a good further education might provide. However, from the perspective of critical mathematics education, this group of students is also in need of developing a deeper conception of what social and environmental justice could mean and of what injustice could include.[3]

Naturally, one needs to be careful and not assume anything like: leftist ideologues are welcomed among low-income groups, while right-wing ideologues flourish in wealthier environments. The situation is much more complex. Around the world, one can observe neo-fascist or neo-liberal ideologies developing in different socio-economic settings. Consequently, critical mathematics education – when confronting totalitarian ideologies, working against oppressions and brutalities, and challenging any form of exploitation – might get into situations where it cannot operate in solidarity with the positions expressed by certain groups of students, but rather needs to confront such positions.

João Luiz Muzinatti (2022) has been working with students around 15 years old, who come from an upper-middle-class neighbourhood in São Paulo. His idea was to address how mathematics-based arguments can help to point out preconceptions broadly assumed among upper-middle-class students: for instance, concerning how much the state spends on welfare systems supporting people with low or no income. He tried to challenge general socio-political assumptions that often make up a traditional upper-middle-class outlook.

Muzinatti addressed a set of such assumptions related to *Bolsa Familia*.[4] This system of family support was introduced during the presidency of Luiz Inácio Lula da Silva (2003–2010) and extended during the presidency of Dilma Rousseff (2011–2016). But *Bolsa Familia* has met much resistance: Why give money to people that do not want to work? Why celebrate laziness in this way? Why give money to unproductive people, when the money could have been used much better in other ways? Why should we use our tax money for people not paying any taxes?

By opening up *Bolsa Familia* as a landscape of investigation, Muzinatti made it possible to formulate other questions like: How much money is involved in this programme? How many families do receive this support? How much of the total tax

---

[3] For a discussion of social justice, see Chap. 1, "Concerns and Hopes". For an introduction of the notion for environmental justice, see Chap. 4, "Sustainability and Risks".

[4] In the following presentation of the project *Bolsa Familia*, I draw on Skovsmose (2016).

paid in Brazil goes to *Bolsa Familia*? How much does the individual taxpayer contribute to *Bolsa Familia*? How much money is necessary for a family to spend on food? How does this number depend on the numbers of members of the family? What is the minimum budget for a families' total expenses? How does this minimum budget relate to the official minimum salary? What is the impact of inflation of food on a rich family compared to a poor family? How great a percentage of its income does a rich family spend on food compared to a poor family?

By working with *Bolsa Familia*, it was revealed that the common upper-middle-class critique of this social programme is not substantiated. Rather, this critique is part of discourses encompassing a range of stereotypes. By putting *Bolsa Familia* on the agenda, the students were invited to consider issues like economic inequalities and redistribution of welfare. With respect to students in comfortable positions, it is extremely important to establish discussions of what can be considered social injustices, as these students might be unaware of or indifferent to many such injustices. An important ambition of critical mathematics education is to challenge such indifference. Students in comfortable positions does not refer to a homogeneous group; it might include a variety of political positions, even of the most radical and progressive form. It is important to provide a space where different conceptions of social and environmental justice are put on the agenda.

15.3 University Students in Mathematics

It is important to critical mathematics education to prevent the formation of a banality of mathematical expertise among university students. By this banality, I refer to an expertise entirely focussed on intrinsic features of mathematics, while being oblivious of the socio-political context within which mathematics might be brought into action. I refer to an expertise that is oblivious to the ways and for what purposes the mathematics might be developed and used.[5]

To many university students, mathematics appears to be a unified and homogeneous structure, which it is their task to understand and to master. This image of mathematics is propagated by textbooks, lectures, lecture notes, and exercises with one and only one correct answer. This general pattern I refer to as the *university mathematic tradition*. Naturally, there are exceptions to this tradition, but it dominates many mathematical departments the world around. As an integral part of the university mathematics tradition, mathematics is presented as being an intrinsically valuable form of knowledge that is in no need of being critically addressed. This doctrine seems to legalise mathematics research and university studies fully concentrating on intrinsic mathematics issues. The studies become fully preoccupied with mathematical notions, theorems, theories, and techniques. This leads to a banality of mathematical expertise.

---

[5] See Chap. 7, "Banality of Mathematical Expertise", for a clarification of my use of this metaphor.

For critical mathematics education, it is crucial to confront the university mathematics tradition, and one way of doing so is to show that mathematics brought into action reflects a range of economic, technological, and political interests. This has implications both for the content of the university studies in mathematics and for the organisation of these studies. In order to make space for critical reflections on mathematics, it is not sufficient just to change the contents of textbooks, lectures notes, and lectures. The whole format of the university study programmes in mathematics needs to be changed. One possibility is to base the studies on project work. Since the middle of the 1970s, this possibility has been tried out at Roskilde University Centre and Aalborg University in Denmark, and over longer periods with success.[6]

As an illustration of what such project work in mathematics could include, let me refer to an example from Aalborg University. Every semester, the mathematics students were engaged in project work, on which they spent half of their study time, while the other half was occupied by course work. Both project work and course work were related to the overall theme of the semester. In their project work, a group of students had decided to investigate the North Sea Model, which is a mathematical model used for maximising the outcome of the fishing in the North Sea. The five students in the project group were in their third year of their mathematics studies, and I was the supervisor for this group.

The North Sea Model was developed by the Danish Institute for Fishery and Marine Research in order to understand the dynamic relationships between species of fish in the North Sea. Previous models can be characterised as one-species models trying to capture how one species does develop. The North Sea Model is a multi-species model trying to grasp the dynamic interactions that might exist between different species. For instance, the size of a species of predators will have an impact on the size of a species of prey.

The North Sea Model has important applications closely linked to the economic interests of the fishing industry in both Denmark and in England. By means of the model it appears possible to carry out experimental forecasting.[7] It seems possible to capture the potential impact of implementing particular systems of fishing quota before any such implementation has taken place. By means of systematic experimental forecasting, it seems to be possible to identify a fishing policy that ensures a maximum yield per year without causing over-fishing of the sea.

The students came to acknowledge the tremendous economic and political interests related to the construction of this mathematical model. The fishing industries in Denmark and England have different profiles. While the English give priority to catching some types of fish, the Danish put priority on other types. This difference has, for instance, an impact on the very construction of the fishing boats, their size, and their equipment. It turned out that the hypothetical quotes for fishing that would

---

[6] See Vithal et al. (1995). For broader presentations of problem-based learning as taking place at Aalborg University, see Jensen et al. (2019); and Kolmos et al. (2004).

[7] For an introduction of the notion of experimental forecasting, see the Chap. 4, "Sustainability and Risks".

maximise the yield from the North Sea – according to the North Sea Model – did fit the profile of the Danish fishing fleet, not the English.

In their project work, the students considered the nature of the quantification associated to the different parameters included in the model. There was much to consider, as the model was formed by a range of difference equations, more than 100 in total. The students classified the parameters into three types, according to how their values were determined. The values of some parameters were based on theoretical biological insight; the values of other parameters were based on empirical investigations; while the values of a third group of parameters were determined in an ad hoc manner. Ad hoc parameters were assigned a value between 0 and 1, referring to the likelihood that a fish from one species would eat a fish from another species when passing each other in the sea. There was no theoretical way nor empirical procedure to determine the values of the ad hoc parameters. Their values were only defined as part of the general adjustment of the model.

Some parameters of the model were identified by the students as being sensitive. These are parameters whose values have a significant impact on the output of the model. A small change in the value of a sensitive parameter will bring about a significant change in the output of the model. It turned out that many of the ad hoc parameters were highly sensitive. In this way, it was revealed how easy it could be to adjust the result of the experimental forecasting for particular interests. The students came to realise that the North Sea Model might be designed so that it came to fit Danish business priorities.

Even though a mathematical model might have a high degree of technicality, it might also include a high degree of arbitrarity. Mathematics is no guarantee for objectivity and neutrality. By studying the North Sea Model, the students recognised the power that can be acted out through a mathematical model, and that technical mathematical knowledge can also serve particular political interest. Such possibilities are not related specifically to the North Sea Model. We are dealing with a general phenomenon, which is important for university students in mathematics to experience in order not to be enrolled in a banality of mathematical expertise.[8]

References

Biotto Filho, D. (2015). *Quem não sonhou em ser um jogador de futebol? Trabalho com projetos para reelaborar foregrounds (Who never dreamed of being a soccer player? Working with projects for elaborating foregrounds)*. Doctoral dissertation. Universidade Estadual Paulista (Unesp).

---

[8] Since the 1970s, critique of mathematical modelling has been part of a critique of mathematics from the 1970s; see Chap. 13, "Initial Formulations of Critical Mathematics Education". Here the connections between a *fagkritik* and *modelkritik* (modelling critique) becomes pointed out. For a recent contribution on elaborating mathematical modelling as a critical activity, see Gibbs and Park (2022).

References

Frankenstein, M. (1983). Critical mathematics education: An application of Paulo Freire's epistemology. *Journal of Education, 165*(4), 315–339.

Freire, P. (2000). *Pedagogy of the oppressed* (With an Introduction by Donaldo Macedo). Blomsbury.

Gibbs, A. M., & Park, J. Y. (2022). Unboxing mathematics: Creating a culture of modelling as critique. *Educational Studies in Mathematics, 110*(1), 167–192.

Gutstein, E. (2006). *Reading and writing the world with mathematics: Toward a pedagogy for social justice*. Routledge.

Jensen, A. A., Stentoft, D., & Ravn, O. (2019). *Interdisciplinarity and problem-based learning in higher education: Research and perspectives form Aalborg University*. Springer.

Kolmos, A., Fink, F. K., & Krogh, L. (Eds.). (2004). *The Aalborg PBL model: Progress, diversity and challenges*. Aalborg University Press.

Muzinatti, J. L. (2022). *Matemática: Verdade apaziguadora (Mathematics: Soothing truth)*. Appris.

Skovsmose, O. (2016). What could critical mathematics education mean for different groups of students? *For the Learning of Mathematics., 36*(1), 2–7.

Vithal, R., Christiansen, I. M., & Skovsmose, O. (1995). Project work in university mathematics education: A Danish experience: Aalborg University. *Educational Studies in Mathematics, 29*(2), 199–223.

Chapter 16
Beyond Stereotypes

Abstract Research in mathematics education is guided by paradigmatic assumptions, concerning amongst other things: neatness, affluence, and placidity. In many research publications, the students appear well behaved, well equipped with the appropriate resources, and ready to work on the tasks set for them. The majority of research publications pay little attention to contexts marked by poverty and lack of resources, even though this is a most common situation. Violence due to criminality, upheavals, and war causes uncertainties and instabilities for many groups of people, but still research publications most often bring to the fore stories taking place in calm and peaceful teaching-learning environments. Neatness, affluence, and placidity appear to be taken as a paradigmatic given, not as issues that need to be critically addressed. Much research in mathematics education focusses on offering descriptions, quantitative or qualitative, of that which is taking place. To critical mathematics education, it is equally important to move beyond such descriptions and to formulate pedagogical imaginations, which conceptualise alternatives to what can be observed and described.

Keywords Research paradigm · Neatness · Affluence · Placidity · Prototypical classroom · Descriptive paradigm · Naturalistic inquiry · Pedagogical imagination

Assumptions, priorities, and paradigmatic stereotypes guide research, also in mathematics education. It is widely accepted that one can look at educational phenomena through different theoretical lenses, something which has been intensively discussed in writings about research methodology. This discussion is also crucial to critical mathematics education.

In order to illustrate what I have in mind when talking about stereotypes in mathematics education research, one need look no further than at articles presenting classroom episodes. When leafing through previous issues of prominent international journals, it is common to come across many scenes from mathematics classrooms and many transcriptions of conversations among students and between teachers and students. We all know that such transcriptions have been selected among many other possibilities, and that the full transcript of what really went on might have been much more extensive. One could assume that the selections and summaries of classroom episodes made for the purpose of publication are rather casual; however, this is not how I see it. I find that such selections and summaries might reflect some profound paradigmatic, but also highly questionable research priorities.

We are most often than not presented with well-behaved students and they seem eager to engage in the mathematics tasks presented to them. We do not often read about teachers being troubled by violent students. We do not listen to screams and raised voices. The classrooms appear as comfortable environments for both students and teachers; we are dealing with adequately resourced classrooms in terms of both materials and human personalities. If computers are needed for students to complete their tasks, they are available. There is no trace of violence in the neighbourhood around the school. We do not hear about fights between different criminal gangs haunting the neighbourhood. We do not hear about any security guards positioned at the school gates. Certainly, the classroom is not located close to any war zone.

These observations bring me to refer to the *prototypical mathematics classroom*: it has well-behaved students, it is well equipped, and it is located in a peaceful environment. The prototypical classroom is *neat, affluent,* and *placid.* Where, then, does this prototypical classroom exist? First of all, it exists in research literature in mathematics education. The expression "prototypical" is meant ironically, and I could as well have talked about the "stereotypical mathematics classroom". Often the prototypical classrooms are in stark contrast to real-life classrooms.

In 2008, I participated in the *Symposium on the Occasion of the 100th Anniversary of ICMI* that took place in Rome.[1] In one of the discussions during the symposium, I presented a conjecture. It was not a conjecture that I could justify, but one that could be refuted.[2] My conjecture was: while 90% of classroom research in mathematics education concentrate on well-behaved students in well-equipped classrooms, apparently located in peaceful environments, only 10% of the research addresses situations with difficult students, or poorly resourced classrooms, or violent environments. It was, in all honesty, just a guess, and it is still a prevalent guess as I have not come across any refutation.

It is a concern of critical mathematics education to look at all possible contexts for teaching and learning mathematics, and not to be trapped by assumptions of neatness, affluence, and placidity. Making careful descriptions is an important feature of research in mathematics education. However, to critical mathematics education, it is important also to move beyond descriptions and to create resources for a pedagogical imagination.

---

[1] See Menghini et al. (2008).
[2] In making my formulation, I kept in mind Popper's formulation of the logic of scientific development as following a pattern of conjectures and refutations.

16.1 Neat

When Helle Alrø and I were organising the empirical material for our book *Dialogue and Learning in Mathematics Education* (Alrø & Skovsmose, 2004), we had to make choices and decisions. We had observations from different classrooms, videos and tape recordings – an awful lot of data, so we needed to decide upon our priorities.

In one classroom, several groups of students had been working on some mathematical tasks. We had a video of what was taking place in the classroom, and a tape recorder had been placed on the tables of each group, so we could transcribe the conversations. We knew in advance that we were not going to use all of the data, but which of the group conversations should be transcribed? We were in a good position: almost all groups had worked enthusiastically on the tasks set for them. In fact, only one group had acted in a problematic way, where they disturbed the other groups, made sexist comments, and behaved in other inappropriate ways.

As anyone who has tried it knows, creating a transcription is a huge task, so we spent some time considering which group conversations we were to concentrate on. We were interested in the dialogues taking place among the students, and also between the teacher and the students. We tried to locate situations where dialogues led to clarifications and to the development of mathematical ideas, and which opened a path for critical reflections.

We simply took as a given that the problematic group could be put aside. No relevant transcriptions could come from that experience. Then we came to think again: Why? We were particularly concentrated on finding examples of dialogue leading to interesting clarifications, but why not make a transcription of the conversation in the problematic group? Yes, why not? We decided to do so and see what we might find. This was a touch-and-go, spur-of-the-moment decision, one that we might well not have made. It might not even have come to our mind that we might pay attention to this group – we could just have concentrated on what is usually considered as "relevant empirical material". We realised that we might easily have been captured by some implicit paradigmatic priorities with respect to the selection of data material. We might have opted for the "neat situations".

The group that initially appeared to create obstacles to mathematical development we came to refer to as the "resistance group" (Brian, Robert, Linda, and Martin, all pseudonyms).[3] When the camera was placed in the classroom, Brian announced with a loud voice: "Look, I can make faces!" Sure he could: he put both forefingers into his nostrils. Robert also looked into the camera and did the same. Then the boys looked at Linda: "Look at her! Pimples on her forehead!" Linda was attacked several times with very unpleasant sexist remarks, but she answered back. Martin remained relatively quiet, but he also contributed to the soundtrack for the tape recorder by swinging around a bunch of keys that clattered on the table in a regular rhythm. During the conversation one can hear remarks like "Take your fat tits with you!" followed by a "Shut up!" The "Shut up!" outburst could be heard

---

[3] The result of the transcription can be seen in Chap. 5 in Alrø and Skovsmose (2004).

several times. At other times, the conversation touched upon the mathematical task that had been set for them; however, nothing was taken seriously, but invariably used as an opportunity for jokes and ironic remarks.

Our book *Dialogue and Learning in Mathematics Education* came to include a careful presentation of what took place in the resistance group. It turned out that this material contributed to our elaboration of the notion of intentionality that – together with the notion of reflection – constitutes our understanding of what a critical activity may include.

With the exception of the "resistance group", provocative and disturbing students are normally not presented in research literature in mathematics education. Looking at books about classroom conversations, culture in the mathematics classroom, and inquiry-based learning, one meets students dedicated to engaging in mathematical reasoning. A similar neatness is found in several handbooks in mathematics education. Mathematics classroom appears to be without disruptive elements. In fact, I am not aware of any other "resistance group" than the one described by Alrø and me being presented in mathematics education research literature.

Sometimes teachers complain that research in mathematics education does not appear relevant. When giving presentations to groups of teachers, I have heard this comment many times. Considering the degree to which the assumption of neatness sets a paradigmatic choice in research, this complaint is not surprising. Contrary to the prototypical classrooms tht dominates research literature, real-life classrooms are filled with difficult students and with resistance groups. Very profound paradigmatic criteria must dominate the selection of data considered appropriate for mathematics education research. We seem to be dealing with a form of censorship, a silencing which often takes the form of self-censorship. This brings about a neatness that separates much research literature from teachers' real-life classroom experiences.

16.2 Affluent

In 1993, I travelled to South Africa for the first time in my life. It could be nothing but my first time, as the international academic boycott had just been lifted. A democratic election in South Africa was in sight. The boycott – as recommended by the African National Congress (ANC) in its struggle against the apartheid regime – was strictly followed by Denmark and many other counties, although not by the UK nor by the USA. I was invited to participate in a conference and give lectures. I was also invited to become the supervisor of a group of black and Indian PhD students that would come to belong to the first generation of post-apartheid doctors in mathematics education. I was honoured by being asked to become their supervisor.

During my numerous subsequent stays in South Africa, I visited many schools. Although the apartheid regime had come to an end in 1994 through the election of Nelson Mandela as president, many leftovers from the apartheid period were still in place, also with respect to the day-to-day realities of the schools. Laws can be changed from 1 day to another, but not the location of poor and rich neighbourhoods, nor the separation between black, Indian, and white neighbourhoods. The whole "topology" of the apartheid remained unchanged. The educational system had been segregated into black, Indian, and white schools. Formally speaking, this segregation belonged to the past, but the schools continued to be located in the same neighbourhoods, which continued to be positioned according to the topology of apartheid. The historically black schools continued to be poor, the historically Indian schools continued to be moderately resourced, and the white schools continued to be well resourced.

Some of the historically black schools that I visited appeared to be miserable places for learning. They were often missing everything: doors, electric installations, and tiles from the roof might have been stolen and used as construction material in the neighbourhood. In a private house near the school, one could, for instance, find a door still with a "4a" nicely painted on it. In some schools, one could not find a functioning toilet, either for the students or for the teachers. If needed, one had to go into the nearby bush. For many reasons, this could be dangerous, so a common solution was to drink as little as possible.

Before getting to South Africa, I knew that under-resourced teaching-learning conditions did exist, but it was here for the first time that I experienced such conditions in their extreme. I was well read in the literature of mathematics education, but I cannot remember ever having seen descriptions of classrooms like the ones I encountered in South Africa. Later on, I thought more systematically about this point, and came to see the omission of extreme poverty from the descriptions in mathematics education research as an example of the dominance of prototypical thinking.

That is not to say that matters have been static. If we look at the proceedings of the Mathematics Education and Society (MES) conferences taking place in 2017 in Volos in Greece (see Chronaki, 2017), in 2019 in Hyderabad in India (see Subramanian, 2019), and in 2021 in Klagenfurt in Austria (online) (see Kollosche, 2021), we see a clear opening towards broader contexts within which mathematics education takes place. Awareness of the relationships between mathematics education and society has brought about studies in poorly resourced classrooms. Research in ethnomathematics has also addressed teaching-learning situations that are far from prototypical. Ethnomathematics and related studies have made significant efforts to broaden the research scope of mathematics education. However, all these important studies do not eliminate my impression that research in mathematics education disproportionately focusses on affluent contexts.

In *Education for All: Statistical Assessment 2000* by UNESCO, we find the traditional distinction between developed countries, less developed countries, and countries in transition.[4] The statistics tell us that, by the turn of the Century, the world's population of children between 6 and 11 years old was distributed in the following way: 10% of the world's population of children came from developed

---

[4] See http://unesdoc.unesco.org/images/0012/001204/1204/120472e.pdf

countries; 86% came from less developed countries, while 4% came from countries in transition. Furthermore, one shocking statistic from the UNESCO report informs us that 16% of the world's population of children did not go to school. This percentage was higher than the total number of children in developed countries.[5]

The distinction between developed countries, less developed countries, and counties in transitions is misleading. This has been clearly pointed out by Manuel Castells (1996, 1997, 1998) who has introduced the notion of the *Fourth World*. In the *End of Millennium*, he defines social exclusion "as the process by which certain individuals and groups are systematically barred from access to positions that would enable them to an autonomous livelihood within the social standards framed by institutions and values in a given context" (Castells, 1998, p. 73).

On several occasions, I have highlighted that processes of ghettoising run parallel with the processes of globalisation. Globalisation means connecting people, but simultaneously it also means disconnecting other people.[6] Castells (1998) summarises the observations in the following way:

> This widespread, multiform process of social exclusion leads to the constitution of what I call, taking the liberty of a cosmic metaphor, the black holes of informational capitalism. These are the regions of society from which, statistically speaking, there is no escape from the pain and destruction inflicted on the human condition of those who, in one way or another, enter these social landscapes (p. 162).

This brings him to coin the notion of the Fourth World:

> The Fourth World comprises large areas of the globe, such as much of Sub-Saharan Africa, and impoverished rural areas of Latin America and Asia. But it is also present in literally every country, and every city, in this new geography of social exclusion. It is formed of American inner-city ghettos, Spanish enclaves of mass youth unemployment, French banlieues warehousing North Africans, Japanese Yoseba quarters, and Asian mega-cities' shanty towns. And it is populated by millions of homeless, incarcerated, prostituted, criminalized, brutalized, stigmatized, sick and illiterate persons. They are the majority in some areas, the minority in others, and a tiny minority in a few privileged contexts. But, everywhere, they are growing in number, and increasingly in visibility, as the selective triage of informational capitalism, and the political breakdown of the welfare state, intensify social exclusion. In the current historical context, the rise of the Fourth World is inseparable from the rise of informational, global capitalism (pp. 164–165).

The Fourth World is not located in a particular part of the world. It runs across the distinction between developed countries, less developed countries, and countries in transition. The Fourth World can be found in any big metropolis in the world as London, Bombay, Paris, New York, and São Paulo.

Let me also mention my notion of *Zero-Point-One World*. Like the Fourth World, the Zero-Point-One World can be found all over the world, but contrary to the Fourth World, it is composed of the world's wealthiest neighbourhoods. The

---

[5] Naturally some of the children not going to school come from the so-called developed countries.

[6] Bauman (1998) makes the following observation: "Globalization divides as much as it unites; it divides as it unites – the causes of division being identical with those which promote the uniformity of the globe." (p.2)

Zero-Point-One World can be found in every metropolis. Like the Fourth World, the Zero-Point-One World runs across the distinction between developed countries, less developed countries, and countries in transition. The Zero-Point-One World can also be found in London, Bombay, Paris, New York, and São Paulo.

In spite of all the initiatives for broadening the scope of research, I find that mainstream research in mathematics education still pays insufficient attention to Fourth World contexts, and I find that the assumption of affluence continues to manifest itself disproportionately in much research literature.

16.3 Placid

When the PhD project in South Africa had started and the students had formulated their study programmes, plans for the data production were created. After the initial steps were taken, one of the PhD students remarked: it must be much easier to do classroom research in Denmark than here in South Africa. The PhD student had made a well-elaborated plan for the data-production, and the data had to be collected in a school far away, in a remote part of the countryside. It was a long journey to reach the school, but the first phase in the data production was completed according to schedule. The next step, however, came to nothing, as local violence had haunted the neighbourhood. Both children and teachers stopped attending the school, and soon it was closed down.

How are researchers to collect data in such unstable environments? All the planning – with pre-observations, educational initiatives, and post-observations – seemed ruined. We needed to give the whole situation much more thought. We were faced with the challenge: in what sense could something be too difficult to research? Considering the very notion of research, it should be possible to research both stable and unstable educational situations. Stability cannot be a pre-condition for researching something. It can only be a condition if one subscribes to some specific assumptions about how research has to be conducted.

This is my preoccupation: Does research in mathematics education, to some extent, assume that the teaching-learning processes take place under some settled conditions? Do dominant methodological patterns imply that, first of all, stable educational settings are researched? Does mathematics education research presuppose placidity as a paradigmatic requirement?

Displacement is a common phenomenon in the world today. According to the UNHCR, the UN High Commissioner for Refugees, by the end of 2018 "70.8 million individuals were forcibly displaced worldwide as a result of persecution, conflict, violence or human rights violations. That was an increase of 2.3 million people over the previous year, and the world's forcibly displaced population remained at a record high".[7] Refugees are found in all parts of the world. In Africa, people have

---

[7] See https://www.unrefugees.org/refugee-facts/statistics/

been fleeing from the Central African Republic to neighbouring countries. In Central America, one finds massive streams of refugees: for instance, those who have emigrated due to the violence in El Salvador, Guatemala, and Honduras. Europe is receiving many refugees from the Middle East, others from Africa, and recently also from Ukraine. In Iraq itself, more than three million people have been displaced. In Southern Sudan, more than three million have been forced to leave their homes. In Syria, one finds 13 million displaced people, which is more than half of the whole population of the country. and what are the living conditions like for the remaining half? Refugees are displaced people, but displacement may take many other forms. Due to certain methods of city planning, huge groups of people – in particular, those from poor neighbourhoods – can be allocated to different neighbourhoods. Uncertainty and instability form the life-conditions for many people around the world.[8]

When we look at the research literature in mathematics education, we do not find many records of any such uncertainties and instabilities. Instead, mathematics education research tells many stories that take place in peaceful teaching-learning environments. I see the assumption of placidity as a problematic bias in mathematics education research. This bias might be maintained by many factors, and some could be methodological to the extent that stability and predictability operate as preconditions for research planning. I do not suggest that research in neat, affluent, and placid contexts should be ignored; critical mathematics education itself needs to explore teaching and learning in all kind of environments. My problem is when neatness, affluence, and placidity become taken as a paradigmatic given.

16.4 Descriptive

Quantitative research has been dedicated to creating descriptions. Such a dedication has been largely cultivated by its adherence to logical positivism, which was deeply inspired by what took place in the natural sciences. According to logical positivism, all sciences (including social sciences and the humanities) should be organised according to the same unifying pattern. This implied that quantitative descriptions were given first priority, and that mathematics came to be considered the language of science. When properly organised, apparently different sciences like physics, chemistry, biology, psychology, sociology, anthropology, history, and literature would all contribute to the same unity of science.

Independent of the type of science, the aim of scientific theories would be the same: to construct descriptions of reality. The aim of any description was to provide reliable explanations, which in turn was identified with making predictions. According to logical positivism, the principal characteristic of a scientific theory is

---

[8] For a discussion of mathematics education and immigrant students, see Carrijo (2023), and Civil (2012, 2020).

16.4 Descriptive

that it can be tested empirically. It has been claimed that it should be possible to verify a theory by outlining (but not necessarily actually carrying out) empirical procedures through which the theory could be verified. It has also been claimed that the defining feature of a scientific theory is to specify conditions for falsifying the theory.

This outlook of logical positivism has inspired a number of research recommendations with respect to mathematics education. It has been pointed out that theory-building in mathematics education should demonstrate a descriptive power. When descriptions are made in a meticulous way, the theory could obtain an explanatory power. This, in turn, can be identified as a predictive power. It has been emphasised that both a principle for verification and a principle for falsification can be applied with respect to theory-building in mathematics education. It has also been emphasised that research results should have replicability, meaning that such results cannot depend on the particular researcher, nor on the particular research approach. Subject factors should be eliminated from scientific research; one way of doing so is to establish multiple sources for obtaining scientific insight, a method which has also been referred to as making triangulations.

Not only quantitative but also qualitative research has been dedicated to constructing descriptions of what does exist. An elaborated formulation of researching "what is" is found in *Naturalistic Inquiry* by Lincoln and Guba (1985). They distance themselves from the positivist assumption that it is possible to provide accurate descriptions of real states of affairs by eliminating subjective factors from the research process. Lincoln and Guba do not assume the existence of an objective reality, which can be properly described. They assume that reality is constructed and that there is a multiplicity of legitimate ways of looking at such constructions. They doubt the existence of any supreme perspective by means of which one can ensure "objectivity"; instead, they find that reality does not exist before it has been constructed. While a positivist inquiry addresses "what is positively given", a naturalist inquiry addresses "what is constructed". While positivist inquiries might give priority to quantitative descriptions, Lincoln and Guba give priority to qualitative descriptions, but both types of inquiries are firmly positioned in the descriptive research paradigm. They concentrate on "what is".

Qualitative research and different expressions of naturalistic inquiries are common in mathematics education research. In much research literature, one finds careful descriptions of what (it is claimed) has taken place in the classrooms. Much research in mathematics education is deeply rooted in the descriptive paradigm.

Critical mathematics education fully recognises the importance of formulating descriptions based on both quantitative and qualitative studies. However, to critical mathematics education it is *also* important to reach beyond the descriptive paradigm. It is not only important to research "what is", but also "what could be". It is important to research alternative educational possibilities, otherwise we might be confined by the limitations imposed by working within the political status quo. This brings me to the notion of *pedagogical imagination*. With this expression, I refer to formulations of alternatives to what is actually taking place in, say, a classroom context. A pedagogical imagination can be based on loose speculations, but it can also be based on systematic studies. To critical mathematics education, it is important that research does not restrict itself to present accounts of what is taking place, but also tries to explore what could come to take place.

In 1996 in a shopping centre in Durban, I had a longer supervision session with the group of South African PhD students. Several of them were concerned about the nature of the data that they planned to produce. On the one hand, they felt that the data should show what was taking place in the classrooms, and that the data should fulfil the classic criteria of validity. On the other hand, they also found that *only* producing descriptions of what was taking place was not adequate in the situation they found themselves in. They wanted *also* to explore alternatives to what was taking place. They were deeply concerned about the continuing post-apartheid separation between black, Indian, and white schools. They wanted to explore multicultural contexts for learning mathematics. But no multicultural classrooms were in sight. We recognised the importance of addressing not only what is taking place, but also what is *not* taking place, but might come to take place. This whole discussion brought us to recognise the importance of forming a pedagogical imagination as integral part of a research approach.[9]

In order to conceptualise alternatives, it is important to try to "look through the data" that is produced. If one identifies and describes processes of exclusion taking place in the mathematics classroom, it is important *also* to try to formulate alternatives to such processes. If one identifies and describes classroom conversations dominated by teachers' expositions of new mathematical topics, it is important also to formulate other ways of student-teacher interaction. If one describes project work concentrated on mathematical problems, it is important also to identify alternative topics to address. It is important, with reference to data about what is currently taking place, simultaneously to "look through" these data. In her study of immigrant students in Brazil and their conditions for learning mathematics, Manuella Carrijo (2023) has used the metaphor of "looking through a window" as a way of "looking through the data". Naturally, one can look at the window, just as one can look at the data. But by "looking through the window/data" one might catch a glimpse of alternatives. In this way, the produced data can resource a pedagogical imagination.

To formulate alternatives to what is taking place is not an individual undertaking. Pedagogical imaginations grow from interactions between researchers, teachers, and students. Doing research from the perspective of critical mathematics education is not principally a process of "looking at" but rather a process of "interacting with". To critical mathematics education, it is important to reach beyond the descriptive paradigm, and not to be captivated by prototypical assumptions.

---

[9] Vithal (2003) elaborates in detail on such a research approach in order to prepare for multicultural settings for leaning mathematics in a South African school context. In Skovsmose and Borba (2004), and in Chap. 9, "Researching Possibilities", in Skovsmose (2014, pp. 111–126), one finds further elaborations of this research approach. In Biotto Filho (2015), this approach has been applied with respect to a Brazilian context. Lima (2022) has provided an extensive empirical study of the formation of pedagogical imagination as part of a mathematics teacher education programme. See also Helliwell and Ng (2022).

To critical mathematics education it is crucial not to be captured by illusions of the prototypical classroom, where one finds well-behaving students, well-resourced setting, and peaceful school environments. However, I am afraid that the conjecture I presented in Rome in 2008 has still not been refuted.

References

Alrø, H., & Skovsmose, O. (2004). *Dialogue and learning in mathematics education: Intention, reflection, critique*. Kluwer Academic Publishers.
Bauman, Z. (1998). *Globalization: The human consequences*. Polity Press.
Biotto Filho, D. (2015). *Quem não sonhou em ser um jogador de futebol? Trabalho com projetos para reelaborar foregrounds (Who never dreamed of being a soccer player? Working with projects for elaborating foregrounds)*. Doctoral Dissertation. Universidade Estadual Paulista (Unesp).
Carrijo, M. (2023). *Beyond borders and racism: Towards an inclusive mathematics education for immigrant students*. Doctoral dissertation. São Universidade Estadual Paulista (Unesp).
Castells, M. (1996). *The information age: Economy, society and culture. Volume I, the rise of the network society*. Blackwell.
Castells, M. (1997). *The information age: Economy, society and culture. Volume II, the power of identity*. Blackwell.
Castells, M. (1998). *The information age: Economy, society and culture. Volume III, end of millennium*. Blackwell.
Chronaki, A. (Ed.). (2017). *Mathematics education in the life and times of crisis: Proceedings of the ninth international mathematics education and society conference (MES 9)*. Mathematics Education and Society.
Civil, M. (2012). Mathematics teaching and learning of immigrant students: An overview of the research field across multiple settings. In O. Skovsmose & B. Greer (Eds.), *Opening the cage: Critique and politics of mathematics education* (pp. 127–142). Sense Publishers.
Civil, M. (2020). Immigrant students in mathematics education. In S. Lerman (Ed.), *Encyclopedia of mathematics education* (2nd ed., pp. 359–365). Springer.
Helliwell, T., & Ng, O.-L. (2022). Imagining possibilities: Innovating mathematics (teacher) education for sustainable futures. *Research in Mathematics Education, 24*(2), 128–149.
Kollosche, D. (Ed.). (2021). *Exploring new ways to connect: Proceedings of the eleventh international mathematics education and society conference (MES 11)*. University of Klagenfurt.
Lincoln, Y. S. & Guba, E. G. (1985). *Naturalistic inquiry*. Sage Publications.
Lima, P. C. (2022). *Imaginação pedagógica e educação inclusiva: Possibilidades para a formação de professores de matemática (Pedagogical imaginations and inclusive education: Possibilities for mathematics teacher education)*. Doctoral dissertation. Universidade Estadual Paulista (Unesp).
Menghini, M., Furinghetti, F., Ciacardi, L., & Arzarello, F. (2008). *The first century of the international commission on mathematical instruction (1908–2008): Reflecting and shaping the world of mathematics education*. iInstituto della Enciclopedia Italiana.
Skovsmose, O. (2014). *Critique as uncertainty*. Information Age Publishing.
Skovsmose, O., & Borba, M. (2004). Research methodology and critical mathematics education. In P. Valero & R. Zevenbergen (Eds.), *Researching the socio-political dimensions of mathematics education: Issues of power in theory and methodology* (pp. 207–226). Kluwer Academic Publishers.
Subramanian, J. (Ed.). (2019). *Proceedings of the tenth international mathematics education and society conference (MES 10)*. Mathematics Education and Society.
Vithal, R. (2003). *In search of a pedagogy of conflict and dialogue for mathematics education*. Kluwer Academic Publishers.

Chapter 17
Social Theorising and the Formatting Power of Mathematics

Abstract In order to interpret and understand a range of social phenomena, it is crucial to consider the social role of mathematics. I address four conceptions which play an important role in social theorising, namely: structuration, risk society, life-world, and violence. I try to illustrate that an adequate understanding of these four social phenomena cannot be obtained without an understanding of the formatting power of mathematics. I find it urgent to integrate this insight into our methods of social theorising.

Keywords Social theorising · Structuration · Risk society · Life-world · Violence · Formatting power of mathematics

With the term "social theorising", I refer to studies that in broad terms try to grasp basic forms of social development. They could be studies developed around the notion of structuration, as undertaken by Anthony Giddens; considerations of the risk society, as initiated by Ulrich Beck; discussions concerning life-worlds initiated by Edmund Husserl and conducted by Alfred Schutz and Thomas Luckmann; and reflections on violence provoked by the capitalist production machinery, as addressed by Slavoj Žižek.

Looking at such studies, one rarely finds specific reference to mathematics, something that I find to be a huge problem for social theorising. I want to show that, in order to interpret and understand many contemporary social phenomena, it is crucial to address the social role of mathematics.

I have talked elsewhere about the formatting power of mathematics in order to highlight the possible social roles of mathematics. At times I have encountered objections to using this expression, because, it is claimed, mathematics by itself does not exercise any power; it is the people *using* mathematics that exercise the power. However, it is an accepted stance to argue that some discourses can themselves constitute a worldview – for instance by being carriers of racist outlooks. I see mathematics as being a discourse, and consequently I continue to talk about the formatting power of mathematics.

I will address four conceptions that play an important role in social theorising, namely: *structuration, risk society, life-world,* and *violence*. I will illustrate that an adequate understanding of these four social phenomena cannot be obtained without an understanding of the formatting power of mathematics.

17.1 Structuration

In *The Constitution of Society: Outline of the Theory of Structuration*, Anthony Giddens (1984) highlights structuration as being a crucial notion for interpreting social phenomena. In this book, he does not propose a straightforward definition of structure; rather, he reveals the complexity of the notion. In order to do this, he pays special attention to the notion of rules, which can operate in many ways. Rules can be explicitly stated by means of legal laws, but they can also be incorporated as part of a cultural tradition. Rules are not static phenomena; some can be explicitly changed; some change over time; and some can change as a result of ill-defined decision making. According to Giddens, the formation of rules constitutes a part of social structuration.

In order to understand the significance of Giddens' theory of structuration, we have to remember the position of logical positivism which for a long period dominated sociology and the social sciences. An important principle of logical positivism was that it should be possible to address any scientific concept through empirical investigation. It was claimed that, when one considers a group of individuals, there is nothing more in the group than the sum of the individuals. Every possible statement about a social group is equivalent to a conjunction of statements about the individual members of the group. This is the principle of reductionism that logical positivism advocated. Notions and statements about society can be reduced to notions and statements about individuals.[1]

When paying particular attention to the phenomenon of structuration, Giddens is up against the principle of reductionism. It is in no way clear to what extent it may be possible to reduce statements about social structuration to statements about individuals. From the perspective of logical positivism, Giddens is introducing a metaphysical notion, since structuration is not explicitly defined in terms of possible empirical observations. I do not subscribe to any principle of reduction, and I fully follow what Giddens is proposing: it is important to study how social structures come to be formed, how they might interact with other structures, how they might be substituted by other structures, and how they might fade away. It is important to study how structurations form our life worlds.

---

[1] The reduction, as advocated by logical positivism, goes further: notions and statements about psychological phenomena can be reduced to notions and statements about biological phenomena; and similarly, biological phenomena can be reduced to chemical phenomena, which in turn can be reduced to physical phenomena.

17.1 Structuration

The point I am going to emphasise is that, today, *mathematics-based structurations* make up an integral part of many forms of social structuration. In all spheres of life, mathematics-based formal structures turn into real-life structures. Such structures might become established through mathematical algorithms put into operation, and through the creation of digital platforms.

In the article "Algorithmic Work: The Impact of Algorithms on Work With Social Media", Livia Norström, Anna Islind, and Ulrika Lundh Snis (2020), highlight:

> A substantial part of people's daily activities is supported by the use of digital platforms. The digital platforms rest on algorithms, making it increasingly important to understand the influence of changes in the algorithms and how to govern them in order to serve both private and work-related purposes and goals. Consequently, ambitions to understand the influence of algorithms is no longer merely a concern for computer scientists and mathematicians. The interest in algorithms and the influence from the way algorithms are designed is advancing beyond computational understanding of optimization and efficiency, towards human, organizational and societal understandings of its applications and effects (p. 1).

The understanding of the functioning of mathematical algorithms and digital platforms cannot remain a concern only for computer scientists and mathematicians. To grasp the structuring of the modern workplace, it is crucial to grasp how mathematical algorithms are developed and brought into operation.

Algorithmic management concerns the detailed distribution of work tasks within a company; it concerns economic decision-making; it concerns the instalments of automatic forms of control and surveillance; and it concerns the selection of new workers for the company. In the ACAS (Advisory, Conciliation and Arbitration Service) study *My Boss the Algorithm: An Ethical Look at Algorithms in the Workplace*, one finds a broad overview of what algorithmic management might include; one finds references to several companies, among them the Amazon, as being a paradigmatic example of algorithmic management brought into practice. As an example of one feature of algorithmic management, let me quote from the ACAS study:

> One of the main ways in which algorithms are both assisting and, in some areas, even replacing human managers is in work allocation; either in allocating workers to particular shifts or teams or in allocating tasks to individual workers. Shift allocation algorithms can automatically match workers to shifts when they are needed and also handle the routine business of allowing workers to swap or change their shifts.[2]

Algorithmic workplaces are structured by a network of rules engraved in the computer systems brought into operation. An algorithmic workplace is an example for a powerful structuration of the reality. The nature of this structuration cannot be captured without a deep awareness of the formatting power for mathematics, in this case executed as rule-producing devices. The mathematical algorithm is such a device which formats workplaces around the world.

In line with what Norström, Islind, and Snis emphasised: the understanding of the way algorithms are designed is advancing beyond computational understanding

---

[2] See *My Boss the Algorithm: An Ethical Look at Algorithms in the Workplace*, Acas.

towards an understanding of social structuring in general. Addressing social structuration without paying attention to the extension and nature of mathematics-based structuration causes a devastating limitation to social theorising.

17.2 Risk Society

The existence of global catastrophic risks scaffolds our living conditions. Nuclear terrorism is a risk linked to the very existence of nuclear weapons, adding to the risks that the nuclear powers might, in fact, use them themselves.[3] By means of nanotechnology, we are able to construct technical devices that operate at the nano level, but how are we to control the functioning of such technologies?[4] Risks emerge due to climate change, as well as the way we try to cope with such change.[5] There are risks related to pandemic situations and to health situations in general, as the whole world recently experienced. Risks are associated with the production of food: biotechnology with unpredicted health implications might be used in farming.[6] Other risks are caused by the limits of food production, as an unequal distribution of what is produced around the globe might lead to hunger crises. Risk might emerge from the very way we try to manage different forms of risk; thus, processes of risk management might by themselves provoke risks.[7]

In his book *Risk Society: Towards a New Modernity*, the original German version of which was published in 1986, Ulrich Beck (1992) creates an awareness of new patterns of risk. Nature provides the basic conditions for human life, but over time nature has also presented devastating threats to human life: eruptions of volcanos, floods, droughts, and pandemics. By talking about the risk society, Beck makes clear that risks do not only emerge from a "hostile" nature; they are also human fabrications due to technological development. Consequently, risks need to be addressed systematically as part of social theorising.

Beck concentrated on demonstrating how the formation of risks has become an integral part of technological development. I find it important, as well, to pay attention to the role of mathematics in the formation and handling of risks.[8]

The very identification of climate change depends upon mathematical modelling, which provides a means for looking into the climatic future. However, such modelling also fabricates risks. A mathematical model might grasp some features of future realities, but might simultaneously ignore other features. Mathematical modelling is not by any means a reliable approach; still, it is the only approach we have

---

[3] See Ackerman and Potter (2008) and Cirincione (2008).
[4] See Phoenix and Treder (2008).
[5] See Frame and Allen (2008).
[6] See Nouri and Chyba (2008).
[7] See Yudkowsky (2008).
[8] For a discussion of risk society, see also Chap. 4, "Sustainability and Risks".

for identifying possible future climate change. The constructions and interpretations of climate models are controversial issues, because different interpretations of the models are possible.[9]

Mathematics is part of the handling as well as the mishandling of different kinds of risks. Mathematics may as well contribute to the very fabrication of risks. Living with risks is an inevitable part of the human condition, and consequently it is important for social theorising to address all kinds of risk phenomena. If the role of mathematics in the formation and handling of risks is ignored, social theorising turns inadequate.

17.3 Life-Worlds

In *The Crisis of European Sciences and Transcendental Phenomenology*, which appeared in German in 1936, Edmund Husserl (1970) coined the notion of life-world (*Lebenswelt*). In 1935, the Nazi government had instituted the so-called Nuremberg Laws that stripped the Jewish population in Germany of all human rights; they no longer counted as German citizens. This had a tremendous impact, also for Husserl. As with other academics with Jewish origins, he lost his academic position and was expelled from the university; he was not even allowed to access the library. Yet during this gloomy period, Husserl completed the book. His health was also precarious, and it was urgent for him to provide a further. and from his own perspective, final clarification of the phenomenological outlook. Husserl coined the notion of life-world at a time when his own life-world lay in ruins. Husserl died in 1938.

According to Husserl, a life-world refers to a person's immediate experiences before they are structured by scientific, philosophical, or other already established conceptual frameworks. An experience of a sunset might be an immediate impression of the moving down of the sun, while recognising a sunset as being due to the rotation of the earth is a theoretical elaboration. A life-world is an original world, existing before the formation of the scientific outlook. Husserl (1970) points out that the "life-world was always there for mankind before sciences" (p. 123). The scientific outlook does not substitute the life-world; instead, the "objective-scientific world is 'grounded' in the self-evidence of the life-world" (p. 130). Husserl states that "the life-world, for us who wakingly live in it, is always already there, existing in advance for us, the 'ground' for all praxis whether theoretical or extrateoretical" (p. 142). A life-world is the foundation for a range of human activities, for any kind of praxis, and for any kind of theoretical construction.

Husserl formulated the notion of life-world in order to ensure a further clarification of the phenomenological outlook. However, based on the idea that a life-world

---

[9] In Chap. 4, "Sustainability and Risks", I refer to Lomborg (2001), who reached the conclusion that climate change is not caused by human inventions. This conclusion was broadly celebrated by industry.

forms all human activity, the notion gets a broader significance. In *The Structures for the Life-World*, Alfred Schutz and Thomas Luckmann (1973) point out:

> By the everyday life-world is to be understood that province of reality which the wide-awake and normal adult simply takes for granted in the attitude of common sense. By this taken-for-grantedness, we designate everything which we experience as unquestionable; every state of affairs is for us unproblematic until further notice (pp. 3–4).

According to Schutz and Luckmann, a life-world is "the taken-for-granted frame in which all the problems which I must overcome are placed" (p. 4). With such interpretations, the notion of life-world becomes crucial, not only for philosophical clarification of phenomenology, but for social theorising in general.

I also find the notion of life-world to be extremely relevant for social theorising. However, my interpretation of the notion differs from what is common in phenomenology. I do not ascribe any epistemic uniqueness to a life-world. I do not try to associate to a life-world any quality of being a "foundation" of, or "grounding" for, scientific knowledge. A lge-world does not have any a priori qualities, but is itself highly structured. There is no genuine life-world which subsequently gets more and more structured. A person's life-world is already structured, and it becomes re-structured again and again. This may be by dominant discourses, cultural practices, and layers of preconceptions.[10]

For structuration and restructuration of life-worlds, mathematical formatting plays a crucial role. Previously I collected the invoices that I received, and once per month I went to the bank in the small city where I was living to have everything paid. It was a monthly event, and on the counter of the bank there was an open box with small cigars so one could leave the bank with everything in order, smoking a cigar. Later on, the box with cigars was removed. Then I got what in Denmark is referred to as a NemID, and by means of this virtual identity I could access from my computer at home not only my bank but public institutions as well. The communication between citizens and public administration became paperless. In the beginning of 2022, my NemID was substituted by MitID, based on extra security precautions. In addition to my computer, I need an app installed on my mobile phone in order to access the bank and public institutions. Any such security devices are developed through extensive and expensive research programmes. This is just one single example of how a province of our life-world can be changed due to mathematical algorithms brought into action.

Life-worlds have many provinces. Apps are brought into operation at workplaces, when we do shopping, when we search for a good offer on the internet, and when we communicate with friends. Some apps we know about, but many mathematical applications are brought into operation in our daily lives, the existence and functioning of which we have no idea. Today, our life-worlds are formatted by mathematical applications, and it is crucial for social theorising to try to also grasp this kind of formatting.

---

[10] For a further clarification of my use of the notion of life-world, see Skovsmose (2014).

17.4 Violence

Violence can be explicit, physical, and brutal. In the beginning of February 2022, when I was drafting this chapter, the TV in Brazil showed the killing of Moïse Kabagambe, a 24-year-old man, who was working in a kiosk named Tropicália at the beach at Barra da Tijuca in Rio de Janeiro. Moïse was beaten to death by three young men using heavy wooden sticks, while a few others looked on. Around 10 years previously, Moïse and his family had fled from Congo due to the violence there.

The security camera at the Tropicália captured the killing, and the scene was shown many times on the TV, watched by everybody in Brazil: young, old, and children. The three man who carried out the killing were arrested; they came from a poor neighbourhood; they were black, like Moïse. As far as I know, the motive for the killing has not been properly clarified, maybe some kind of revenge due the conflicting interests connected to selling on the beach. The police did state that they had not found indications of the involvement of *militias*, which are rather more interested in drug dealing.

Violence need not only be explicit and physical, but can be structural, and as such can still be extremely brutal. In the article "Affirmative Actions in Terms of Special Rights: Confronting Structural Violence in Brazilian Higher Education", Guilherme da Silva and I (2019) characterise structural violence by highlighting the following: (1) Structural violence is acted out through a fabric of power: it does not have any specific agent. Neither a particular person nor any well-defined institution might exercise this violence. (2) Structural violence is exercised through forms of discrimination and exclusion. Discrimination can, for instance, concern access to food, healthcare, and welfare in general. Exclusion can be exercised through, for instance, racism and sexism. (3) Structural violence is accompanied by legitimising discourses. The function of such discourses is to make the violence appear natural and unquestionable. For instance, colonial systems have sought legitimisation through discourses celebrating white supremacy.

Structural violence can have several expressions, one being in *structural racism*. This notion is explored profoundly in *Racismo Estrutural* by Silvio Almeida (2019). Nazi Germany, the apartheid South Africa, and the Jim Crow States in the USA are examples of state-implemented racism; however, racism can be acted out even when the formal laws are completely democratic and the same for everybody. Behind the surface of we-are-all-in-it-together, a brutal form of structural racism might be operating. Through the notion of structural racism, one tries to draw attention to the fact that racism need not be expressed through physical violence, as in the case of Moïse, but can find many other forms of violence.

In the book *Violence*, Slavoj Žižek (2008) uses the expression "systemic violence" as referring to the brutal way a capitalist economy might exploit workers. Back in time, industrialisation was accompanied by an explicit and violent abuse of workers and their families; strikes were confronted by police violence; and health problems were ignored. Today, it might be more difficult to identify general explicit violence in the workplace, but if we look at workers' living conditions in poor environments producing goods to be consumed in the wealthy parts of the world, we find cases of brutal exploitation. The logic of colonialism, to extract value from the colonies, has been prolonged into an almost identical logic: to extract value from parts of the world rich in natural resources with a cheap labour force and to transfer it to the privileged part of the world. We are dealing with a crushing form of structural violence.

Violence is not some extraordinary social phenomena, but constitutes a part of how societies are operating today. This makes it necessary for social theorising to address different forms of violence, including structural violence.

Mathematics has a profound but ambiguous relationship with violence. In Brazil, structural racism can be revealed by looking at statistics telling us about the likelihood that a black person will take up a course in further education, come to live in an affluent environment, get a good job, come to live in poverty, get involved in crimes, and finally end up in jail. Mathematics can function as a principal instrument for identifying and documenting cases of structural racism, as well as of other forms of structural violence.

Can mathematics contribute to structural violence? In order to fit smoothly into semi-automatised processes, workers need to demonstrate a prescription readiness. Such a readiness can be a violent imposition: workers have to submit themselves to predefined operations. Mathematics education, as formed by the school mathematics tradition, can be seen as a state-organised preparation for prescription readiness. A basic principle of this tradition is that mathematical competence can be developed by students solving pre-formulated exercises with one, and only one, correct answer. Does the school mathematics tradition bring about mathematics knowledge and understanding? Hardly. But certainly this tradition helps to exercise prescription readiness, which is an important component of a submissive working force. Exercising prescription readiness through mathematics education might still be an example of a form of structural violence, preparing for the structural violence that workers subsequently become subjected to.[11]

Can mathematics itself be violent? Both the First and the Second World War were dominated by advances in war technologies that made it possible to kill at an industrial scale. However, since the Second World War a principal new development has taken place: mathematics has come to play a crucial role, not only in the design and construction of war technology, but also in the very functioning of this technology. We can observe the firing of a missile and also the explosion when it reaches its target, but we do not see all the mathematical algorithms that form part of the steering of the missile and the timing of the explosion. Mathematical algorithms are brought into operation everywhere in modern war technology.

Violence in its many manifestations needs to be addressed by social theorising, something that was also recognised by Žižek. But we need to add that mathematics plays different and contradictory roles with respect to violence. Mathematics can contribute to the construction and functioning of violent technologies and to the

---

[11] For a discussion of prescription readiness, see Chap. 2, "Landscapes and Racism".

implementation, preservation, and escalation of structural violence. Simultaneously, mathematics can also play a role in addressing structural violence. For social theorising, it is crucial to grasp this double role of mathematics.

References

Ackerman, G., & Potter, W. C. (2008). Catastrophic nuclear terrorism: A preventable peril. In N. Bostrom & M. M. Cirkovic (Eds.), *Global catastrophic risks* (pp. 402–449). Oxford University Press.
Almeida, S. (2019). *Racismo estrutural (Structural racism)*. Editora Jandaíra.
Beck, U. (1992). *Risk society: Towards a new modernity*. SAGE Publications.
Cirincione, J. (2008). The continuing threat of nuclear war. In N. Bostrom & M. M. Cirkovic (Eds.), *Global catastrophic risks* (pp. 381–401). Oxford University Press.
Frame, D., & Allen, M. R. (2008). Climate change and global risk. In N. Bostrom & M. M. Cirkovic (Eds.), *Global catastrophic risks* (pp. 265–286). Oxford University Press.
Giddens, A. (1984). *The constitution of society: Outline of the theory of structuration*. University of California Press.
Husserl, E. (1970). *The crisis of European sciences and transcendental phenomenology* (Translated with an introduction by David Car). Northwestern University Press.
Lomborg, B. (2001). *The skeptical environmentalist: Measuring the real state of the world*. Cambridge University Press.
Norström, L., Islind, A. S., & Lundh Snis, U. (2020). Algorithmic work: The impact of algorithms on work with social media. Conference: European Conference on Information Systems (ECIS 2020). *ECIS Research Papers, 185*. https://aisel.aisnet.org/ecis2020_rp/185
Nouri, A., & Chyba, C. F. (2008). Biotechnology and biosecurity. In N. Bostrom & M. M. Cirkovic (Eds.), *Global catastrophic risks* (pp. 481–503). Oxford University Press.
Phoenix, C., & Treder, M. (2008). Nanotechnology as global catastrophic risk. In N. Bostrom & M. M. Cirkovic (Eds.), *Global catastrophic risks* (pp. 450–480). Oxford University Press.
Schutz, A., & Luckmann, T. (1973). *The structures of the life-world* (Vol. 1). Northwestern University Press.
Skovsmose, O. (2014). *Foregrounds: Opaque stories about learning*. Sense Publishers.
Yudkowsky, E. (2008). Artificial intelligence as a positive and negative factor in global risk. In N. Bostrom & M. M. Cirkovic (Eds.), *Global catastrophic risks* (pp. 308–345). Oxford University Press.
Žižek, S. (2008). *Violence*. Picador.

Chapter 18
A Philosophy of Critical Mathematics Education

Abstract I outline a philosophy of critical mathematics education by addressing conceptions of social justice, environmental justice, mathematics in action, foregrounds, dialogue, and critique. For critical mathematics education, dialogical processes resource critical activities. It is important to be critical with respect to mathematics, which means to be ready to question any form of assumed mathematical knowledge and to question ways of bringing mathematics into action. It is also important to be critical by means of mathematics, which means to be ready to draw on mathematical resources when questioning cases of social and environmental injustice. Like dialogue, critique is open-ended. No critique can reach definite validity. Critique continues tentatively and preliminarily. This observation has implications for any critical enterprise, including critical mathematics education.

Keywords Philosophy of critical mathematics education · Social justice · Environmental justice · Mathematics in action · Foregrounds · Dialogue · Critique

In 1994, I published the book *Towards a Philosophy of Critical Mathematics Education*. The very word "towards" in the title indicates that something more clarified could be expected in the future. I find that the time has come for me to try to write a text with the title *A Philosophy of Critical Mathematics Education*, even though I still do not feel quite ready to do so.

A philosophy of mathematics education is different from practices of mathematics education and also from theories of mathematics education. However, contrary to many positions in philosophy, I am not concerned about keeping any clear demarcation between, on the one hand, a "philosophy of" and, on the other hand, a "practice of" and a "theory of". Still, the focus of a philosophy is different from that of a theory and of a practice. A philosophy of education elaborates, in particular, on concepts and notions relevant for interpreting and understanding educational phenomena, for formulating educational theories, and for changing educational practices. A philosophy might create openings towards new interpretations, new theories, and new practices.

Also a philosophy of critical mathematics education should pay attention to conceptual ideas, and in the following I address the concepts of *social justice, environmental justice, mathematics in action, foregrounds, dialogue,* and *critique.* Social justice and environmental justice embody visions with respect to the socio-political context within which learning takes place. Mathematics in action refers to possible social roles played by mathematics, and to how mathematics and power might be interrelated. Foregrounds signify features of the students' and teachers' life-conditions. And dialogue and critique capture qualities of the very process of learning, and how this process might lead to political positionings.

These concepts are woven together in a texture including many other notions like citizenship, democracy, and landscapes of investigation. I do not claim that the concepts that I refer to are the only ones that are important for mapping out a philosophy of critical mathematics education, but they help me to express concerns and hopes of this education.

18.1 Social Justice

Critical mathematics education recognises that education has a socio-political role to play. Education can be part of a struggle against oppression, atrocities, and exclusions. It can work for social justice. Mathematics education for social justice is a label that is currently used quite broadly, and I do not try to make any distinction between this type of education and critical mathematics education.

The notion of social justice is far from being straightforward and transparent. It needs to be handled with care, as it might be stuffed with common-sense interpretations, assumptions, and preconceptions. Social justice does not have any pre-given interpretation, but rather is a notion under construction. It is a contested construction. In all circumstances, social justice is an important notion for critical mathematics education to operate with.

Clarifying what "justice" could mean is a classic preoccupation of western philosophy. In the dialogue *The Republic*, Plato addressed the notion, and according to him, the very idea of justice is located in the world of ideas, and one needs to *discover* what justice really means by exploring this world.[1] According to Plato and to Platonism in general, this is a process quite similar to discovering mathematical truths, as the real mathematical objects also belong to the world of ideas. The notion of justice has been addressed by many religious thinkers, such as Augustine and Thomas Aquinas. According to religious dogmatism, it is possible to discover what justice means by carefully studying the Bible.

In 1848, several countries in Europe experienced social upheavals and revolutions. In Palermo, Italy, revolutionary manifestations took place in January of that year. In France, we speak about the February Revolution, and in several cities

---

[1] See Plato, *The Republic.* Translated by Benjamin Jowett. *The Internet Classics Archive.* http://www.constitution.org/pla/republic.htm

groups of radical workers clashed with other political groups. Many German states experienced a March revolution. The Austrian empire faced severe problems, and in Vienna revolutionary groups raised barricades. In Denmark, absolute monarchism was abandoned, and a new, more democratic constitution was implemented. Karl Marx and Friedrich Engels published *The Communist Manifesto*, which not only reflected what was taking place, but also radicalised the revolutionary trends. In 1848, the very notion of "social justice" appeared for the first time, namely in the title of the book *Costituzione secondo la giustizia sociale* (*The Constitution Under Social Justice*) written by Antonio Rosmini-Serbati (2007). In this book, social justice was related to both religious conceptions and to political visions. In his writing, Marx did not use the expression of social justice, nor did the expression become part of orthodox Marxist terminology. Only later did social justice develop as a commonly used expression indicating a leftist position.

To critical mathematics education, it is important *not* to consider the identification of what social justice could mean as a process of discovery, but rather as a process of construction. I find that ethical conceptions – like equality, fairness, and social justice – do not emerge from philosophical principles, ethical assumptions, or religious dogma. Nor do I assume that ethical claims are grounded in empirical observations, as assumed by utilitarian positions. I find that ethical claims emerge through social constructions in real-life contexts, and one such context is the classroom.[2]

I do not see a mathematics education for social justice as a set of lessons about social justice, but as an education that engages students and teachers in formulating what social justice might mean. It engages students and teachers in expressing what could be considered social oppression, economic inequalities, structural poverty, social exclusion, sexism, and racism. By exploring such issues, the notion of social justice might be reformulated and further developed. It is important that such reformulations are not dominated by the teacher or any other person, through informing the class about what social justice could be, but instead should be guided by dialogic interactions and shared concern for providing clarification.

It seems straightforward to claim that mathematics education for social justice is an education for citizenship. However, behind this apparently straightforward statement hide huge controversies. We need to ask questions like: What could citizenship mean in a country structured according to the capitalist mode of production and consumption? In a country permeated by religious convictions? In a country still suffering the impact of colonial oppression? In a country governed by totalitarian regimes? In a country assuming a caste system? In a country occupied by other forces, like Palestine? For people entering another country as refugees?

It seems important to make a distinction between a mathematics education that tends to bring about functional and adapted citizens within a given economic and political order, and a mathematics education that stimulates the formation of citizens ready to question the given order. Such a distinction, however, might give the impression that citizenship is a bipolar notion, stretching out between critical and

---

[2] For a discussion of social justice as construction, see Chap. 1, "Concerns an Hopes".

uncritical forms. In contrast, I see citizenship as a multi-dimensional notion with a diversity of possible interpretations. What kind of citizenship we should work towards and what conception of social justice we should engage with is a principal concern of critical mathematics education.

Conceptions of social justice also relate to conceptions of democracy. One needs to contest the use of the notion of democracy by asking questions like: Could England be considered an old democracy, while simultaneously being an empire that embraced slavery? Could a country with huge differences between poor and rich people be a democracy? Do democratic ideals resonate with neo-liberal ideologies? Certainly, the notion of democracy is as difficult to handle as the notions of citizenship and social justice. Such notions might include a variety of ambiguities, but simultaneously they might include resources for expressing concerns and hopes for the future.

Critical mathematics education engages students and teachers in formulating what social justice, citizenship, and democracy could mean. Such overall formulations are directly related to addressing specific cases of injustice.

18.2 Environmental Justice

How much more of the remaining rainforest can be cut down before living conditions on earth become intolerable? How much more plastic can be dumped into rivers and oceans before clean water becomes a myth? How much more can the air be polluted before breathing turns into a health problem?

Today such questions are urgent, but they only emerged as such after the Second World. In line with the ideas of Karl Marx and Critical Theory, the critical projects had economic structures and human relationships as their focus, not relationships between humanity and nature. According to classic Marxism, as well as to the modern outlook in general, nature is an infinite resource for material progress, and in order to ensure welfare, nature needs to be dominated and used efficiently.

The concern about sustainability developed in steps. One such step brought the Second World War to a close: the atomic bombs dropped over Hiroshima and Nagasaki. In the most brutal way, it became obvious that humanity had reached a tremendous capacity of destruction. An atomic war could extinguish human life on earth. Soon it was recognised that not only war technology, but also apparently "peaceful" technology could have devastating implications for both humanity and nature. The construction and operation of atomic power plants not only served the production of energy, but simultaneously brought about dramatic new risk structures. A risk need not become a reality, but in this case it did: the Three Mile Island accident took place in 1979, the Chernobyl accident in 1986, and the Fukushima-Daiichi accident in 2011. Many other forms of risks have become integral parts of technological innovations. What long-term effects could the production and consumption of genetically manipulated food have? Would the number of cancer cases increase dramatically in, say, thirty years' time due to pollution? Would new and until now unknown illnesses and pandemics emerge? No well-defined laboratory experiments can answer such questions. We have to wait and see. As Ulrich Beck (1998) has observed: our whole society has turned into an experimental laboratory, where nobody is in charge.

Environmental issues need to be critically addressed in mathematics education, and sustainability is a concern of critical mathematics education. In 2021, Lisa Steffensen published the study *Critical Mathematics Education and Climate Change: A Teaching and Research Partnership in Lower-Secondary School*, which represents an important further development of critical mathematics education. Originally, critical mathematics education focussed on problems of social justice, while concerns about sustainability point in new directions. I find it important that critical mathematics education operates with sustainability as one of its concerns (naturally, without ignoring any of its other concerns).

There exists a close relationship between social justice and environmental justice. Environmental justice concerns our relationships to nature, but simultaneously it concerns social relationships[3]. Colonialism was a brutal scheme of extracting natural resources from the colonies to be used by the colonisers. The colonised were forced to do the very extraction of the resources, while the colonisers did the monitoring and controlling. Capitalist industrial production has incorporated many elements of colonialism, including using resources and cheap labour forces from poorer parts of the world to extract resources and produce goods to be consumed in richer parts of the world. Both social justice and environmental justice are deep concerns of critical mathematics education.

Environmental justice concerns access to natural resources. Like economic resources, this access is extremely unequally distributed the world around. Access to clean water is one example. Access to clean air is another. Some industries are causing more pollution than others, and highly polluting industries are often placed in poorer neighbourhoods. Access to natural resources is far from being the same to everybody. The extraction of natural resources is dominated by economic priorities. What to extract – and for whom – reflects profit-generating opportunities, and not human needs in general. Hunger is a huge problem in Brazil today, not because Brazil does not produce food in sufficient quantities, but because exporting food creates more profit for the agricultural industry. Environmental injustice is a global phenomenon closely integrated with social injustice.

18.3 Mathematics in Action

To critical mathematics education, it is crucial to engage in a critique of mathematics. It is important to address the question: What socio-political and economic roles might be acted out through mathematic? By raising this question, critical mathematics education distances itself from any view of mathematics as harmless and innocent.

---

[3] For an introduction of the notion of environmental justice, see Chap. 4, "Sustainability and Risks".

Mathematics put into action can operate in powerful ways. Mathematics is part of all kinds of automatised production processes. Robots and robot-like machines only function via algorithms. Work processes can be observed and controlled via mathematics, and the modern version of a panopticon is a mathematical construct. A panopticon can take the form of an ever-expanding database that includes all information about workers' performances in a production process. In today's informational capitalist society, economic transactions are mathematics based. This applies to everyday actions of doing shopping, it applies to the functioning of the stock market, and it applies to large-scale business transactions. Today, mathematical techniques have become an integral part of the capitalist profit-generating machinery.

Mathematics plays a crucial role in identifying new technological possibilities, which might include fascinating opportunities as well as devastating risks. Today, *technological imagination* is most often based on mathematics. As an example, aeronautical research investigates the possibility of making the human pilot superfluous. The removal of the pilot can take place in steps. One is to substitute two pilots in an aircraft with one pilot and a virtual pilot. A further step is to let the human pilot remain on the ground, and from there to direct the aircraft alongside the virtual pilot. A final step is for the virtual pilot to take over completely. Such technological imaginations are based on mathematical resources. The virtual pilot is a mathematical construct. The further construction of mathematical algorithms is a prerequisite for manufacturing this construct. By establishing conditions for carrying out experimental forecasting, mathematics plays a crucial role in conducting *hypothetical reasoning*. Empirical investigations of establishing virtual piloting cannot be carried out with real airplanes; this could be catastrophic. All initial experiments are carried out in mathematics-created virtual realities. Hypothetical reasoning might concern questions like: How would the automatics pilot handle a situation when the airplane cannot land at the designated airport, say due to weather problems? For a company to engage in a project of creating virtual piloting, different lines of *justification* need to be formulated. One important argument concerns the possibility of saving money. A significant factor is the use of petrol, which depends on the weight of the aircraft. If we consider smaller aircrafts, like those for carrying parcels or taxiing people, the weight of the pilot adds significantly to the total weight of the aircraft. To elaborate on relevant cost-benefit analyses, mathematics is brought into action. When mathematical algorithms are implemented in the shape of virtual pilots, parts of our reality get changed. In this sense, mathematics becomes *real*. When the human pilot has made space for the virtual pilot, the reality will be changed for the passengers. Going by airplane carries some risk, and such risk structures will take a different format due to automatic piloting. The reality will also change for the airplane companies. They might save money, but they will have to deal with new kinds of danger. Accidents might happen, but who can be held responsible? The company? The managers that argued for saving money? The computer expert that designed the automatic pilot? The mathematicians that constructed the necessary algorithms? When mathematics is brought into action, a *dissolution of responsibility* often seems to take place.

18.3 Mathematics in Action

By paying attention to technological imagination, hypothetical reasoning, justification, realisation, and dissolution of responsibility, I have tried to show connections between mathematics and power. Mathematics provides resources that can be brought into action for any kinds of purpose and with different kinds of social impact.[4]

To critical mathematics education, it is important not only to address mathematics through critique, but also to conduct critiques by means of mathematics. Mathematics can be brought into action in order to identify cases of social and environmental injustice. Mathematics can be used for pointing out economic inequalities. The Gini index is one such tool, and explorations by means of this index can, for instance, be part of project work in secondary schools. The mathematical details of the Gini Index reach beyond the secondary-school level, but the mathematical principles associated with the index can be clarified for secondary-school students.[5] Students can address economic conditions as they might be experienced by a poor family. Students might come to observe that inflation might hit poor families much harder than it will hit well-off families. The identification of social injustice is not only for students suffering such injustice; it is for all groups of students.[6] Problems related to racism can be exposed in terms of numbers.[7] All students can be engaged in addressing environmental injustice. By means of simple mathematics, primary-school students can investigate the use of water in a household. The use of water and other environmental issues can also by university students in mathematics. Thus, the discussion of climate change is rooted in applications of advanced mathematical models.[8]

Critical mathematics education recognises that the relationship between mathematics and power is indefinite. Mathematics can be brought into action for any kinds of technological, economic, ideological, and political purpose. It can also be brought into action to empower students. Mathematics and power make up a contested relationship.

---

[4] See Chap. 7, "Banality of Mathematical Expertise", for a discussion of technological imagination, hypothetical reasoning, justification, realisation, and dissolution of responsibility. See Chap. 8, "Mathematics and Ethics", for a different way of addressing mathematics in action.

[5] See Chap. 3, "Democracy and Erosions", for a presentation of the Gini Index.

[6] See Chap. 15, "All Students", for an example of the ways in which upper-middle-class students can address living conditions of poor families.

[7] See Chap. 2, "Landscapes and Racism", for discussions of how racism can be addressed in a classroom context.

[8] See Chap. 4, "Sustainability and Risk", for a discussion of how environmental issues can be addressed in primary education, as well as in mathematics teacher education.

18.4 Foregrounds

For a philosophy of critical mathematics education, it is important to consider the students and teachers being the principal agents in learning processes. I do not make a particular distinction between learning and teaching, as we are dealing with features of the same integral process. I choose to refer to this process as a learning process.

Several educational theories have assumed rather simplified conceptions of students. Behaviourism is a classic example, seeing learners as stimulus-response systems. Through his genetic epistemology, Jean Piaget provides a different simplification of learners, finding them to be subject to processes of assimilation and accommodation.[9] For critical mathematics education, it is crucial not to assume any simplified interpretation of students, but to see them as holistic human beings situated in the complexities of their life-worlds, formatted by economic conditions, cultural traditions, religious convictions, political powers, and dominant discourses.

An important province of a student's life-world is the *student's foreground*.[10] With this expression, I refer to the opportunities in life that are available to the student. These opportunities form the horizon towards which students may direct their actions. The opportunities can be very different for different groups of students. Statistics show that the likelihood that students undertake further education, come to enjoy adequate economic security, get access to healthcare (and so on) depends on factors such as gender, ethnicity, and economic resources. Such factors bring about a social conditioning of foregrounds.

In trying to understand students' motives – or lack of motives – for learning, it is important to consider their foregrounds and what they consider to be their possibilities in life. Teachers might find that the students do not want to engage in the mathematical activities that they have prepared so carefully. The students may not care, even though (by all standards) the activities appear fascinating. The students might even try to obstruct the teaching. Possible explanations need not be located within the classroom itself. They need not have anything to do with the teachers, their preparations, or the nature of the proposed activities. It could be that the students are subjected to a devastating social conditioning of their foregrounds. It could be that they have acknowledged that they, realistically speaking, cannot dream of any further education, and certainly not any further education where they can make use of mathematics. The students might feel that they are living in a hopeless situation. Hopelessness is a devastating obstruction for learning.

The notion of foreground is part of a circle of concepts, by means of which I try to capture the social complexity that the learning of mathematics takes place in.[11] The circle includes the following links: *learning and action, action and intention,*

---

[9] For a brief summary of genetic epistemology, see Piaget (1970).

[10] See Chap. 5, "To Learn or Not to Learn", for a discussion of students' foregrounds. See also Skovsmose (2014).

[11] See Chap. 5, "To Learn or Not to Learn", for a presentation of such a conceptual circle.

18.4 Foregrounds

intention and foreground, foreground and motive, motive and mathematics, mathematics and life-world, and *life-world and learning*. Here, let me make some comments about the two links including the notion of foreground.

Intention and *foreground* are closely related concepts. One can try to characterise actions by means of intentions. Taking a walk is something that I see as an action, while stumbling over a stone when walking I refer to as a physical behaviour. I try to differentiate between a physical behaviour and an action by means of the notion of intention. One has the intention of taking a walk, but not of stumbling over a stone. An intention gives direction to an action. I see learning as action, and therefore I pay particular attention to the possible intentions that students might bring into their learning activities. The foregrounds establish horizons of the future towards which the students might direct their intentions. Some features of a foreground might be illusionary, some might be realistic. Some groups of students might experience their foregrounds being ruined. With this expression, I refer to situations where many future possibilities are eliminated: for instance, due to poverty. Facing amputated foregrounds, students might come to withdraw their intentions for learning. There is little to hope for.

Foreground and *motive* are related, as motives for learning are formed by what students perceive as their possibilities. It is difficult for a student to find motives for learning mathematics, if there are no visible possibilities in life where a mastering of mathematics appears relevant. Ruined foregrounds annihilate destroy motives for learning. They bring about devastating learning obstacles. To interpret students' achievements in mathematics, I find it crucial to consider how their foregrounds are formatted, maybe ruined, due to social factors. I see low levels of achievement in mathematics, not as a consequence of some students' personal inabilities, but rather as an expression of brutal processes of social marginalisation. Groups of students get dumped into hopeless situations. This is a painful learning obstacle.

Like students, teachers have been subjected to simplification through educational theorising. The Modern Mathematics Movement, as developed throughout the 1960s and 1970s, represents one such example. According to this movement, the educational organisation of mathematics should be similar to its structural organisation, in particular as presented by Bourbaki. Like mathematics in general, mathematics education should start out with set theory. It was recognised that teachers might not master modern mathematics, but new textbooks were written, and they were accompanied by detailed teacher manuals. If teachers followed the prescriptions carefully, the reform would be implemented with success. This was the assumption. The role of the teacher was to function as a mouthpiece, through which the reformers could talk directly to the students.

Contrary to any such conception of teachers, critical mathematics education finds it important to see teachers as an integral part of learning processes. Such processes are formed through meetings between students and teachers submerged in the complexities of life. Like students, teachers have aspirations and hopes for the future. However, teachers are operating within a restricted set of possibilities contrived by regulations, curriculum demands, and test requirements. In this sense, teachers' foregrounds might be, if not amputated, then highly structured. Teachers are also operating within the labour market, controlled by political and economic factors. Teacher are not only educators but also salaried workers. This dual role might lead to contradictory demands, e.g. with respect to how much time to use for preparation. Teachers have families and private obligations that might also conflict with the demand of being well prepared.

In educational theories, teachers' working conditions have normally not been addressed. Issues about salaries, working hours, holidays, time for preparation, time for collaboration with colleagues, and so on have been considered trade union issues. However, I suggest that we consider such issues as being educational as well.

In critical mathematics education literature, one finds descriptions of educational practices where the teachers play an impressive role in engaging students in challenging activities, conducting profound socio-political investigations, and engaging students in activist programmes. I admire the work of these super-teachers. However, I do not want to impose the role of a super-teacher as a requirement for engaging with critical mathematics education. I find that any teacher could come to share the concern of critical mathematics education, but also that the conditions for acting depend on the teachers' work circumstances and life situations in general.

18.5 Dialogue and Critique

To think of learning in terms of dialogue is far from a common conception. Neither behaviourism, nor Piaget, nor the Modern Mathematics Movements paid much attention to dialogic processes. The school mathematics tradition can also be seen as an obstruction for establishing mathematical learning as dialogue; according to this tradition, the quality of the students' leaning depends first of all on the quality of the teachers' presentations.

To critical mathematics education, learning is a process of interaction. Learning takes place through meetings between students and teachers with aspirations, ambitions, frustrations, worries, and intentions. It is a meeting between people with different foregrounds, located in complex life-worlds. Diverse patterns of interaction are possible, and to critical mathematics education, dialogue plays a crucial role. Inspiration for assigning a special role to dialogue comes from Paulo Freire (2000), who saw learning as a political process addressing issues of oppression and socioeconomic injustice. In order to play this role, education cannot be reduced to informative delivery.

Dialogic interactions provide resources for critical activities. This observation makes it important to create learning environments that stimulate dialogic interactions, and not to maintain the teacher's exposition as the primary component of a learning process. The idea of creating landscapes of investigation and inviting students to explore such landscapes is to make space for dialogic interactions. Such interaction is vital for any form of critical activity.

By means of a dialogic theory of learning mathematics, I have tried to show the intimate relationship between dialogue and critique. The dialogic theory is

18.5 Dialogue and Critique

condensed in terms of six learning interacts: *getting in contact, exploring, positioning, foregrounding, externalising,* and *doubting*.[12] Here, let me make some comments about three of these interacts, namely positioning, externalising, and doubting.

Positioning becomes an important learning interact when one acknowledges that there are no absolute truths. This also applies to mathematics. Positioning refers to the act of relating one kind of knowledge to another kind of knowledge. Whenever I talk about knowledge, I think of it as referring to what students or other groups of people assume to be known. I do not embark on any idealised conception of knowledge as capturing truths with certainty. Knowledge is a temporary and local phenomenon. When knowledge is considered in this way, positioning turns into an important learning interact. When students come to learn something new, it must be positioned to what they already know. Is there a resonance between these two kinds of knowledge? Does one add to the other? Are there some kinds of tension between the two kinds of knowledge? Do they contradict each other? As part of positioning, it is also important to relate knowledge to what one can refer to as community knowledge. This is the knowledge that is generally adopted in the cultural, economic, and political setting to which the students belong. Positioning might reveal resonance or tension between students' newly obtained knowledge and their parents' knowledge. It is also important that teachers position themselves: this could be with respect to mathematical as well as socio-political issues. Positioning is a way of questioning; it prevents one kind of knowledge from assuming an undisputable status. Positioning is a dialogic interact. It is established through meetings between different positions. Simultaneously, positioning is a critical interact. It creates resources for questioning any taken-for-granted position.

Externalising is a different form of internalising. By "internalising", I mean the individual act of incorporating new knowledge into what a person already knows. In many theories of learning, internalising has been assigned a principal role for the way we come to know, but for a dialogic theory of learning mathematics, externalising plays an even more important role. By "externalising", I mean the communication and application of what is learned. Externalising is a way of bringing knowledge into action. Externalising can take place as part of students' group work, where some students tell the others how they would address a certain problem. A class might have been working on a certain project, and the students might present their results to the other students from the class. Such a process does not simply consist of presenting obtained knowledge; it includes a further elaboration of knowledge. The group that makes the presentation might come to recognise that what they present is only a preliminary insight. They might also be surprised by some questions being raised by other students. They might recognise that they have, at some point, been mistaken in their own beliefs. Students might present the results of a project to other students from the school and, in this way, bring externalisation beyond the walls of the classroom. Externalisation might reach beyond the walls of the school,

---

[12] See Chap. 14, "A Dialogical Theory of Learning Mathematics?", for a presentation of the learning interacts.

when results are communicated to parents, families, or the community. Externalising is a process of communicating something to others, but it is not a one-way process. Whenever one communicates something, one gets feedback. Externalisation is a learning interact that brings dialogue and critique together. Eventually, externalising can turn into political activism.

Doubting is related to questioning, by means of which one can try to clarify an uncertainty. One can get more information, and one can try to check the validity of one's insight. However, I find doubting to reach deeper than questioning. Doubting concerns the status of what might be assumed to count as knowledge. As part of daily-life practices, one takes many things for granted. If one looks back in time, it is easier to acknowledge the questionable character of much taken-for-granted knowledge. It has been "common knowledge" that men are more rational than women and, consequently, that women could not have the right to vote. Many versions of racism have been part of "common knowledge": it has been taken as evident that some groups of people were inferior compared to other groups of people. It has been "common knowledge" that poor people are the principal cause of their own misery. Today, we might still experience leftovers from historical taken-for-granted knowledge, and we might have added a portion of new questionable presumptions to what we today consider to be knowledge. In the future, much of our taken-for-granted knowledge might be revealed as being highly questionable. Coming to doubt is an important feature of any learning process. Whatever new knowledge one obtains, it might raise doubts about already established knowledge; and any new knowledge also needs to be doubted. Nothing can be taken for granted.

I have presented doubting as a learning interact. However, doubting represents an existential experience of uncertainty. Many philosophical trends have been in search of certainties: How can we establish knowledge with certainty? How can we locate ethical guidelines that we can rely on? How can we identify valid political positions? However, I find that any such search for certainty is going to fail. The dream of reaching certainty is a deception. Our life-worlds are submerged in uncertainty. Uncertainty constitutes a human condition. Critical mathematics education is not accompanied by any well-defined operational approaches, but needs to permanetly be a search for what to do. Critical mathematics education is an expression of concerns and hopes.

Note

This chapter includes some pieces of text from "A Philosophy of Critical Mathematics Education" published in 2021 in the *Philosophy of Mathematics Education Journal*, (37). I thank the editor for their kind permission to include these pieces.

References

Beck, U. (1998). Politics of risk society. In J. Franklin (Ed.), *The politics of risk society* (pp. 9–22). Polity Press.

Freire, P. (2000). *Pedagogy of the oppressed* (With an Introduction by Donaldo Macedo). Bloomsbury.

References

Piaget, J. (1970). *Genetic epistemology*. Columbia University Press.
Rosmini-Serbati, A. (2007). *The constitution under social justice* (Translated by Albetto Mingardi). Lexington Books.
Skovsmose, O. (2014). *Foregrounds: Opaque stories about learning*. Sense Publishers.

Composed Reference List

Abtahi, Y., Gøtze, P., Steffensen, L., Hauge, K. H., & Barwell, R. (2017). Teaching climate change in mathematics classroom: An ethical responsibility. *Philosophy of Mathematics Education Journal, 32*.

Ackerman, G., & Potter, W. C. (2008). Catastrophic nuclear terrorism: A preventable peril. In N. Bostrom & M. M. Cirkovic (Eds.), *Global catastrophic risks* (pp. 402–449). Oxford University Press.

Adorno, T. W. (1971). *Erziehung zur Mündigkeit (Education for authority)*. Suhrkamp.

Aguirre, J. M., & Zavala, M. R. (2013). When equal isn't fair. In E. Gutstein & B. Peterson (Eds.), *Rethinking mathematics: Teaching social justice by the numbers* (2nd ed., pp. 115–121). Rethinking Schools.

Albertazzi, L. (2006). *Immanent realism: An introduction to Brentano*. Springer.

Almeida, S. (2019). *Racismo estrutural (Structural racism)*. Editora Jandaíra.

Alrø, H., & Skovsmose, O. (2004). *Dialogue and learning in mathematics education: Intention, reflection, critique*. Kluwer Academic Publishers.

Andersson, A., & Barwell, R. (Eds.). (2021). *Appling critical mathematics education*. Brill.

Apple, M. W. (1982). *Education and power*. Routledge and Kegan Paul.

Apple, M., Au, W., & Gandin, L. A. (Eds.). (2009). *The Routledge international handbook of critical education*. Routledge.

Araújo, J. L. (2007). *Educação matemática crítica (Critical mathematics education)*. Argumentum Editora.

Araújo, J. L. (2009). Uma abordagem sócio-crítica da modelagem matemática: A perspectiva da educação matemática crítica (A socio-critical approach to mathematical modeling: A critical mathematics education perspective). *Alexandria, 2*, 55–68.

Arendt, H. (1951). *The origins of totalitarianism*. Schocken Books.

Arendt, H. (1977). *Eichmann in Jerusalem: A report on the banality of evil*. Penguin Books.

Au, W. (2011). Teaching under the new Taylorism: High-stakes testing and the standardization of the 21st century curriculum. *Journal of Curriculum Studies, 43*(1), 25–45.

Austin, J. L. (1962). *How to do things with words*. Oxford University Press.

Avci, B. (2018). *Critical mathematics education: Can democratic mathematics education survive under neoliberal regime?* Brill and Sense Publishers.

Avigad, J. (2008). Computers in mathematical inquiry. In P. Mancosu (Ed.), *The philosophy of mathematical practice* (pp. 302–316). Oxford University Press.

Ayer, A. J. (Ed.). (1959). *Logical positivism*. The Free Press.

Azevedo, S. S. (2019). *Educação financeira nos livros didáticos de matemática nos anos finais do ensino fundamental (Financial education in mathematics textbooks in the final years of elementary school)*. Master's thesis. Federal University of Pernambuco (UFPE).

Baber, S. A. (2007). *Interplay of citizenship, education and mathematics: Formation of foregrounds of Pakistani immigrants in Denmark*. Doctoral Dissertation. Aalborg University.

Baber, S. (2012). Learning of mathematics among Pakistani immigrant children in Barcelona: A socio-cultural perspective. In O. Skovsmose & B. Greer (Eds.), *Opening the cage: Critique and politics of mathematics education* (pp. 144–166). Sense Publishers.

Bales, K. (2012). *Disposable people: New slavery in the global economy* (3rd ed.). University of California Press.

Baroni, A. K. C., Hartmann, A. L. B., & Carvalho, C. C. S. (Eds.). (2021). *Uma abordagem crítica da educação financeira na formação do professor da matemática (A critical approach to financial education in mathematics teacher education)*. Editora Appris.

Barros, D. D. (2021). *Leitura e escrita de mundo com a matemática e a comunidade LGBT+: As lutas e a representatividade de um movimento social (Reading and writing the world with mathematics and LGBT+ commmunity: The struggles and representation of a social movement)*. Doctoral Dissertation. Universidade Estadual Paulista (Unesp).

Barwell, R. (2013). The mathematical formatting of climate change: Critical mathematics education and post-normal science. *Research in Mathematics Education, 15*(1), 1–16.

Barwell, R. (2018). Some thoughts on a mathematics education for environmental sustainability. In P. Ernest (Ed.), *The philosophy of mathematics education today* (pp. 145–160). Springer.

Barwell, R., & Hauge, K. H. (2021). A critical mathematics education for climate change. In A. Andersson & R. Barwell (Eds.), *Appling critical mathematics education* (pp. 166–184). Brill.

Bauman, Z. (1998). *Globalization: The human consequences*. Polity Press.

Beck, U. (1992). *Risk society: Towards a new modernity*. Sage Publications.

Beck, U. (1998). Politics of risk society. In J. Franklin (Ed.), *The politics of risk society* (pp. 9–22). Polity Press.

Beck, U. (1999). *World risk society*. Polity Press.

Bell, D. (1987). *And we are not saved: The elusive ques for racial justice*. Basic Books.

Bell, D. (2002). *Ethical ambition: Living a life of meaning and worth*. Bloomsbury.

Bell, D. (2004). *Silent covenants*. Oxford University Press.

Benjamin, W. (1999). *The arcades project*. The Belknap Press of Haward University Press.

Bernacerraf, P., & Putnam, H. (Eds.). (1964). *Philosophy of mathematics*. Prentice-Hall.

Bernier, M. (2015). *The task of hope in Kierkegaard*. Oxford University Press.

Beth, E. W., & Piaget, J. (1966). *Mathematics, epistemology and psychology*. Reidel.

Biotto Filho, D. (2015). *Quem não sonhou em ser um jogador de futebol? Trabalho com projetos para reelaborar foregrounds (Who never dreamed of being a soccer player? Working with projects for elaborating foregrounds)*. Doctoral Dissertation. Universidade Estadual Paulista (Unesp).

Birsch, D., & Fielder, J. H. (Eds.). (1994). *The Ford Pinto case: A study in applied ethics, business, and technology*. State University of New York.

Bishop, A. (1988). *Mathematical enculturation*. Kluwer Academic Publishers.

Bloch, E. (1995). *The principle of hope, I-III*. MIT Press.

Blum, W., & Niss, M. (1991). Applied mathematical problem solving, modelling applications, and link to other subjects: State, trends and issues in mathematics education. *Educational Studies in Mathematics, 22*(1), 37–68.

Blum, W., et al. (Eds.). (1989). *Applications and modelling in learning and teaching mathematics*. Ellis Horwood.

Bohm, D. (1996). *On dialogue*. Routledge.

Bohman, J., & Rehg, W. (Eds.). (1997). *Deliberative democracy: Essays on reason and politics*. The MIT Press.

Bostock, D. (2009). *Philosophy of mathematics: An introduction*. Wiley Blackwell.

Bostrom, N., & Cirkovic, M. M. (2008). *Global catastrophic risks*. Oxford University Press.

Bourbaki, N. (1950). The architecture of mathematics. *The American Mathematical Monthly, 57*(4), 221–232.
Bourdieu, P. (1991). *Language and symbolic power* (Edited with an introduction by John B. Thompson). Polity Press.
Bourdieu, P. (1996). *The state nobility: Elite schools in the field of power*. Polity Press.
Brams, S. J. (2008). *Mathematics and democracy: Designing better voting and fair-division procedures*. Princeton University Press.
Brauer, F., Castillo-Chavez, C., & Feng, Z. (2019). Mathematical models in epidemiology. *Texts in applied mathematics, 69*. Springer.
Brentano, F. (1995a). *Psychology from an empirical standpoint* (Translated by A. C. Rancurello, D. B. Terrell, & L. L. McAlister; with an introduction by Peter Simons). Routledge.
Brentano, F. (1995b). *Descriptive psychology* (Translated and edited by Benito Müller). Routledge.
Bridges, K. M. (2019). *Critical race theory: A primer*. Foundation Press.
Britto, R. (2013). Educação matemática e democracia: Mídia e racismo. *Anais do VII CIBEM*, 3355–3362.
Britto, R. (2022). Media and Racism. In M. G. Penteado & O. Skovsmose (Eds.), *Landscapes of investigation: Contributions to critical mathematics education* (pp. 39–56). Open Book Publishers.
Brouwer, L. E. J. (1913). Intuitionism and formalism. *Bulletin of the America Mathematical Society, 20*(2), 81–96.
Brouwer, L. E. J. (1996). Life, art, and mysticism. *Notre Dame Journal of Formal Logic, 37*(3), 389–429.
Brown, D. (1970). *Bury my heart at Wounded Knee*. Henry Holt and Company.
Brown, J. R. (2008). *Philosophy of mathematics: A contemporary introduction to the world of proofs and pictures* (2nd ed.). Routledge.
Buber, M. (1996). *I and thou*. A Touchstone Book, Simon and Schuster.
Bueno, O., & Linnebo, Ø. (Eds.). (2009). *New waves in philosophy of mathematics*. Palgrave Macmillan.
Burton, L. (2004). *Mathematicians as enquirers: Learning about learning mathematics*. Kluwer Academic Publishers.
Calibaba, J. (2017). *The road to vision zero: Zero traffic fatalities and serious injuries* (2nd ed.). CreateSpace Independent Publishing Platform.
Campos, I. S. (2013). *Alunos em ambientes de modelagem matemática: Aaracterização do envolvimento a partir da relação com o background e o foreground (Students in mathematical modeling environments: Characterisation of involvement from the relationship with background and foreground)*. Master's thesis. Universidade Federal de Minas Gerais (UFMG).
Campos, I. S. (2018). *A divisão do trabalho no ambiente de aprendizagem de modelagem matemática segundo a educação matemática crítica (The division of labor in the mathematical modeling learning environment according to critical mathematics education)*. Doctoral dissertation. Belo Horizonte Universidade Federal de Minas Gerais (UFMG).
Capasso, V. (2008). Mathematical structures of epidemic systems. *Lecture Notes in Biomathematica, 97*. Springer.
Carrijo, M. H. S. (2023). *Beyond borders and racism: Towards an inclusive mathematics education for immigrant students*. Doctoral Dissertation. Universidade Estadual Paulista (Unesp).
Carter, J. (2014). Mathematics dealing with 'hypothetical states of things'. *Philosophia Mathematica, 22*(3), 209–230.
Carter, J. (2019). Philosophy of mathematical practice: Motivations, themes and prospects. *Philosophia Mathematica, 27*(1), 1–32.
Castells, M. (1996). *The information age: Economy, society and culture. Volume I, The rise of the network society*. Blackwell.
Castells, M. (1997). *The information age: Economy, society and culture. Volume II, The power of identity*. Blackwell.
Castells, M. (1998). *The information age: Economy, society and culture. Volume III, End of millennium*. Blackwell.

Chang, R. (2013). Toward an Asian American legal scholarship: Critical race theory, poststructuralism, and narrative space. In R. Delgado & J. Stefancic (Eds.), *Critical race theory: The cutting edge* (3rd ed., pp. 466–478). Tempe University Press.

Chassapis, D. (2017). "Numbers have the power" or the key role of numerical discourse in establishing a regime of truth about crisis in Greece. In A. Chronaki (Ed.), *Mathematics education and life at times of crisis: Proceedings of the Ninth International Mathematics Education and Society Conference* (pp. 45–55). Volos, Greece.

Chriss, N. A. (1997). *Black-Scholes and beyond: Option pricing models*. McGraw-Hill.

Chronaki, A. (Ed.). (2017). *Mathematics education and life at times of crisis: Proceedings of the Ninth International Mathematics Education and Society Conference*. Volos, Greece.

Cirincione, J. (2008). The continuing threat of nuclear war. In N. Bostrom & M. M. Cirkovic (Eds.), *Global catastrophic risks* (pp. 381–401). Oxford University Press.

Civiero, P. A. G. (2016). *Educação matemática crítica e as implicações sociais da ciência e da tecnologia no processo civilizatório contemporâneo: Embates para formação de professores de matemática (Critical education and the social implications of science and technology in the contemporary civilizing process: Conflicts for mathematics teachers education)*. Doctoral dissertation. Universidade Federal de Santa Catarina.

Civiero, P. A. G., & Oliveira, F. P. Z. (2020). Landscapes of investigation and scientific initiation: Possibilities in civilizatory equation. *Revista Acta Scientiae, 22*, 165–185.

Civiero, P. A. G., Milani, R., Lima, A. S., & Lima, A. S. (Eds.). (2022a). *Educação matemática crítica: Múltiplas possibilidades na formação de professores que ensinam matemática (Critical mathematics education: Multiple possibilities in the education of teachers who teach mathematics)*. Editora da SBEM, Brasília.

Civiero, P. A. G., Milani, R., Lima, A. S. & Miranda, F. O. (Eds.). (2022b). *Alçando voos com a educação matemática crítica: Discussões sobre a formação de professores que ensinam matemática (Get flying with critical mathematics education: Discussions about the education of teachers who teach mathematics)*. Editora IFC.

Civil, M. (2012). Mathematics teaching and learning of immigrant students: An overview of the research field across multiple settings. In O. Skovsmose & B. Greer (Eds.), *Opening the cage: Critique and politics of mathematics education* (pp. 127–142). Sense Publishers.

Civil, M. (2020). Immigrant students in mathematics education. In S. Lerman (Ed.), *Encyclopedia of mathematics education* (2nd ed., pp. 359–365). Springer.

Clark, M. (1990). *Nietzsche on truth and philosophy*. Cambridge University Press.

Coiffier, J. (2011). *Fundamentals of numerical weather prediction*. Cambridge University Press.

Coles, A. (2016). *Mathematics education in the anthropocene. Open Access*. University of Bristol.

Coles, A., Barwell, R., Cotton, T., Winter, J., & Brown, L. (2013). *Teaching secondary mathematics as if the planet matters*. Routledge.

Cortella, M. S. (2015). *Educação, convivência e ética: Audácia e esperança! (Education, coexistence and ethics: Audacity and hope!)*. Cortez Editora.

Cortella, M. S. (2017). *A escola e o conhecimento: Fundamentos epistemológicos e políticos (School and knowledge: Epistemological and political foundations)*. Cortez Editora.

Cozzens, P. (2016). *The earth is weeping: The epic story of the Indian wars for the American West*. Vintage Books.

Crenshaw, K., Gotanda, N., Peller, G., & Thomas, K. (Eds.). (1995). *Critical race theory: The key writing that formed the movement*. The New Press.

Crutzen, P. J., & Stoermer, E. F. (2000). The "anthropocene". *Global Change Newsletter, 41*(17), 17–18.

Curry, H. B. (1970). *Outlines of a formalist philosophy of mathematics*. North-Holland Publishing Company.

Damerow, P., Elwitz, U., Keitel, C., & Zimmer, J. (1974). *Elementarmatematik: Lernen für die praxis (Elementary mathematics: Learning for the practice)*. Ernst Klett.

Davis, P. C. (2013). Law of microaggression. In R. Delgado & J. Stefancic (Eds.), *Critical race theory: The cutting edge* (pp. 187–196). Tempe University Press.

Davis, P. J., & Hersh, R. (1988). *Descartes' dream: The world according to mathematics.* Penguin Books.
Deleuze, G. (1986). *Nietzsche and philosophy.* Continuum.
Delgado, R., & Stefancic, J. (Eds.). (2005). *The Derrick Bell reader.* New York University Press.
Delgado, R., & Stefancic, J. (Eds.). (2013). *Critical race theory: The cutting edge* (3rd ed.). Tempe University Press.
Descartes, R. (1993). *Meditations on first philosophy.* Routledge.
Descartes, R. (1998). *The world and other writings.* Cambridge University Press.
Dewey, J. (1966). *Democracy and education.* The Free Press.
Dewey, J. (1974). *On education* (Selected writings, edited and with an introduction by Reginald D. Archambault). The University of Chicago Press.
Dieudonné, J. A. (1970). The work of Nicholas Bourbaki. *The American Mathematical Monthly, 77,* 134–145.
Diffie, W., & Hellman, M. E. (1976). New direction in cryptography. *IEEE Transactions on Information Theory, 22,* 644–654.
Dilthey, W. (2002). *The formation of the historical world in the human sciences* (Selected works, Vol. 3). Princeton University Press.
Drummond, M. F., Sculpher, M., Claxton, K., Stoddart, G., & Torrance, G. (2015). *Methods for the economic evaluation of health care programmes* (4th ed.). Oxford University Press.
Duggan, J. (2016). *System dynamic modelling with R.* Springer.
Eisensee, T., & Strömberg, D. (2007). News droughts, news floods, and US disaster relief. *The Quarterly Journal of Economics, 122*(2), 693–728.
Eknoyan, G. (2008). Adolphe Quetelet (1796–1974): The average man and indices of obesity. *Nephrology Dialysis Transplantation, 23*(1), 47–51.
Engels, F. (1993). *The condition of the working class in England* (Edited with an introduction and notes by Davis Mclellan). Oxford University Press.
Ernest, P. (1998). *Social constructivism as a philosophy of mathematics.* State University of New York Press.
Ernest, P. (2016). A dialogue on the ethics of mathematics. *The Mathematical Intelligencer, 38*(3), 69–77.
Ernest, P. (Ed.). (2018). *The philosophy of mathematics education today.* Springer.
Ernest, P. (2018b). The ethics of mathematics: Is mathematics harmful? In P. Ernest (Ed.), *The philosophy of mathematics education today* (pp. 187–216). Springer.
Ernest, P. (2020). The ethics of mathematical practice. In B. Sriraman (Ed.), *Handbook of the history and philosophy of mathematical practice.* Springer.
Ernest, P., Sriraman, B., & Ernest, N. (Eds.). (2015). *Critical mathematics education: Theory, praxis, and reality.* Information Age Publishing.
Ernest, P., Skovsmose, O., Bendegem, J.-P., Bicudo, M., Miarka, R., Kvasz, L., & Möller, R. (Eds.). (2016). *The philosophy of mathematics education: ICME-13 Topical Surveys.* Springer.
Fanon, F. (2004). *The wretched on the earth.* Grove Press.
Fanon, F. (2008). *Black skin, white masks.* Grove Press.
Faustino, A. C. (2018). *"Como você chegou a esse resultado?": O processo de dialogar nas aulas de matemática dos anos iniciais do Ensino Fundamental. ("How did you get to this result?": The dialogue in mathematics classes of the early years of Elementary School).* Doctoral dissertation. Universidade Estadual Paulista (Unesp).
Faustino, A. C. (2021). Trabalho com projetos nos anos inicias do ensino fundamental: Dialogando para ler e escrever o mundo com matemática (Work with projects in the early years of elementary school: Dialogue in order to read and write the world with mathematics). In G. H. G. Silva, I. M. S. Lima, & F. A. G. Rodríguez (Eds.), *Educação matemática crítica e a (in)justiça social: Práticas pedagógicas e formação de professores (Critical mathematics education and social (in)justice: Pedagogical practices and teacher education)* (pp. 259–292). Mercado de Letras.
Faustino, A. C., & Skovsmose, O. (2020). Dialogic and non-dialogic acts in learning mathematics. *For the Learning of Mathematics, 40*(1), 9–14.

Ferreirós, J. (2016). *Mathematical knowledge and the interplay of practices*. Princeton University Press.

Feuerbach, L. (1989). *The essence of Christianity*. Prometheus Books.

Forrester, T. (1981). *The microelectronic revolution*. The MIT Press.

Foucault, M. (1989). *The archaeology of knowledge*. Routledge.

Foucault, M. (1994). *The order of things: An archaeology of the human sciences*. Vintage Books.

Foucault, M. (2000). *Power*. The New Press.

Frame, D., & Allen, M. R. (2008). Climate change and global risk. In N. Bostrom & M. M. Cirkovic (Eds.), *Global catastrophic risks* (pp. 265–286). Oxford University Press.

Frankenstein, M. (1983). Critical mathematics education: An application of Paulo Freire's epistemology. *Journal of Education, 165*(4), 315–339.

Frankenstein, M. (1989). *Relearning mathematics: A different third R – Radical maths*. Free Association Books.

Frankenstein, M. (1998). Reading the world with math: Goals for a critical mathematical literacy curriculum. In E. Lee, D. Menkart, & M. Okazawa-Rey (Eds.), *Beyond heroes and holidays: A practical guide to K-12 anti-racist, multicultural education and staff development*. Network of Educators on the Americas.

Frankenstein, M. (2012). Beyond math content and process: Proposals for underlying aspects of social justice education. In A. A. Wager & D. W. Stinson (Eds.), *Teaching mathematics for social justice: Conversations with mathematics educators* (pp. 49–62). NCTM, National Council of Mathematics Teachers, USA.

Frege, G. (1893/1903). *Grundgesetze der Arithmetik I-II*. H. Poble.

Frege, G. (1967). Begriffsschrift: A formula language, modelled upon that of arithmetic, for pure thought. In J. van Hiejenoort (Ed.), *From Frege to Gödel: A source book in mathematical logic, 1879–1931* (pp. 1–82). Harvard University Press.

Frege, G. (1978). *Die Grundlagen der Arithmetik/The foundations of arithmetic*. Blackwell.

Freire, P. (2000). *Pedagogy of the oppressed* (With an Introduction by Donaldo Macedo). Bloomsbury.

Freire, P. (2014). *Pedagogy of hope: Reliving pedagogy of the oppressed* (With notes by Ana Maria Araújo Freire; translated by Robert R. Barr). Bloomsbury.

Freire, P. (2021). *Pedagogia do Oprimodos, 80ª Edição (Pedagogy of the oppressed, 80th Edition)*. Paz e Terra.

Fremstedal, R. (2012). Kierkegaard on the metaphysics of hope. *The Heythrop Journal, 53*(1), 51–60.

Freud, S. (1997). *The interpretations of dreams* (Translation by A. A. Brill, Introduction by Stephen Wilson). Wordsworth Classics.

Galilei, G. (1957). The assayer. In G. Galilei (Ed.), *Discoveries and opinions of Galileo* (pp. 229–281). Anchor Books.

George, A., & Velleman, D. J. (2002). *Philosophies of mathematics*. Blackwell.

Giaquinto, M. (2008). Visualizing in mathematics. In P. Mancosu (Ed.), *The philosophy of mathematical practice* (pp. 22–42). Oxford University Press.

Gibbs, A. M., & Park, J. Y. (2022). Unboxing mathematics: Creating a culture of modelling as critique. *Educational Studies in Mathematics, 110*(1), 167–192.

Giddens, A. (1984). *The constitution of society: Outline of the theory of structuration*. University of California Press.

Giroux, H. A. (1989). *Schooling for democracy: Critical pedagogy in the modern age*. Routledge.

Gödel, K. (1962). *On formally undecidable propositions of Principia Mathematica and related systems*. Basic Book.

Grabiner, J. V. (1974). Is mathematical truth time-dependent? *The American Mathematical Monthly, 81*(4), 354–365.

Greer, B., Gutiérrez, R., Gutstein, E., Mukhopadhyay, S., & Rampal, A. (2017). Majority counts: What mathematics for life, to deal with crises? In A. Chronaki (Ed.), *Mathematics education* and life at times of crisis: Proceedings of the Ninth International Mathematics Education and Society Conference (pp. 159–163). Volos, Greece.

Gutstein, E. (2003). Teaching and learning mathematics for social justice in an urban, Latino school. *Journal for Research in Mathematics Education, 34*, 37–73.

Gutstein, E. (2006). *Reading and writing the world with mathematics: Toward a pedagogy for social justice*. Routledge.

Gutstein, E. (2009). Possibilities and challenges in teaching mathematics for social justice. In P. Ernest, B. Greer, & B. Sriraman (Eds.), *Critical issues in mathematics education* (pp. 351–373). Information Age Publishing.

Gutstein, E. (2012). Reflections on teaching and learning mathematics for social justice in urban schools. In A. A. Wager & D. W. Stinson (Eds.), *Teaching mathematics for social justice: Conversations with mathematics educators* (pp. 63–78). NCTM, National Council of Mathematics Teachers, USA.

Gutstein, E. (2016). "Our issue, our people – Math as our weapon": Critical mathematics in a Chicago neighborhood high school. *Journal for Research in Mathematics Education, 47*(5), 454–504.

Gutstein, E. (2018). The struggle is pedagogical: Learning to teach critical mathematics. In P. Ernest (Ed.), *The Philosophy of Mathematics Education Today* (pp. 131–143). Springer.

Habermas, J. (1971). *Knowledge and human interests*. Beacon Press.

Habermas, J. (1984/1987). *The theory of communicative action I–II*. Heinemann/Polity Press.

Hacking, I. (2014). *Why is there philosophy of mathematics at all?* Cambridge University Press.

Hales, S. D., & Welshon, R. (2000). *Nietzsche's perspectivism*. University of Illinois Press.

Hall, J., & Barwell, R. (2021). The mathematical formatting of obesity in public health discourse. In A. Andersson & R. Barwell (Eds.), *Applying critical mathematics education* (pp. 210–228). Brill and Sense Publishers.

Hannaford, C. (1998). Mathematics teaching *is* democratic education. *Zentralblatt für Didaktik der Mathematik, 98*, 181–187.

Hardy, G. H. (1967). *A mathematician's apology* (With a foreword by C. P. Snow). Cambridge University Press.

Harford, T. (2012). *Black Scholes: The maths formula linked to the financial crash*. https://www.bbc.com/news/magazine-17866646

Hauge, K. H., Gøtze, P., Hansen, R., & Lisa Steffensen, L. (2017). Categories of critical mathematics based reflection on climate change. In A. Chronaki (Ed.), *Mathematics education and life at times of crisis: Proceedings of the Ninth International Mathematics Education and Society Conference* (pp. 524–532). Volos, Greece.

Hawkins, A. J. (2019). *Deadly Boeing crashes raise questions about airplane automatisation*. https://www.theverge.com/2019/3/15/18267365/boeing-737-max-8-crash-autopilot-automation

Helliwell, T., & Ng, O.-L. (2022). Imagining possibilities: Innovating mathematics (teacher) education for sustainable futures. *Research in Mathematics Education, 24*(2), 128–149.

Hempel, C. G. (1959). The empiricist criterion of meaning. In A. Ayer (Ed.), *Logical positivism* (pp. 108–129). The Free Press.

Hempel, C. G. (1965). *Aspects of scientific explanation*. The Free Press.

Hersh, R. (1979). Some proposals for reviving the philosophy of mathematics. *Advances in Mathematics, 31*, 31–50.

Hersh, R. (1990). Mathematics and ethics. *Humanistic Mathematics Network Journal, 5*(9), 20–23.

Hersh, R. (1997). *What is mathematics, really?* Oxford University Press.

Hilbert, D. (1922). Neubegründung der Mathematik: Erste Mitteilung. *Abhandlungen aus dem Seminar der Hamburgischen Universität (Refoundation of Mathematics: First Communication), 1*, 157–177.

Hilbert, D. (1950). *Foundations of geometry*. The Open Court Publishing Company.

Hilbert, D., & Ackermann, W. (1958). *Principles of mathematical logic*. Chelsea Publishing Company.

Hofflander, A. E. (1966). The human life value: A theoretical model. *The Journal of Risk and Insurance, 33*(4), 529–536.
Hood, K. (2017). The science of value: Economic expertise and the valuation of human life in US federal regulatory agencies. *Social Studies of Science, 47*(4), 441–456.
Horkheimer, M. (2002). *Critical theory: Selected essays*. Continuum.
Horkheimer, M., & Adorno, T. W. (2002). *Dialectic of enlightenment*. Continuum.
Husserl, E. (1970). *The crisis of European sciences and transcendental phenomenology* (Translated with an introduction by David Car). Northwestern University Press.
Husserl, E. (1998). *Ideas pertaining to a pure phenomenology and to a phenomenological philosophy: First book*. Kluwer Academic Publishers.
Husserl, E. (1999). *Cartesian meditations: An introduction to phenomenology* (Translated by Dorion Cairns). Kluwer Academic Publishers.
Husserl, E. (2001). *Logical investigations: Volume 1–2* (Translated by J. N. Findlay). Routledge.
Imperial College COVID-19 Response Team. (2020). *Impact of non-pharmaceutical interventions (NPIs) to reduce COVID19 mortality and healthcare demand*. Imperial College.
Ince, J. (in progress). *Belief in African Americans as learners in mathematics*. Doctoral dissertation. Universidade Estadual Paulista (Unesp).
Jablonka, E., & Gellert, U. (2007). Mathematisation – demathematisation. In U. Gellert & E. Jablonka (Eds.), *Mathematization and de-mathematization: Social, philosophical and educational ramifications* (pp. 1–18). Sense Publishers.
Jacobini, O. R. A. (2004). *Modelagem matemática como instrumento de ação política na sala de aula (Mathematical modeling as an instrument of political action in the classroom)*. Doctoral dissertation. Universidade Estadual Paulista (Unesp).
Jacquette, D. (Ed.). (2002). *Philosophy of mathematics: An anthology*. Blackwell.
Jacquette, D. (Ed.). (2004). *The Cambridge companion to Brentano*. Cambridge University Press.
Jane, B., Fleer, M., & Gipps, J. F. (2007). Changing children's views of science and scientists through school-based teaching. *Asia Pacific Forum on Science Learning and Teaching, 8*(1), 1–21.
Jaworski, B. (2006). Theory and practice in mathematics teaching development: Critical inquiry as a mode of learning in teaching. *Journal of Mathematics Teacher Education, 9*(2), 187–211.
Jensen, A. A., Stentoft, D., & Ravn, O. (2019). *Interdisciplinarity and problem-based learning in higher education: Research and perspectives form Aalborg University*. Springer.
Johnson, B. (2010). *Algorithmic trading and DMA: An introduction to direct access trading strategies*. 4Myeloma Press.
Jørgensen, M. W., & Phillips, L. (2002). *Discourse analysis as theory and method*. Sage Publications.
Joseph, G. G. (2000). *The crest of the peacock: The non-European roots of mathematics*. Princeton University Press.
Kant, I. (1929). *Critique of pure reason* (Translated by Norman Kemp Smith). MacMillan.
Kant, I. (2007). *Critique of judgment*. Oxford University Press.
Kant, I. (2010). *Critique of practical reason*. Classic Books International.
Kiss, I. Z., Miller, J., & Simon, P. L. (2017). *Mathematics of epidemics on networks: From exact to approximate models*. Springer.
Klein, N. (2014). *This changes everything: Capitalism vs. the climate*. Simon and Schuster.
Knijnik, G. (1996). *Exclução e resistência: Educação matemática e legitimidade cultural (Exclusion and resistance: Mathematics education and cultural legitimacy)*. Artes Médicas.
Knijnik, G. (2007). Mathematics education and the Brazilian landless movement: Three different mathematics in the context for the struggle for social justice. *Philosophy of Mathematics Education Journal, 21*, 1–18.
Knijnik, G., & Wanderer, F. (2015). Mathematics education in Brazilian rural areas: An analysis of the public policy and the Landless Movement Pedagogy. *Open Review of Educational Research, 2*, 143–154.

Knijnik, G., & Wanderer, F. (2018). Broadening school mathematics curriculum: the complexity of teaching mathematical language games of different forms of life. In K. Yasukawa, A. Rogers, A. K. Jackson, & B. Street (Eds.), *Numeracy as social practice: Global and local perspectives* (pp. 121–132). Routledge.

Kollosche, D. (2017). Auto-exclusion in mathematics education. *Revista Paranaense de Educação Matemática, 6*(12), 38–63.

Kollosche, D. (Ed.). (2021). *Exploring new ways to connect: Proceedings of the Eleventh International Mathematics Education and Society Conference. 3 Volumes*. University of Klagenfurt.

Kolmos, A., Fink, F. K., & Krogh, L. (Eds.). (2004). *The Aalborg PBL model: Progress, diversity and challenges*. Aalborg University Press.

Körner, S. (1968). *The philosophy of mathematics*. Hutchinson University Library.

Lakatos, I. (1976). *Proofs and refutations*. Cambridge University Press.

Langville, A. N., & Meyer, C. D. (2012). *Google's PageRank and beyond: The science of search engine rankings*. Princeton University Press.

Lanigan, C. (2007). From assembly line to just-in-time: Preparing a capable workforce for the knowledge economy. *CIO*. https://www.cio.com/article/2439128/from-assembly-line-to-just-in-time%2D%2Dpreparing-a-capable-workforce-for-the-knowledge-economy.html

Latour, B. (2018). *Down to earth: Politics in the new climate regime*. Polity Press.

Leggett, C. (1999). *The Ford Pinto case: The valuation of life as it applies to the negligence efficiency argument*. https://users.wfu.edu/palmitar/Law&Valuation/Papers/1999/Leggett-pinto.html

Leibniz, G. (1703). *Explanation of binary arithmetic*. http://www.leibniz-translations.com/binary.htm

Leitgeib, H. (2009). On formal and informal provability. In O. Bueno & Ø. Linnebo (Eds.), *New waves in philosophy of mathematics* (pp. 263–299). Palgrave Macmillan.

Lempert, W. (1971). *Leistungsprinzip und Emanzipation (Performance and emancipation)*. Suhrkamp.

Lerman, S. (2000). The social turn in mathematics education research. In J. Boaler (Ed.), *Multiple perspectives on mathematics teaching and learning* (pp. 19–44). Ablex.

Lima, L. F. (2015). *Conversas sobre matemática com pessoas idosas viabilizadas por uma ação de extensão universitária (Conversations about mathematics with elderly people made possible by a university extension action)*. Doctoral dissertation. Universidade Estadual Paulista (Unesp).

Lima, A. S. (2018). *A relação entre conteúdos matemáticos e o campesinato na formação de professores de matemática (The relationship between mathematical content and the peasantry in the training of mathematics teachers)*. Doctoral dissertation. Universidade Federal de Pernambuco (UFPE).

Lima. P. C. (2022). *Imaginação pedagógica e educação inclusiva: Possibilidades para a formação de professores de matemática (Pedagogical imaginations and inclusive education: Possibilities for mathematics teacher education)*. Doctoral dissertation. Universidade Estadual Paulista (Unesp).

Lima, A. S., & Lima, I. M. S. (2019). Diálogo, investigação e criticidade em um curso de licenciatura em educação do campo (Dialogue, investigation and critique in a degree course in rural education). *Revista de Matemática, Ensino e Cultura, 14*, 67–79.

Lincoln, Y. S. & Guba, E. G. (1985). *Naturalistic inquiry*. Sage Publications.

Linnebo, Ø. (2017). *Philosophy of mathematics*. Princeton University Press.

Locke, J. (1988). *Two treatises of government* (Edited by P. Laslett). Cambridge University Press.

Locke, J. (1997). *An essay concerning human understanding*. Penguin Books.

Lomborg, B. (2001). *The sceptical environmentalist: Measuring the real state of the world*. Cambridge University Press.

Lynn, M., & Dixon, A. D. (Eds.). (2022). *Handbook of critical race theory in education* (2nd ed.). Routledge.

Madley, B. (2017). *An American genocide: The United States and the California Indian catastrophe, 1846–1873.* Yale University Press.

Madsen, K. B. (1974). *Modern theories of motivation.* Munksgaard.

Makarenko, A. S. (2001). *The road to life: An epic of education* (Vol. 1). University Press of the Pacific.

Malheiros, A. P. (2002). *A produção matemática dos alunos em ambiente de modelagem (Students' mathematical production in a modeling environment).* Master's thesis. Universidade Estadual Paulista (Unesp).

Mancosu, P. (Ed.). (2008). *The philosophy of mathematical practice.* Oxford University Press.

Manders, K. (2008). Diagram-based geometric practice. In P. Mancosu (Ed.), *The philosophy of mathematical practice* (pp. 65–79). Oxford University Press.

Marcone, R. (2015). *Deficiencialismo: A invenção da deficiência pela normalidade. (Deficiencialism: The invention of disability by normality).* Doctoral dissertation. Universidade Estadual Paulista (Unesp).

Marcuse, H. (1964). *One-dimensional man: Studies in the ideology of advanced industrial society.* Beacon Press.

Marx, K. (1844). *Economic and philosophic manuscripts of 1844.* https://www.marxists.org/archive/marx/works/download/pdf/Economic-Philosophic-Manuscripts-1844.pdf

Marx, K. (1970). *A contribution to the critique of political economy.* International Publishers.

Marx, K. (1992/1993). *Capital: A critique of political economy, I–III.* Penguin Classics.

Marx, K. (1998). *The German ideology.* Prometheus Books.

McKenzie, D. (2007). *Mathematics of climate change: A new discipline for an uncertain century.* Mathematical Sciences Research Institute.

McLaren, P. (1997). *Revolutionary multiculturalism: Pedagogies of dissent for the new millennium.* Westview Press.

McLaren, P. (2016). *Pedagogy of insurrection: From resurrection to revolution.* Peter Lang.

Mehlberg, H. (Ed.). (1960). The present situation in the philosophy of mathematics. *Synthese, 12*(4), 380–414. Reprinted in D. Jacquette (Ed.) (2002), *Philosophy of mathematics: An anthology* (pp. 65–82). Oxford: Blackwell.

Mellin-Olsen, S. (1977). *Læring som sosial prosess (Learning as a social process).* Gyldendal.

Mellin-Olsen, S. (1981). Instrumentalism as an educational concept. *Educational Studies in Mathematics, 12*, 351–367.

Mellin-Olsen, S. (1987). *The politics of mathematics education.* Reidel Publishing Company.

Menghini, M., Furinghetti, F., Ciacardi, L., & Arzarello, F. (2008). *The first century of the international commission on mathematical instruction (1908–2008): Reflecting and shaping the world of mathematics education.* Instituto della Enciclopedia Italiana.

Merton, R. (1968). *Social theory and social structure.* Free Press.

Milani, R. (2015). *O processo de aprender a dialogar por futuros professores de matemática com seus alunos no estágio supervisionado (The process of prospective mathematics teachers learning to dialogue with their students during their supervised internship).* Doctoral dissertation. Universidade Estadual Paulista (Unesp).

Milani, R., & Skovsmose, O. (2014). Inquiry gestures. In O. Skovsmose (Ed.), *Critique as uncertainty* (pp. 45–56). Information Age Publishing.

Milani, R., Civiero, P. A. G., Soares, D. A. & Lima, A. S. (2017). O diálogo nos ambientes de aprendizagem nas aulas de matemática (Dialogue in learning environments in mathematics classrooms). *Revista Paranaense de Educação Matemática, 6*(12), 221–245.

Miller, M. B. (2014). *Mathematics and statistics for financial risk management* (2nd ed.). Wiley.

Miranda, F. O. (2015). *A inserção da educação matemática crítica na escola pública: Aberturas, tensões e potencialidades (The insertion of critical mathematics education in the public school: Openings, tensions and potentialities).* Doctoral dissertation. Universidade Estadual (Unesp).

Moddleton, J. (2020). Motivation in mathematical learning. In S. Lerman (Ed.), *Encyclopaedia of mathematics education* (pp. 635–637). Springer.

Moe, T. M. (2003). Politics, control, and the future of school accountability. In P. E. Peterson & M. R. West (Eds.), *No child left behind? The politics and practice of school accountability* (pp. 80–106). Brookings Institution Press.

Mollenhauer, K. (1973). *Erziehung und Emanzipation (Education and emancipation)*. Juventa Verlag.

Mollenhauer, K. (2013). *Forgotten connections: On culture and upbringing*. Routledge.

Moran, D. (2005). *Husserl: Founder of phenomenology*. Polity Press.

Morgenthau, H. (1918). *Ambassador Morgenthau's story: A personal account of the Armenian genocide*. Doubleday, Page and Company.

Moses, R. B., & Cobb, C. E. (2001). *Radical equations: Civil rights from Mississippi to the Algebra Project*. Beacon Press.

Moura, A. C. (2020). *O Encontro entre surdos e ouvintes em cenários para investigação: Das incertezas às possibilidades nas aulas de Matemática (Meeting amongst deaf and hearing student in landscapes of investigation: From uncertainties to possibilities in mathematics educations)*. Doctoral dissertation. Universidade Estadual Paulista (Unesp).

Münzinger, W. (Ed.). (1977). *Projektorientierter Mathematikunterricht (Project-oriented mathematics education)*. Urbau and Schwarzenberg.

Muzinatti, J. L. (2018). *A "verdade" apaziguadora na educação matemática: Como a argumentação de estudantes de classe média pode revelar sus visão acerca da injustiça social ("Truth" as a pacifier in mathematics education: How the argumentation of middle-class students can reveal their views on social injustice)*. Doctoral dissertation. Universidade Estadual Paulista (Unesp).

Muzinatti, J. L. (2022). *Matemática: Verdade apaziguadora (Mathematics: Soothing truth)*. Appris.

Nagel, E. (1961). *The structure of science: Problems in the logic of scientific explanations*. Hardcourt, Brace and World.

Negt, O. (1968). *Soziologische Phantasie und exemplarisches Lernen. Zur Theorie der Arbeiterbildung (Sociological Imagination and exemplary learning: On the theory of workers' education)*. Europäische Verlagsanstalt.

Neumann, F. (2009). *Behemoth: The structure and practice of National Socialism*. Roman and Littlefield.

Newton, I. (1999). *The principia: Mathematical principles of natural philosophy*. University of California Press.

Nietzsche, F. (1968). *The will to power*. Vintage Books.

Nietzsche, F. (1974). *The gay science*. Vintage Books.

Nietzsche, F. (1986). *Human, all too human: A book for free spirits* (Translated by Marion Faber with Stephen Lehmann). University of Nebraska Press.

Nietzsche, F. (1998a). *Beyond good and evil*. Oxford University Press.

Nietzsche, F. (1998b). *On the genealogy of morality*. Hackett Publishing Company.

Nietzsche, F. (2003). *Twilight of the idols and the Anti-Christ*. Penguin Books.

Nietzsche, F. (2007). *Ecce homo: How to become what you are* (Translated with an Introduction and Notes by Duncan Large). Oxford University Press.

Nietzsche, F. (2009). *Beyond good and evil: Prelude to a philosophy of the future* (Translated by Ian Johnson). Richer Resources Publications.

Nietzsche, F. (2010). On truth and lie in a nonmoral sense. In F. Nietzsche (Ed.), *On truth and untruth: Selected writings* (pp. 17–49). Harperperennial and Modern Thought.

Niss, M. (1989). Aims and scope of applications and modelling in mathematics curricula. In W. Blum et al. (Eds.), *Applications and modelling in learning and teaching mathematics* (pp. 22–31). Ellis Horwood.

Norström, L., Islind, A. S., & Lundh Snis, U. (2020). Algorithmic work: The impact of algorithms on work with social media. Conference: European Conference on Information Systems (ECIS 2020). *ECIS Research Papers, 185*. https://aisel.aisnet.org/ecis2020_rp/185

Nouri, A., & Chyba, C. F. (2008). Biotechnology and biosecurity. In N. Bostrom & M. M. Cirkovic (Eds.), *Global catastrophic risks* (pp. 481–503). Oxford University Press.

O'Neil, C. (2016). *Weapons of math destruction: How big date increase inequality and threatens democracy*. Broadway Books.

OEEC. (1961). *New thinking in school mathematics*. Organisation for European Economic Co-operation.

Oliveira, M. (2015). *Laboratório de investigação matemática e educação financeira (Landscapes of investigaion and financial education)*. Master's Thesis. Universidade Federal de Juiz de Fora (UFJF).

Oliveira, L. P. F. (in progress). *Ler e escrever o mundo com matemática: explorando possibilidades com estudantes dos anos finais do Ensino Fundamental (Reading and writing the world with mathematics: exploring possibilities with middle school students)*. Doctoral dissertation. Universidade Estadual Paulista (Unesp).

Orwell, G. (1937). *The road to Wigan Pier*. Victor Gollancz.

Pais, A. (2012). A critical approach to equity in mathematics education. In O. Skovsmose & B. Greer (Eds.), *Opening the cage: Critique and politics of mathematics education* (pp. 49–91). Sense Publishers.

Parenti, C. (2001). Big brother's corporate cousin: High-tech workplace surveillance is the hallmark of a new digital Taylorism. *The Nation, 273*(5), 26–30.

Parra, A. I. S. (2018). *Curupira's walk: Prowling ethnomathematics theory through decoloniality*. Doctoral dissertation. Aalborg University Press.

Parra, A., Bose, A., Alshwaikh, J., Gonzáles, M., Marcone, R., & D'Souza, R. (2017). "Crisis" and the interface with mathematics education research and practice: An everyday issue. In A. Chronaki (Ed.), *Mathematics education and life at times of crisis: Proceedings of the Ninth International Mathematics Education and Society Conference* (pp. 174–178). Volos, Greece.

Paul, S. (2021). *Philosophy of action: A contemporary introduction*. Routledge.

Peirce, C. S. (1960). *The simplest mathematics: Collected papers. Volume IV*. The Belknap Press of Harvard University Press.

Penteado, M. G., & Skovsmose, O. (2015). Mathemacy in a mathematized and demathematized world. In B. S. D'Ambrosio & C. E. Lopes (Eds.), *Creative insubordination in Brazilian mathematics education research* (pp. 107–117). Lulu Press.

Penteado, M. G., & Skovsmose, O. (Eds.). (2022). *Landscapes of investigation: Contributions to critical mathematics education*. Open Book Publishers.

Phillips, C. (2011). Institutional racism and ethnic inequalities: An expanded multilevel framework. *International Journal of Sociology and Social Policy, 40*(1), 173–192.

Phoenix, C., & Treder, M. (2008). Nanotechnology as global catastrophic risk. In N. Bostrom & M. M. Cirkovic (Eds.), *Global catastrophic risks* (pp. 450–480). Oxford University Press.

Piaget, J. (1970). *Genetic epistemology*. Columbia University Press.

Pólya, G. (1957). *How to solve it*. Doubleday.

Popper, K. R. (1965). *The logic of scientific discovery*. Harper and Row.

Powell, J. A. (2008). Structural racism: Building upon the insights of John Calmore. *North Carolina Law Review, 86*, 791–816.

Powell, A., & Frankenstein, M. (Eds.). (1997). *Ethnomathematics: Challenging Eurocentrism in mathematics education*. State University of New York Press.

Ravn, O., & Skovsmose, O. (2019). *Connecting humans to equations: A reinterpretation of the philosophy of mathematics*. Springer.

Rawls, J. (1999). *A theory of justice*. Oxford University Press.

Reed, A. L. (2022). *The south: Jim Crow and its afterlives*. Verso.

Restivo, S. (1992). *Mathematics in society and history*. Kluwer Academic Publishers.

Restivo, S., van Bendegem, J. P., & Fisher, R. (Eds.). (1993). *Math worlds: Philosophical and social studies of mathematics and mathematics education*. State University of New York Press.

Roger, C. R., & Farson, R. E. (1969). Active listening. In R. C. Huseman, C. M. Logue, & D. L. Freshley (Eds.), *Readings in interpersonal and organizational communication* (pp. 480–496). Holbrook.

Roncato, C. R. (2021). *Significado em educação matemática e estudantes com deficiências: Possibilidades de encontros de conceitos (Meaning in mathematics education and students with disabilities: Possibilities and meaning between concepts)*. Doctoral dissertation. Universidade Estadual Paulista (Unesp).

Rosa, M., D'Ambrosio, U., Orey, D. C., Shirley, L., Alangui, W. V., Palhares, P., & Gavarrete, M. E. (Eds.). (2016). *Current and future perspectives of ethnomathematics as a program. CME-13 Topical Surveys*. Springer.

Rosmini-Serbati, A. (2007). *The constitution under social justice* (Translated by Albetto Mingardi). Lexington Books.

Rousseau, J.-J. (1968). *The social contract*. Penguin Books.

Rousseau, J.-J. (1992). *Discourse on the origin and basis of inequality among men*. Indian Hackett Publishing.

Russell, B. (1905). On denoting. *Mind, 14*(56), 479–493.

Said, E. (1979). *Orientalism*. Vintage Books.

Sampaio, L. O. (2010). *Educação estatística crítica: Uma possibilidade? (Critical statistical education: A possibility?)* Master's thesis. Universidade Estadual Paulista (Unesp).

Scagion, M. P. (in progress). *Ler e escrever o mundo com matemática: A condição de vida de pessoas idosas e a educação matemática (Reading and writing the world with mathematics: Elderly peoples' life conditions and mathematics education)*. Doctoral dissertation. Universidade Estadual Paulista (Unesp).

Schneier, B. (1996). *Applied cryptography*. John Wiley and Sons.

Schoenfeld, A. H. (2000). Purposes and methods of research in mathematics education. *Notices*. American Mathematical Society.

Schumpeter, J. A. (1987). *Capitalism, socialism and democracy*. Unwin Paperbacks.

Schutz, A., & Luckmann, T. (1973). *The structures of the life-world, I–II*. Northwestern University Press.

Searle, J. (1969). *Speech acts*. Cambridge University Press.

Searle, J. (1983). *Intentionality*. Cambridge University Press.

Shapiro, S. (2000). *Thinking about mathematics: The philosophy of mathematics*. Oxford University Press.

Sieg, W. (1994). Mechanical procedures and mathematical experience. In A. George (Ed.), *Mathematics and mind* (pp. 71–117). Oxford University Press. Reprinted in D. Jacquette (Ed.) (2002), *Philosophy of mathematics: An anthology* (pp. 226–258). Oxford: Blackwell.

Silva, A. G. O. (2005). *Modelagem matemática: Uma perspectiva voltada para a educação matemática crítica (Mathematical modeling: A perspective of critical mathematics education)*. Master's thesis. Universidad estadule de Londrina.

Silva, A. A. (2013). *Os artefatos e mentefatos nos ritos e ceremonias do danhono (The artifacts and mental facts in the rites and ceremonies of the danhono)*. Doctoral dissertation. Universidade Estadual Paulista (Unesp).

Silva, E. B. (2014). *O diálogo entre diferentes sujeitos que aprendem e ensinam matemática no contexto escolar dos anos finais do ensino fundamental (The dialogue between people who learn and teach mathematics in the final years of elementary school)*. Doctoral dissertation, Universidade Federal de Brasilia (UNB).

Silva, G. H. G. (2016). *Equidade no acesso e permanência no ensino superior: O papel da educação matemática frente ás politicas de ações afirmativas para grupos sub-representados (Equity in access and permanence in higher education: The role of mathematics education in relation to affirmative action policies for underrepresented groups)*. Doctoral dissertation. Universidade Estadual Paulista (Unesp).

Silva, G. H. G., & Powell, A. B. (2016). Microagressoes no ensino superior nas vias da Educacão Matemática (Microaggressions in higher education occurring in mathematics education). *Revista Latinoamericana de Etnomatemática: Perspectivas Socioculturales de la Educacion Matemática, 9*(3), 44–76.

Silva, A. A., & Severinho Filho, J. (Eds.). (2021). *Etnomatemática: Os múltiplos olhares para os saberes locais (Ethnomathematics: The multiple viewa at local knowledge)*. Editora Tangará.

Silva, G. H. G., & Skovsmose, O. (2019). Affirmative actions in terms of special rights: Confronting structural violence in Brazilian higher education. *Power and Education, 11*(2), 204–220.

Silva, G. H. G., Lima, I. M. S., & Rodríguez, F. A. G. (Eds.). (2021). *Educação matemática crítica e a (in)justiça social: Práticas pedagógicas e formação de professores (Critical mathematics education and social (in)justice: Pedagogical practices and teacher education)*. Mercado de Letras.

Skinner, B. F. (1957). *Verbal behaviour*. Appleton-Century-Crofts.

Skovsmose, O. (1980). *Forandringer i matematikundervisningen (Changes in mathematics education)*. Gyldendal.

Skovsmose, O. (1981a). *Matematikundervisning og kritisk pædagogik (Mathematics education and critical pedagogy)*. Gyldendal.

Skovsmose, O. (1981b). *Alternativer og matematikundervisningen (Alternaitves in mathematics education)*. Gyldendal.

Skovsmose, O. (1984). *Kritik, undervisning og matematik (Critique, education, and mathematics)*. Lærerforeningernes Materialeudvalg.

Skovsmose, O. (1990). Perspectives in curriculum development in mathematics education. *Scandinavian Journal of Educational Research, 34*(2), 151–168.

Skovsmose, O. (1994). *Towards a philosophy of critical mathematics education*. Kluwer Academic Publishers.

Skovsmose, O. (2001). Landscapes of investigation. *Zentralblatt für Didaktik der Mathematik, 33*(4), 123–132. Reprinted as Chapter 1 in Skovsmose, O. (2014), *Critique as uncertainty*, (pp. 3–20). Information Age Publishing.

Skovsmose, O. (2004). Mathematics in action: A challenge for social theorising. *Philosophy of Mathematics Education Journal, 18*.

Skovsmose, O. (2005a). *Travelling through education: Uncertainty, mathematics, responsibility*. Sense Publishers.

Skovsmose, O. (2005b). Meaning in mathematics education. In J. Kilpatrick, C. Hoyles, & O. Skovsmose, in collaboration with Valero, P. (Eds.), *Meaning in mathematics education* (pp. 85–102). Springer.

Skovsmose, O. (2007). Foregrounds and politics of learning obstacles. In U. Gellert & E. Jablonka (Eds.), *Mathematisation – Demathematisation: Social, philosophical, sociological and educational ramifications* (pp. 81–94). Sense Publishers.

Skovsmose, O. (2008). Mathematics education in a knowledge market. In E. de Freitas & K. Nolan (Eds.), *Opening the research text: Critical insights and in(ter)ventions into mathematics education* (pp. 159–174). Springer.

Skovsmose, O. (2012). Justice, foregrounds and possibilities. In T. Cotton (Ed.), *Towards an education for social justice: Ethics applied to education* (pp. 41–62). Peter Lang.

Skovsmose, O. (2014a). *Foregrounds: Opaque stories about learning*. Sense Publishers.

Skovsmose, O. (2014b). *Critique as uncertainty*. Information Age Publishing.

Skovsmose, O. (2016a). An intentionality-interpretation of meaning in mathematics education. *Educational Studies in Mathematics, 92*(3), 411–424.

Skovsmose, O. (2016b). What could critical mathematics education mean for different groups of students? *For the Learning of Mathematics., 36*(1), 2–7.

Skovsmose, O. (2018a). Students' foregrounds and politics of meaning in mathematics education. In P. Ernest (Ed.), *The philosophy of mathematics education today* (pp. 115–130). Springer.

Skovsmose, O. (2018b). Critical constructivism: Interpreting mathematics education for social justice. *For the Learning of Mathematics, 38*(1), 38–44.

Skovsmose, O. (2019). Crisis, critique and mathematics. *Philosophy of Mathematics Education Journal, 35*.

Skovsmose, O. (2020). Three narratives about mathematics education. *For the Learning of Mathematics, 40*(1), 47–51.

Skovsmose, O., & Borba, M. (2004). Research methodology and critical mathematics education. In P. Valero & R. Zevenbergen (Eds.), *Researching the socio-political dimensions of mathemat-*

ics education: Issues of power in theory and methodology (pp. 207–226). Kluwer Academic Publishers.

Skovsmose, O., & Penteado, M. G. (2016). Mathematics education and democracy: An open landscape of tensions, uncertainties, and challenges. In L. D. English & D. Kirshner (Eds.), *Handbook of International research in mathematics education* (3rd ed., pp. 359–373). Routledge.

Skovsmose, O., & Yasukawa, K. (2009). Formatting power of 'mathematics in a package': A challenge for social theorising? In P. Ernest, B. Greer, & B. Sriraman (Eds.), *Critical issues in mathematics education* (pp. 255–281). Information Age Publishing.

Skovsmose, O., Alrø, H., & Valero, P. in collaboration with Silvério, A. P. & Scandiuzzi, P. P. (2007). "Before you divide you have to add": Interviewing Indian students' foregrounds. In B. Sriraman (Ed.), *International perspectives on social justice in mathematics education. The Montana mathematics enthusiast*, Monograph 1 (pp. 151–167). The University of Montana Press.

Smidt, D. W. (2007). *Husserl*. Routledge.

Smith, B. (1994). *Australian philosophy: The legacy of Franz Brentano*. Open Court.

Smith, A. (2009). *The wealth of nations*. Thrifty Books.

Soares, D. A. (2008). Educação matemática crítica (Cricla mathematics educaion). *XII Encontro Brasileiro de Estudantes de Pós-graduação em Matematica*. Universidade Estadual Paulista (Unesp).

Soares, D. A. (2013). *O Ensino de matemática em um perspective crítica: Dimensões teóricas e acadêmicas (Teaching mathematics in a critical perspective: Theoretical and academic dimensions)*. Novas Edições Acadêmicos.

Soares, D. A. (2022). *Sonhos de adolescentes em desvantagem social: Vida, escola e educação matemática (Dreams of socially disadvantaged teenagers: Life, school and mathematics education)*. Doctoral dissertation. Universidade Estadual Paulista (Unesp).

Solnit, R. (2016). *Hope in the dark*. Haymarket Books.

Solórzano, D. G., & Huber, L. P. (2020). *Racial microaggressions: Using critical race theory to respond to everyday racism*. Teachers College Press.

Souza, M. H. R. (2021). *A Matemática em ação no curso técnico de nível médio em agroecologia do SERTA (Mathematics in action in the high-level technical course in agroecology at SERTA)*. Master's thesis. Universidade Federal de Pernambuco (UFPE).

Souza-Carneiro, D. V. (2021). *A matemática em ação no ensino superior: Possibilidades por meio do problem-based learning (Mathematics in action in higher education: Possibilities through problem-based learning)*. Doctoral dissertation. Universidade Estadual Paulista (Unesp).

Spinoza, B. (2005). *Ethics*. Penguin Classics.

Sriraman, B. (Ed.). (2020). *Handbook of the history and philosophy of mathematical practice*. Springer.

Stallings, W. (2017). *Cryptography and network security: Principle and practice* (7th ed.). Pearson.

Steele, L. (2013). Sweatshop accounting. In E. Gutstein & B. Peterson (Eds.), *Rethinking mathematics: Teaching social justice by the numbers* (2nd ed., pp. 78–94). Rethinking Schools.

Steffensen, L (2021). *Critical mathematics education and climate change: A teaching and research partnership in lower-secondary school*. Doctoral Dissertation. Western Norway University of Applied Sciences.

Steffensens, L., Herheim, R., & Rangnes, T. E. (2021). The mathematical formatting of how climate change is perceived. In A. Andersson & R. Barwell (Eds.), *Appling critical mathematics education* (pp. 185–209). Brill.

Steiner, H.-G. (1988). Mathematization of voting systems: Some classroom experiences. *International Journal of Mathematical Education in Science and Technology, 19*(2), 199–213.

Strauss, D. (2019). *The life of Jesus: Critically examined* (Vol. I). HardPress.

Streck, D. R., Redin, E., & Zitkoski, J. J. (2018). *Dicionário Paulo Freire* (4th ed.). Autêntica.

Subramanian, J. (Ed.). (2019). *Proceedings of the tenth international mathematics education and society conference (MES 10)*. Mathematics Education and Society.

Sweeney, T. (2016). Hope against hope: Søren Kierkegaard on the breath of eternal possibility. *Philosophy and Theology, 28*(1), 165–184.
Swetz, F. J. (1987). *Capitalism and arithmetic: The new math of the 15th century*. Open Court Publishing Company.
Taylor, F. W. (2006). *The principles of scientific management*. Cosimo Classics.
Thomas, P. (2018). Calculating the value of human life: Safety decisions that can be trusted. *Policy Report 25*. University of Bristol. https://www.bristol.ac.uk/media-library/sites/policybristol/PolicyBristol-Report-April-2018-value-human-life.pdf
Thomas, J. P., & Vaughan, G. J. (2015). Testing the validity if the "value of a prevented fatality" (VPF) used to assess UK safety measures. *Process Safety and Environmental Protection, 93*, 293–298.
Torfing, J. (1999). *New theories of discourse: Laclau, Mouffe and Žižek*. Wiley-Blackwell.
Turing, A. M. (1937). On computable numbers, with an application to the *Entscheidungsproblem*. *Proceedings of the London Mathematical Society, s2-42*(1), 230–265.
Tymoczko, T. (1979). The four-color problem and its philosophical significance. *The Journal of Philosophy, 76*(2), 57–83.
Tymoczko, T. (Ed.). (1986). *New directions in the philosophy of mathematics*. Birkhäuser.
Vithal, R. (2003). *In search of a pedagogy of conflict and dialogue for mathematics education*. Kluwer academic Publishers.
Vithal, R., Christiansen, I. M., & Skovsmose, O. (1995). Project work in university mathematics education: A Danish experience: Aalborg University. *Educational Studies in Mathematics, 29*(2), 199–223.
Volk, D. (1975). Plädoyer für einen problemorientierten Mathematikunterricht im emanzipatorischer Absicht (Plea for a problem-oriented mathematics education with an emancipatory interest). In M. Ewers (Ed.), *Naturwissenschaftliche Didaktik zwischen Kritik und Konstruktion (Natural science education between critique and construction)* (pp. 203–234). Belz Verlag.
Volk, D. (Ed.). (1979a). *Kritische Stichwörter zum Mathematikunterricht (Critical keywords for mathematics education)*. Wilhelm Fink Verlag.
Volk, D. (1979b). *Handlungorientierende Unterrichtslehre am Beispiel Mathematikunterricht. Band A (Action-oriented teaching using the example of mathematics education. Volume A)*. Päd Extra Buchverlag.
Volk, D. (1980). *Zur Wissenschaftstheorie der Mathematik: Handlungorientierende Unterrichtslehre. Band B (On the philosophy mathematics: Action-oriented teaching. Volume B)*. Päd Extra Buchverlag.
von Glasersfeld, E. (Ed.). (1991). *Radical constructivism in mathematics education*. Kluwer Academic Publishers.
von Glasersfeld, E. (1995). *Radical constructivism: A way of knowing and learning*. Falmer.
Wagenschein, M. (1965/1970). *Ursprüngliches Verstehen und exaktes Denken, I–II (Original understanding and exact thinking, I–II)*. Ernst Klett Verlag.
Wallis, W. D. (2014). *The mathematics of elections and voting*. Springer.
Warner, T. (2011). *Numerical weather and climate prediction*. Cambridge University Press.
Watson, J. B. (1913). Psychology as the behaviourists view is. *Psychological Bulletin, 31*(10), 755–776.
Whitehead, A., & Russell, B. (1910–1913). *Principia mathematica I–III*. Cambridge University Press.
Whitt, L., & Clarke, A. W. (2019). *North American genocides: Indigenous nations, settler colonialism, and international law*. Cambridge University Press.
Wilkerson, I. (2020). *Caste: The origins of our discontents*. Random House.
Wilson, G. (1989). *The intentionality of human action*. Stanford University Press.
Wittgenstein, L. (1953). *Philosophical investigations*. Blackwell.
Wittgenstein, L. (1978). *Remarks on the foundations of mathematics*. Basil Blackwell.
Wittgenstein, L. (2002). *Tractatus logico-philosophicus*. Routledge.
Wright Mills, C. (1959). *The sociological imagination*. Oxford University Press.
Wright Mills, C. (1960). Letter to the new left. *New Left Review, 5*.

Wundt, W. (1904). *Principles of physiological psychology*. Macmillan.
Yasukawa, K., Skovsmose, O., & Ravn, O. (2012). Mathematics as a technology of rationality: Exploring the significance of mathematics for social theorising. In O. Skovsmose & B. Greer (Eds.), *Opening the cage: Critique and politics of mathematics education* (pp. 265–284). Sense Publishers.
Yasukawa, K., Skovsmose, O., & Ravn, O. (2016). Scripting the world in mathematics and its ethical implications. In P. Ernest, B. Sriraman, & N. Ernest (Eds.), *Critical mathematics education: Theory, praxis, and reality* (pp. 81–98). Information Age Publishing.
Young, R. E. (1989). *A critical theory of education: Habermas and our children's future*. Harvester Wheatsheaf.
Young, R. J. C. (2001). *Postcolonialism: A historical introduction*. Blackwell.
Yudkowsky, E. (2008). Artificial intelligence as a positive and negative factor in global risk. In N. Bostrom & M. M. Cirkovic (Eds.), *Global catastrophic risks* (pp. 308–345). Oxford University Press.
Žižek, S. (2008). *Violence*. Picador.
Zola, É. (1954). *Germinal*. Penguin Books.
Zweig, S. (2000). *Brazil: A land of the future*. Ariadne Press.

Name Index

A
Abtahi, Y., 53
Ackerman, G., 226
Ackermann, W., 107
Adorno, T.W., 83, 84, 167, 170, 172, 175
Aguirre, J.M., 42
Albertazzi, L., 64
Allen, M.R., 226
Almeida, S., 21, 229
Alrø, H., 17, 68, 187–189, 191, 213, 214
Alshwaikh, J., 80, 128
Andersson, A., 180
Apple, M.W., 172, 173
Aquinas, T., 234
Araújo, J.L., 179
Arendt, H., 34, 96, 97
Arzarello, F., 212
Au, W., 111
Augustine, 234
Austin, J.L., 138, 139
Avci, B., 180
Avigad, J., 115
Ayer, A.J., 93
Azevedo, S.S., 179

B
Baber, S.A., 69
Bales, K., 129
Baroni, A.K.C., 179
Barros, D.D., 25, 43, 178
Barwell, R., 49–51, 53, 126, 127, 140, 141, 180
Bauman, Z., 216

Beck, U., 48, 49, 223, 226, 237
Bell, D., 20
Bendegem, J.P. van, 91
Benjamin, W., 83, 84, 234
Bernacerraf, P., 92, 103
Bernier, M., 7
Beth, E.W., 62
Biotto Filho, D., 178, 202, 203, 220
Birsch, D., 107
Bishop, A., 150
Bloch, E., 7–9
Blum, W., 135
Bohm, D., 17, 162, 164
Bohman, J., 37
Bolsonaro, J., 40
Borba, M., 220
Bourdieu, P., 68, 139
Bose, A., 80, 128
Bostock, D., 103
Bostrom, N., 119
Bourbaki, N., 89, 241
Bourdieu, P., 68, 139
Brams, S.J., 41
Brauer, F., 121
Brentano, F., 63, 64
Bridges, K.M., 20, 21, 84
Britto, R., 22, 23
Brouwer, L.E.J., 92, 147
Brown, D., 20
Brown, J.R., 103
Brown, L., 49
Bruno, G., 146
Buber, M., 163, 189
Bueno, O., 104
Burton, L., 88, 95, 96

C

Calibaba, J., 107
Campos, I.S., 179
Capasso, V., 121
Carrijo, M., 21, 72, 178, 218, 220
Carter, J., 90, 94, 96, 115
Carvalho, C.C.S., 179
Castells, M., 216
Castillo-Chavez, C., 121
Chang, R., 20
Chassapis, D., 80, 128
Chriss, N.A., 124
Christiansen, I.M., 175, 207
Chronaki, A., 79, 128, 215
Chyba, C.F., 226
Ciacardi, L., 212
Cirincione, J., 226
Cirkovic, M.M., 119
Civiero, P.A.G., 179
Clark, M., 160
Clarke, A.W., 20
Claxton, K., 128
Cobb, C.E., 69
Coiffier, J., 50, 126
Coles, A., 49, 80, 119
Copernicus, N., 90, 146
Cortella, M.S., 9
Cotton, T., 49
Cozzens, P., 20
Crenshaw, K., 20, 84
Crutzen P.J., 79, 119
Curry, H.B., 134, 135

D

D'Ambrosio, U., vii
Damerow, P., 173
Darwin, C., 63
Davis, P.C., 21
Davis, P.J., 94
Deleuze, G., 159
Delgado, R., 20, 21, 84
Descartes, R., 6, 7, 146, 155–159
Dewey, J., 168, 189
Diderot, D., 157
Dieudonné, J.A., 89
Diffie, W., 151
Dilthey, W., 169, 171
Dixon, A.D., 22, 84
Drummond, M.F., 128
D'Souza, R., 80, 128
Duggan, J., 122
Dutschke, R., 170

E

Eichmann, A., 96, 97
Einstein, A., 148, 151
Eisensee, T., 128
Eknoyan, G., 141
Elwitz U., 173
Engels, F., 5, 82, 235
Ernest, N., 180
Ernest, P., 90, 91, 96, 104, 180

F

Fanon, F., 5
Farson, R.E., 189
Faustino, A.C., 17, 53, 178, 188
Feng, Z., 121
Ferreirós, J., 88, 96, 115
Feuerbach, L., 82
Fielder, J.H., 107
Fink, F.K., 207
Forrester, T., 108
Fleer, M., 65
Foucault, M., 112, 162
Frame, D., 226
Frankenstein, M., 69, 91, 175, 202
Frege, G., 92, 147
Freire, P., 8, 9, 67, 69, 141, 162–164, 168, 175, 177, 186, 202, 242
Fremstedal, R., 7
Freud, S., 63
Freudenthal, H., 175
Furinghetti, F., 212

G

Galilei, G., 90, 146, 160, 162
Gandin, L.A., 173
Gavarrete, M.E., 91
Gellert, U., 71
George, A., 103
Giaquinto, M., 115
Giddens, A., 223, 224
Gini, G., 41
Gipps, J.F., 65
Giroux, H.A., 172
Glasersfeld, E. von, 62
Gödel, K., 107, 134, 135
Gonzáles, M, 80
Gotanda, N., 20, 21, 84
Gøtze, P., 52, 53
Grabiner, J.V., 91, 150
Greer, B., 80, 128
Guba, E.G., 219
Gutiérrez, R., 80, 128

Name Index

Gutstein, E., 69, 73, 80, 98, 128, 141, 192, 202

H
Habermas, J., 70, 171, 172, 176
Hacking, I., 90, 104
Hales, S.D., 160
Hall, J., 140, 141
Hannaford, C., 44
Hansen, R., 52
Hardy, G.H., 68, 70, 91, 148, 149, 151
Harford, T., 124
Hartmann, A.L.B., 179
Hauge, K.H., 52, 53
Hawkins, A.J., 125
Hegel, G.W.F., 82, 83
Hellman, M.E., 151
Hempel, C.G., 63, 186
Hersh, R., 90, 94, 95, 104, 114
Hilbert, D., 92, 107, 134, 147
Hitler, A., 39, 168
Hofflander, A.E., 105
Hood, K., 105, 129
Horkheimer, M., 83, 84
Huber, L.P., 21
Husserl, E., 64, 70, 73, 223, 227

I
Ince, J., 72, 178
Islind, A.S., 225

J
Jablonka, E., 71
Jacobini, O.R.A., 180
Jacquette, D., 64, 103
Jane, B., 65
Jaworski, B., 44
Jensen A.A., 207
Johnson, B., 122, 125
Jørgensen, M.W., 139
Joseph, G.G., 151

K
Kant, I., 7, 81, 82, 90, 105, 111–113, 155–158, 160
Keitel, C., 173
Kepler, J., 146
Kierkegaard, S., 7, 8
King, M.L., 170
Kipling, R., 19
Kiss, I.Z., 122

Klein, N., 120
Knijnik, G., 179
Kollosche, D., 22, 70, 215
Kolmos, A., 207
Körner, S., 92, 103
Krogh, L., 207
Kvasz, L., 91

L
Lakatos, I., 149
Langville, A.N., 112
Lanigan, C., 110
Latour, B., 120
Leggett, C., 107
Leibniz, G.W., 108
Leitgeib, H., 115
Lempert, W., 172
Lerman, S., 185
Lima, A.S, 179
Lima, I.M.S., 179, 187, 220
Lima, L.F., 179
Lima, P.C., 179, 187, 220
Linnebo, Ø., 103
Locke, J., 34, 160
Lomborg, B., 50, 227
Lorenz, M.O., 41, 42
Luckmann, T., 70, 223, 228
Ludendorff, E., 39
Lula, L.I.S., 205
Lundh Snis, U., 225
Lynn, M., 22, 84

M
Madley, B., 20
Madsen, K.B., 68
Makarenko, A.S., 83
Malcolm X, 170
Malheiros, A.P., 180
Mancosu, P., 90, 96, 115
Manders, K., 115
Marcone, R., 80, 128, 179
Marcuse, H., 83, 84
Marx, K., 4, 5, 47, 82, 83, 85, 109, 155, 156, 158, 235, 236
McKenzie, D., 49
McLaren, P., 83
Mehlberg, H., 103
Mellin-Olsen, S., 68, 174
Menghini, M., 212
Merton, R., 89
Meyer, C.D., 112
Miarka, R., 91

Milani, R., 17, 179, 188
Mill, J.S., 10
Miller, J., 122
Miller, M.B., 125
Miranda, F.O., 179
Moddleton, J., 67
Moe, T.M., 111
Mollenhauer, K., 172
Möller, R., 91
Moran, D., 70
Morgenthau, H., 19
Moses, R.B., 69
Moura, A.C., 42, 179
Mukhopadhyay, S., 80, 128
Münzinger, W., 174
Muzinatti, J.L., 178, 205

N
Nagel, E., 186
Negt, O., 171, 172, 176
Neumann, F., 39
Newton, I., 90, 146, 148, 158
Nietzsche, F., 7, 8, 155, 159–162
Niss, M., 135
Norström, L., 225
Nouri, A., 226

O
Oliveira, F.P.Z., 178, 179
Oliveira, M., 179
O'Neil, C., 94, 124, 125, 130
Orey, D., 91
Orwell, G., 5

P
Pais, A., 15
Palhares, P., 91
Papen, F., 39
Parenti, C., 110
Parra, A.I.S., 37, 80, 128
Paul, S., 62
Peirce, C.S., 94
Peller, G., 20, 21, 84
Penteado, M.G., 16, 35, 42, 54, 71, 98, 178, 194
Phillips, C., 21
Phillips, L., 139
Phoenix, C., 226
Piaget, J., 61, 62, 173, 174, 185, 240, 242
Plato, 6, 90, 125, 159, 164, 234

Pólya, G., 17
Popper, K.R., 186, 212
Potter, W.C., 226
Powell, A.B., 21, 91
Powell, J.A., 21
Putnam, H., 92, 103

R
Rampal, A., 80, 128
Ravn, O., 90, 92, 93, 103, 104, 112, 140, 207
Rawls, J., 10
Redin, E., 8
Reed, A.L., 18
Rehg, W., 37
Restivo, S., 91
Rodríguez, F.A.G., 179
Roger, C.R., 189
Roncato, C.R., 65, 179
Rosa, M., 91
Rosmini-Serbati, A., 10, 235
Rousseau, J.-J., 34, 161
Rousseff, D., 205
Russell, B., 92, 136, 137, 139, 147

S
Said, E.W., 5, 19
Sampaio, L.O., 179
Scandiuzzi, P.P., 68
Scagion, M. P., 179
Schneier, B., 124
Schoenfeld, A.H., 186, 197
Schumpeter, J.A., 35, 36
Schutz, A., 70, 223, 228
Sculpher, M., 128
Searle, J., 62
Severinho Filho, J., 179
Shirley, L., 92
Sieg, W., 108
Silva, A.A., 179
Silva, A.G.O., 179
Silva, E.B., 179
Silva, G.H.G., 21, 65
Silvério, A.P., 68
Simon, P.L., 122
Skinner, B.F., 61
Smith, A., 70, 109
Smith, B., 64
Soares, D.A., 9, 67, 72, 178, 180
Solnit, R., 9
Solórzano, D.G., 21

Souza, M.H.R., 179
Souza-Carneiro, D.V., 179
Spinoza, B., 145
Sriraman, B., 96, 180
Stallings, W., 124
Steele, L., 11
Stefancic, J., 20, 21, 84
Steffensen, L., 52, 53, 180, 237
Steiner, H.-G., 40, 41
Steiner, R., 169
Stentoft, D., 207
Stoddart, G., 128
Stoermer E.F., 79, 119
Strauss, D., 82
Streck, D.R., 8
Strömberg, D., 128
Subramanian, J., 215
Sweeney, T., 7
Swetz, F.J., 151, 152

T
Taylor, F.W., 109, 110
Thomas, K., 20, 21, 84
Thomas, P., 106
Thunberg, G., 56
Torfing, J., 139
Torrance, G., 128
Treder, M., 226
Trump, D., 40, 122
Turing, A., 107
Tymoczko, T., 90, 115, 150

V
Valero, P., 68
Vaughan, G.J., 106
Velleman, D.J., 103
Vithal, R., 97, 174, 207
Volk, D., 173, 174

Vygotsky, L., 185

W
Wagenschein, M., 169, 176
Wallis, W.D., 41
Wanderer, F., 179
Warner, T.T., 50, 126
Watson, J.B., 61
Welshon, R., 160
Whitehead, A.N., 92, 137, 147
Whitt, L., 20
Wilhelm II, 38
Wilkerson, I., 20
Wilson, G., 62
Winter, J., 49
Wittgenstein, L., 59, 91, 121, 122, 133,
    136–139, 155
Wright Mills, C., 113, 176, 186
Wundt, W., 63

Y
Yasukawa, K., 93, 94, 124, 140, 151
Young, R.E., 172
Young, R.J.C., 84
Yudkowsky, E., 226

Z
Zavala, M.R., 42
Zimmer, J., 173
Zitkoski, J.J., 8
Žižek, S., 139, 223, 229, 230
Zola, É., 5
Zweig, S., 177

Subject Index

A
Absolutism, 147, 148
Advocating, 34, 187, 188, 190, 191
Affluence, 212, 217, 218
Algorithm, 52, 55, 94, 95, 120, 124, 125, 152, 153, 197, 225, 238
Anthropocene, 49, 51, 79, 80
Apartheid, 35, 37, 214, 215, 229
Armenian genocide, 20
Axiom, 107, 134, 135, 145, 147, 157
Axiomatic system, 90, 137, 145, 147

B
Banality of mathematical expertise, 87–99, 116, 140, 206, 208, 239
Behaviourism, 61, 240, 242
Bias Index, 16, 25–29, 43
Body Mass index (BMI), 140, 141
Bolsa Familia, 205, 206

C
Capitalism, 35, 83, 84, 151–153, 216
Categorisation, 158
Citizenship, 4, 40, 234–236
Climate changes, 16, 48–51, 53, 55, 84, 85, 114, 119, 126, 127, 180, 226, 227, 237, 239
Colonialism, 5, 19, 84, 230, 237
Completeness, 134
Concern, 3–12, 16, 18, 25, 33, 37, 43, 44, 49, 52–56, 65, 85, 90–92, 104, 109, 125, 135, 146, 148, 160, 164, 169, 173–175, 179, 180, 186, 187, 191–193, 204, 212, 225, 229, 234–238, 242, 244
Consistency, 134
Crisis
  hierarchy of crises, 80–83
  multiplicity of crises, 83–85
  politics of, 120
Crisis and mathematics
  mathematics constituting a crises, 123–126
  mathematics formatting a crises, 126–127
  mathematics picturing a crises, 121–123
Critical education, 40, 47, 167, 168, 170–172, 176, 178
Critical race theory, 20–22, 84
Critical situation, 80, 81, 99, 120, 125, 127, 129, 130
Critical theory, 21, 47, 83, 84, 172, 173, 236
Critique
  of critique, 17, 47, 81–83, 85, 155, 158, 162
  as dialogue, 155, 162
  as hammering, 155, 162
  of mathematics, 90, 112, 145, 148, 149, 153, 237
  as methodology, 155, 162
Cryptography, 149, 151, 153

D
Deliberation, 35–38, 43–44
Democracy
  deliberations, 44
  equity, 36, 43
  fairness, 35–37
  voting, 35–37

Descriptive paradigm, 219, 220
Dialogic act, 187–189, 191
Dialogic theory of learning mathematics, 163, 185–197, 242, 243
 doubting, 196, 197, 243, 244
 exploring, 5, 11, 12, 16, 36, 47, 69, 82, 83, 85, 108, 109, 141, 153, 156, 158, 161, 162, 168, 170, 172, 192, 205, 230
 externalising, 194–197, 243, 244
 foregrounding, 192–194, 196, 197, 243
 getting in contact, 187, 189, 197, 243
 positioning, 164, 191, 192, 196, 197, 234, 243
Dialogue, 6, 10, 16, 17, 23, 24, 156, 162–164, 170, 178, 179, 186–188, 194, 213, 214, 234, 242–244
Digitalisation, 71, 93, 108, 109
Digital revolution, 108, 109, 113
Digital Taylorism, 110–111
Discourse, 6, 10, 15, 16, 19, 21, 33, 68, 80, 121, 122, 128, 139–141, 159, 161, 177, 179, 206, 223, 228, 229, 240
Discursive theory, 139
Dissolution of responsibility, 93–95, 99, 140, 238, 239
Doubting, 196, 197, 243, 244

E
Enlightenment, 81, 157, 161
Entscheidungsproblem, 107, 109
Environmental justice, 48, 56, 205, 206, 234, 236–237
Epistemological dimension, 81, 92, 95, 103, 114, 136
Epistemology, 7, 174, 175, 196
Equity, 17, 35–37, 43, 44
Erosion of democracy, 33, 34, 38–44
Ethical dimension, 88, 91–95, 97, 104
Ethical vacuum, 88, 93, 96, 99, 114–116
Ethics, 7, 93, 103–116, 145, 152
Ethnomathematics, 179, 215
Euclidean geometry, 145, 150, 157
Exemplarity, 177
Exemplary learning, 171, 176, 177
Exercise paradigm, 4, 179
Experimental forecasting, 48–51, 207, 208, 238
Exploring, 5, 11, 12, 16, 36, 47, 69, 82, 83, 85, 108, 109, 141, 153, 156, 158, 161, 162, 168, 170, 172, 192, 205, 230
Externalising, 194–197, 243, 244

F
Fachkritik, 171
Fagkritik, 171, 174, 208
Fairness, 36, 41–44, 69, 235
Ford Pinto case, 106
Foreground
 students, 65, 72, 193, 241
 teachers, 193
Foregrounding, 192–194, 196, 197, 243
Formalised mathematical theory
 alphabet, 134
 axiom, 90, 107, 134, 135, 137, 145, 147, 157
 formula, 88, 91, 107, 124, 134, 151
 proof, 41, 134, 135, 145, 148–150, 191
 rule of deduction, 90, 92, 135
 theorem, 41, 91, 107, 134, 135, 137, 147, 176, 179, 206
Formalism, 90, 92, 95, 96, 107, 114, 115, 133–135, 147, 148
Formatting power of mathematics, 140, 223–231
Four-colour theorem, 150
Four-dimensional philosophy of mathematics
 epistemological dimension, 90, 92
 ethical dimension, 91, 92
 ontological dimension, 90, 92
 sociological dimension, 91, 92
Fourth world, 216, 217

G
Geisteswissenschaftliche Pädagogik, 168, 169, 171
Geisteswissenschaft, 168
Genetic epistemology, 61, 62, 185, 240
Genocide
 Armenian genocide, 19, 20, 81
 North American genocide, 20, 81
Getting in contact, 187, 189, 197, 243
Ghettoising, 216
Gini index, 41, 42, 239
Globalisation, 216
Glocksee School, 172

H
Hierarchy of crises, 80–83, 85
Holocaust, 169
Hope, 3–12, 67, 160, 178, 180, 191, 193, 203, 204, 234, 236, 241, 244
Hypothetical reasoning, 93–95, 98, 238, 239

Subject Index

I
Illocutionary force, 138
Imagination
  pedagogical, 179, 187, 212, 219, 220
  sociological, 113, 171, 176–177, 179, 186
  technological, 238
Inclusive mathematics education, 179
Index
  Bias Index, 16, 25–29, 43
  Body-Mass Index, 140
  Gini Index, 41, 42, 239
Industrial revolution, 108, 109, 113
Infinitesimal, 150
Intentionality, 62–67, 214
Intuitionism, 90–92, 96, 114, 147, 148

J
Justice
  environmental, 48, 56, 205, 206, 234, 236–237
  social, 205, 206, 234, 237
Justification, 5, 19, 51, 93–95, 105, 148, 156, 174, 238, 239

L
Landscapes of investigation, 16, 17, 29, 178, 234, 242
Language as performing, 138, 139
Language as picturing, 121, 133, 136–138
Language game, 59, 138
Learning as action, 60–62, 64, 67, 73, 241
Learning interact
  doubting, 196–197
  exploring, 189–191
  externalising, 194–196
  foregrounding, 192–194
  getting in contact, 189
  positioning, 192
Life-world, 60, 70–73, 93, 109, 111, 114, 126, 164, 223, 224, 227, 228, 240–242, 244
Listening, 96, 189–191, 194
Locutionary content, 138
Logical positivism, 63, 83, 92, 93, 121, 139, 171, 176, 186, 187, 218, 219, 224
Logicism, 90–92, 96, 114, 147, 148
Lorenz curve, 41, 42

M
Marxism, 8, 47, 83–85, 236

Mathematical algorithm, 71, 124, 125, 151–153, 225, 228, 230, 238
Mathematics
  as glorious, vii, 89, 146–148
  as harmless and innocent, 237
  as indefinite, 149–153
  as performing, 140–142
  as picturing, 134, 135
  and power, 52, 145, 234, 239
Mathematics and crises
  mathematics constituting a crisis, 123–126
  mathematics formatting a crisis, 126–127
  mathematics picturing a crisis, 121–123
Mathematics in action
  categorisation, 111–113, 158
  digitalisation, 71, 93, 107–109
  dissolution of responsibility, 93, 94, 99, 140, 238, 239
  hypothetical reasoning, 93–95, 98, 140, 238, 239
  imagination, 70, 113, 114
  justification, 93–95, 140, 239
  quantification, 105, 106, 140
  realisation, 93–95, 140, 239
  serialising, 104, 109–111, 116, 140, 152
  technological imagination, 93, 95, 113, 114, 140, 238, 239
Metamathematics
  completeness, 134
  consistency, 134
  Entscheidungsproblem, 107, 109
Modernity, 47, 150, 155, 156, 159–162, 164, 226
Modern Mathematics Movement, 89, 173–175, 241, 242
Motivation, 68
Motive, 60, 61, 66–70, 72, 73, 178, 193, 194, 229, 240, 241
Mündigkeit, 167
Multiplicity of crises, 83–85

N
Naturalistic inquiry, 219
Neatness, 212, 214, 218
North American genocide, 20
North Sea Model, 207, 208
Nuremberg Laws, 18, 39, 40, 73, 227

O
Ontology, 94, 125, 126

Oppression, 4, 5, 20, 36, 69, 71, 73, 108, 109, 141, 161, 168, 171–173, 180, 205, 234, 235, 242

P
Pedagogical imagination, 179, 187, 212, 219, 220
Performing, 16, 52, 72, 133–142, 150
Perlocutionary force, 138
Phenomenology, 70, 227, 228
Philosophy of mathematics
   The *epistemological dimension*
      addresses, 90
   ethical dimension, 116
   ontological dimension, 90
   sociological dimension, 104
Philosophy of mathematics education, 85, 116, 233, 244
Picture theory, 121, 133, 136–138
Placidity, 212, 217, 218
Platonism, 90, 94, 95, 234
Politics of crises, 128–130
Positioning, 164, 191, 192, 196, 197, 234, 243
Post-modernism, 155, 159
Power, 4, 5, 7, 35, 39–42, 48, 51, 55, 83, 84, 108, 124, 128, 129, 138, 139, 151, 153, 160, 161, 167, 168, 170, 186, 197, 202, 208, 219, 223, 225, 226, 229, 236, 239, 240
Prescription readiness, 16, 230
Presenting, 16, 54, 84, 128, 146, 162, 175, 190, 191, 211, 243
Problem-based learning, 17, 179
Problem orientation, 97
Production line, 11, 12
Project work, 16, 44, 66, 174, 177–179, 189, 190, 194, 203, 207, 208, 220, 239
Proofs and refutations, 149, 150
Prototypical mathematics classroom, 212

Q
Quantification, 22, 25, 105–107, 140, 208
Questioning, 52, 187, 190, 191, 196, 243, 244

R
Racism
   structural, 37, 229
Reading the world with mathematics, 141
Realisation, 93–95, 99, 239
Research paradigm
   affluence, 212, 217, 218
   descriptive paradigm, 219, 220
   naturalistic inquiry, 219
   neatness, 212, 214, 218
   placidity, 212, 217, 218
Risk society, 48, 49, 51, 52, 223, 224, 226, 227
Role model, 88–92, 97, 99

S
School mathematics tradition, 16, 98, 188, 230, 242
Scientific management, 109, 110
Scientific revolution, 62, 90, 105, 111, 140, 146
Serialising, 104, 109–111, 114, 116, 140, 152
Set theory, 89, 173, 241
Similarity gap, 51, 123
Social justice, 4, 10–12, 44, 56, 98, 156, 159, 161, 162, 180, 234–237
Social theorising
   life-world, 227–228
   risk society, 226–227
   structuration, 224–226
   violence, 230
Sociological dimension, 104
Sociological imagination, 113, 171, 176, 177, 179, 186
Speech act
   illocutionary force, 138
   locutionary content, 138
   perlocutionary force, 138
Structural racism, 16, 21, 22, 24, 29, 37, 229, 230
Structuration, 65, 70, 223–226, 228
Students at social risks, 4, 178, 202–204
Students in conformable positions, 202
Students' foregrounds, 67–70, 72, 73, 240
Students' movement, 170, 171
Sustainability, 47–56, 93, 123, 197, 207, 236, 237
Symbolic violence, 139

T
Taylorism
   digital, 110, 111
Teacher's foregrounds, 234, 241
Technological imagination, 93, 95, 113, 114, 238, 239
Topology of apartheid, 215
Turing machine, 107

Two-dimensional philosophy of
 mathematics, 103

U
Uncertainty, 49, 113, 114, 164, 218, 244
University students in mathematics, 4, 175,
 202, 206–208, 239

V
Violence

symbolic, 139
Voting, 35–36, 40–41, 44

W
Weimar Republic, 34, 38, 40, 168–170, 178
Writing the world with mathematics, 69, 73,
 142, 178

Z
Zero-Point-One World, 216, 217

GPSR Compliance
The European Union's (EU) General Product Safety Regulation (GPSR) is a set
of rules that requires consumer products to be safe and our obligations to
ensure this.

If you have any concerns about our products, you can contact us on

ProductSafety@springernature.com

In case Publisher is established outside the EU, the EU authorized
representative is:

Springer Nature Customer Service Center GmbH
Europaplatz 3
69115 Heidelberg, Germany

www.ingramcontent.com/pod-product-compliance
Lightning Source LLC
LaVergne TN
LVHW012248070526
838201LV00091B/154